IT'S ABOUT TIME

SCIENCE HARMONIZED WITH RELIGION
WHERE WE ARE IN GOD'S TIME

What People are Saying

A truly great book!! I read it with enjoyment. I recommend it, as it is written for lay people. — *Fernando Lins*

• • •

You are a global treasure! And so is your precious book, *It's About Time*.

I ordered it online in 2015, and immediately devoured it. As a university graduate in the natural sciences, I had expected to find a mostly scientific book, loaded with difficult factual information and formulas.

To my surprise, and delight, it is mostly about factual religion, and the reality of how God intelligently, purposefully and precisely organized the elements and forces existing in what we call space, into not only this earth – specifically for us His children – but coordinated it all so incredibly among His existing creations in space. All we experience in mortality testifies of Him and His love for us.

The contents of your book blend so beautifully with revealed principles of both religion and science, that they significantly enhance my understanding and conviction of both, particularly man's mortal agency. I feel honored to know you and meet you in a public occasion. Your humility to present your historic material in such humble circumstances, and your open oneness with others of similar spirit, is truly a touch of the divine. Thank you for being you, and for being a tool in God's hands for truth and true progress.

— *Bob Webster (in a personal letter to the author)*

• • •

Amazing insights into the science and religion question. World famous scientist, shares his understanding and experiences, as he defends his testimony of God, as DEFENDED by scientific investigation. Great read…challenges your understanding. I would definitely refer others to this great work.

— *Michael (author)*

• • •

Fortunately, David Allan offers another postulate and another physical model. It leads to recognition of who we really are and what we might become. It offers growth of greatly expanded consciousness, perhaps even telepathic connectivity for humanity. It offers solutions for society via empathy by recognizing our true spiritual nature. It can even offer a new energy source from the orthogonal interaction of the zero-point field, and thus solve mankind's physical needs with abundant energy. It could truly usher in a golden age for humanity. It's About Time!

— *Moray B. King (research scientist)*

Is there a great hidden mystery to the meaning of life when we understand God's Time?

How is man's time different from God's time?

Am I a child of God or am I evolved from simpler life forms?

What is the science behind my prayer channel to God?

If science is true and religion is true, must they not harmonize, since truth cannot contradict truth?

David Allan desires to share with you answers to these questions. He is one of the world's foremost time metrologists, and has been measuring time for over half a century as an atomic clock physicist. He wrote the algorithm for generating time in the United States. He has sliced the second into billionths, quadrillionths, and even smaller segments. Such precision has enabled developments like GPS, cellular communication, and many other major benefits to society. Mr. Allan also reveals the problems that several false traditions have caused to body, mind, and spirit. He also shares their solutions.

The author's devotion to learning and the pursuit of truth has revealed a surprisingly strong connection between things spiritual, scientific, physical, political, historical and mental. It's about time we became aware of the ideas that bind us to the *eternal*. This is the most exciting time in history!

© 2016 Copyright David W. Allan

All rights reserved. No part of this book may be reproduced in any form whatsoever, whether by graphic, visual, filming, microfilming, tape recording, or any other means without the written permission of the publisher, except in the case of brief passages embodied in critical reviews and articles where the title, author and publisher accompany such review or article.
Third Edition Printing: February 2016

ISBN: 978-1-937735-75-3

Digital Legend Press
(An imprint of Legends Library Publishing, Inc.
Rochester, NY) Salt lake City, UT

For inquiries write to: info@digitalegend.com
Or phone: 877-222-1960
Website: www.DigitaLegend.com

Cover design by Rachel Oliveri
Book layout and design by Kevn Lambson

IT'S ABOUT TIME

SCIENCE HARMONIZED WITH RELIGION

WHERE WE ARE IN GOD'S TIME

David W. Allan

Legends
LIBRARY

New York

Dedication

This book is Dedicated to the God of the Universe, And to my
Sweetheart and wife, Edna
And to our precious children: Shelli, Karie, Sterling, Jeannette, Celeste,
McKaylee, and Nathan,
And to my wonderful extended family and friends,
And to my brilliant colleagues, who have inspired me so.
I am grateful from the bottom of my heart for ALL of you.
This book would not be without you.

Acknowledgments

While the acknowledgments are hard to write, they are easy to write. It is hard because there are so many who have contributed to the contents of this book. I could write pages and pages of those who have helped me along the way. As I will say later, I feel like I have been standing on the shoulders of giants.

It is easy to write this acknowledgment because my heart is in deep gratitude for the awesome help I have had in writing this book. Many members of my family have helped and all have been loving and supportive. In particular, my wife, Edna, is a very good reader; our oldest son, Sterling, is an excellent writer and editor. Both of their comments have been very substantive and helpful. Our awesome and very talented grand-daughter, Rachel Oliveri, did the art work for the cover. Fred Kessler, our dear friend from Mesa, Arizona — a retired English teacher — has spent many hours turning my American-physicist writing into English! Professor Chauncey Riddle has become a dear friend through the book, as indicated in the Foreword, and his suggestions, corrections, and guidance have influenced the book greatly. David Tuttle, brother to our publisher, has taken the time to give some very meaningful edits. Our son-in-law, Kevn Lambson, a talented graphic artist, kindly volunteered and did the very challenging job of formatting the book, preparing the index, and assisting with the cover design, as well as helping with many helpful edits. My family support has been truly awesome. The cooperation and support from Boyd Tuttle, our publisher, is deeply appreciated. I have to thank Lee Nelson from my hometown of Mapleton, Utah for encouraging me to write the book in the first place.

As I have dedicated this book to Him, I believe He has helped me make it be the book for the new millennium — challenging traditional thought, while harmonizing science with religion. He has helped me to know where we are in God's Time. It has been a very exciting book to write. My hope and prayer is that you, the reader, will be greatly blessed as I reach out in love to you both in spirit and with what I have written. Thank you for taking the time to ponder and read what I have felt to share. My hope is that you will find many useful and enlightening truths to help you, and that we may work together to make this planet we share a better place.

Foreword by Emeritus Professor Chauncey C. Riddle

Chauncey Cazier Riddle graduated from Brigham Young University and received his MA and PhD degrees in philosophy from Columbia University in the city of New York. He taught at BYU for 40 years, serving as Professor of Philosophy, Chairman of the Department of Graduate Studies in Religious Instruction, Dean of the Graduate School, and Assistant Academic Vice President.

Foreword: Dear Reader

You are in for a special treat. David W. Allan has two special qualifications to write this book: He is a world-renowned scientist and he is a thoroughly grounded Christian.

In this work he initiates a rapprochement of science and religion.

The science he discusses is centered on the study and measurement of time, the area in which he spent his entire professional life. His scientific prowess resulted in great improvements in the measurement of time and the many devices by which time is measured and compared. Unbeknownst to the average layman, the measurement of time is crucial to much technology in our modern world, such as GPS.

David Allan correctly recognizes that science is a work in progress. He recognizes that some scientists currently entertain some questionable hypotheses, but that science itself is a very worthwhile enterprise. It is worthwhile especially because it can help humanity solve the technical problems which abound in our complex human world. He also values religion and sees it as the practical application of good sense to the solution of everyday personal problems. He believes that when one has found true wisdom (true religion), personal happiness is the assured result. He posits an approach to religion that is practical for mind, body and spirit, seeing mental health, spiritual health and body health as a single package.

His science is sound. Please do not be put off by formulas and Greek letter notation. As you read and reread the technical passages you will find them to be keys to understanding science. They enable precision in thinking and acting, which valuables are at the core of all technical excellence in applying scientific thought to our human practical problems.

And please take advantage of the many references he makes to other works, especially those on his own website, which will add much to your understanding of the matters of time, science and religion.

As you immerse yourself in conversation with David Allan by reading his words, please take time to savor carefully what he says. Then challenge what he says. Challenge it by applying the religion to your own life and problems, and by applying the science in some technical way if you are thus skilled. Only in these twin applications will you come to full appreciation of what he says and to be able to improve on what he says. Improving on what he says is the thing he most desires from each of his readers.

Happy reading, careful thinking, and improved living to each of you through your adventure with this book.

Table of Contents

Foreword iii
Author's Background xi
Introduction xxv

Sections and Chapters

Section I

Expanded Scientific Method Brings Harmony with Religion and Answers Many of Life's Most Challenging Questions

Chapter 1 3
It's About Time to Put God Back into the Scientific Method.

Chapter 2 23
It's About Time We Tackle Some of Life's Most Difficult Questions.

Section II

Applying Our New Basis for Scientific Thinking to Several False Traditions

Chapter 3 49
What Is God's Word?

Chapter 4 73
What is the True Nature of Love?

Chapter 5 131
Mary's Miracle; God Is Closer Than You Think; Documented Miracles Showing Evidence of God

Chapter 6 147
The Theory of Evolution Is Pervasive. Have We Been Deceived?

Chapter 7 163
Food, Water, Toxins, and Nourishment To Body, Mind and Spirit.

Table of Contents

Chapter 8 201
Can All Addictions Be Overcome?

Chapter 9 211
Can All Cancer Be Cured?

Chapter 10 213
Should We Have "Mind Over Medicine?"

Chapter 11 217
The "Biology of Belief" and Healing Modalities

Chapter 12 221
Stress, the Killer of Body and Brain? How Can We Best Deal with It?

Chapter 13 223
Growing the Brain and Fullness of Joy

Chapter 14 235
About Energy

Chapter 15 239
Is Our Defense for Our Defenses?

Chapter 16 245
New World Order Infiltration of the Internet

Chapter 17 249
Are We a Christian Nation?

Chapter 18 257
How Are We Saved?

Chapter 19 265
The Ideal Society and the First Law of Heaven

Section III

Man's Time and GPS, God's Time, Transcendental Time, Where We Are in God's Time, and the Wheel of Time

Chapter 20 271
It's About Man's Time, How GPS Works, and How to Improve Them Both

Chapter 21 299
It's About the Fifth-Dimension Beyond Man's Time and Our Mortal Sphere of Existence, Where God and Angels Dwell

Chapter 22 321
It's About How Transcendental Time Couples to Both Mortality and the Eternity-Domain

Chapter 23 335
It's About Where We Are in God's Time

Chapter 24 363
It's About the Wheel of Time — One Eternal Round

Chapter 25 369
It's About Time!

Conclusion 377

Glossary 385

Index 389

The following articles are located at *www.ItsAboutTimeBook.com*:

- *Seven Events in the Seven Steps of the Atonement of Jesus the Christ*
- *Christ will Gather as a Hen Gathers Her Chickens*
- *So You Want To Be One With God?*
- *The Time Table Of The Lord*
- *Don't bypass "And it came to pass"*
- *And It Came To Pass=El=God*
- *And It Shall Come To Pass =Jehovah*
- *"Shall is Sure"*
- *A Third Witness of Grand Scriptural Harmony*
- *As Old As Adam*
- *Where Did The Book Of Mormon Really Take Place and Does It Matter?*
- *The Book That Influenced the World More Than Any Other*

Table of Contents

- *Left To Tell* (Book Report)
- *Six Magical Skills In the Art of Listening*
- *A Super Food That is Too Good To Be True*
- *The Magical Benefits of Vitamin K2* (Interview)
- *Four Simple Secrets For Living To Be 100*
- *Protect Yourself During A Pandemic*
- *Natures Natural Noise Processes – Flicker-Noise and Timing Errors in Clocks*
- *The Science of Keeping Time*
- *Letter to Key Coordinators of GPS Program*

It's About Time

About the Author

Author's Background

This is not a biography, yet it is — synoptically. As I consider the content of what I have written, it contains most of the major Truths that have been opened to me and have been life-changing. I feel greatly blessed that these Truths have come into my life from family and friends, from colleagues, from talks I have heard from a variety of folks, from books that I have read (especially the scriptures, of course), from God's incredible creations, and most importantly from inspiration I have received from On High. I love to read and research for Truth. Surprisingly, I have even found some Truths on the internet! How do I know what is true? We will discuss that in detail.

It is my desire in this book to help you be filled with hope and understanding about what is time and about the time in which we live. It's about time, since I have spent most of my career (fifty-four years) with the study of time and the measurement of time as well as where we are in God's time. It is the most exciting time in the history of the Earth. You have heard the phrase, "Live and learn!" I would like to turn that phrase around: "Learn and live" the abundant life promised by God.[1] And in the process of learning we will see exciting harmony being built between science and religion.

When I say, "the measurement of time," it is actually a misstatement unless you talk about God's time. You cannot measure time here in mortality as we commonly think of it; it is not something you can extract from the universe and determine its value. What is measured is the time difference between two clocks, which we can do to much better than 1 million-millionth of a second (picosecond). Time as we have it in the world is a man-made construct. The Global Positioning System (GPS) is one of many applications that would not work without manmade constructed timing.

For twenty-four of my thirty-two years in Boulder, Colorado, my group was responsible for generating official time for the United States of America, which is the responsibility of the National Bureau of Standards (later renamed the National Institute of Standards and Technology). Extracurricular to that process, I have studied God's Time as part of my religious pursuits, which takes us beyond mortal time into what we call the Eternity-Domain.

---

1) John 10:10

It's About Time

Throughout the book, we will share some very revealing and exciting data (experimental evidence) delving into that domain and addressing the topics on the title page as well as those in the table of contents. I use the word "We" because I have received help from lots of different folks and different directions for this work. I especially thank my Sweetheart, Edna Love Ramsay Allan, who is also my wife of fifty-five years. I am most grateful to God, whose help I feel continuously.

I believe in the "simple power of Truth." This book is a result of my quest for Truth with a capital "T." It is also my desire to share what has brought great joy, peace, and love into my life along with significant and unique insights during my seventy-seven years' sojourn on this beautiful planet, and the quest has been exciting and still is. I give the glory to God for His tender mercies in revealing these Truths to His children. I definitely feel "It's about time" to do so, and I hope I can share with clarity and in a way to touch your heart.

If you have a desire to learn the Truth, then let us learn together and we will be on the same page. You may have heard the phrase, "A man convinced against his will is of the same opinion still!" Many people have a problem with selective listening; they can hear only what fits their paradigm. I believe being open to the Truth from wherever it comes is a most rewarding experience and can cause meaningful paradigm shifts in our lives.

If you laugh at me or with me, that's good too. You have heard that laughter is the best medicine. It has been shown that laughter strengthens your immune system, boosts your energy, diminishes your pain, resolves conflicts, and protects you against the devastating effects of the main killer in our lives — stress. Best of all, this priceless medicine is free and fun.

My family is known for enjoying puns; we find it helps to lighten up and enjoy a good laugh. Before I get into the book, I have to tell you a good one. I was talking to our youngest daughter, McKaylee, across the kitchen counter in Boulder, Colorado, where we lived at the time. There was a sack on the counter with holes in it. I told McKaylee, "This is a religious sack; it's holy!" She quickly responded, "Oh Dad, don't be sacrilegious!"

My resources for "Truth" are wherever I find them. As a scientist, I will use an expanded version of the scientific method. It adds a great richness to the standard approach, to substantiate those Truths that I share as developed in Chapter 1. As much as possible, I like to rely on trustworthy data — experimental information — and broaden the meaning of experimental as well.

We hear a lot about the conflict between science and religion, which is ironic because if you go to their definitions, they are metaphysical; they only exist only in our minds. So any conflict is not real, but only in our minds. Noah Webster defines science as follows: "1) In a general sense, knowledge, or certain knowledge; the comprehension or understanding of Truth or facts by the mind." (Italics added) Then we may apply it to different disciplines — like the science of music. What we typically call science is the putting those comprehensions or understandings into practice — like developing laws and models of our understanding of the physics of natural phenomena. The Law of Gravity is an example. We practice science and call it science. We will follow that colloquial standard in this book, since it is the common practice.

A similar argument follows for religion. Religion is our metaphysical beliefs, which everyone has — even the atheist. The practice of religion is almost always associated with a belief in God and the practice of the moral ethics associated with that religious belief system. The practice of moral ethics we call "religion" — following some belief system. This is the typical use of the word "religion," and this practice will be our usage in this book, since that is the common practice. Science, so called, is typically without God. Religion, so called, is typically with God.

Science and religion, during the last century have largely been in conflict. In contrast, there have been advanced-ancient civilizations that have science and religion integrated.[2] One of my goals is to help bring that integration about. Truth does not contradict Truth. If the Truths of science are true and the Truths of religion are true, then they will harmonize, even if contemporary atheists don't want them to. And sometimes religionists have a similar problem — when they refuse to look at or ignore reliable scientific data. I believe that that harmony between science and religion will bring about an enormous blessing to humanity and to the Earth in the days and years ahead. So, I move forward to share the Truths that have been given to me with the Lord's help as best I can.

---

2) Look at the following as an example: *http://www.allanstime.com/Spiritual/Hopewell_Civilization_Great_Octagon/*

I was born and raised on a farm in Mapleton, Utah, into a very supportive family: my father, Sylvester (Smuss) Allan; my mother, Florence; my older sister and brother, Beverly (Malheiro) and Dean; and my younger sister, Jeanne (Strong). As I rode around town on the graveled roads on my bicycle, I knew every dog by his bite and his bark. Sometimes I would have only one pedal, after breaking off the other from climbing steep hills. My dad and my brother, Dean, were my male role models growing up. Dean was five years older, so when Dean would let me tag along with him and my older cousin, Collin Allan, that was a big deal. During my later teens, Dean served in the Air Force during the Korean War, so I missed him during that important time. Fortunately, I had some excellent teachers and very good friends during my school years, as well as when I went to Brigham Young University.

Dad was very creative, and invented many things. A lot of our farm equipment was made with special features. He had a shop, a welder and an acetylene torch, and could weld my pedals back on my bicycle until I broke them off too many times. And that happened also. Often, if some farmer had something break, they would come to Smuss to fix it. I don't think I ever saw him not "fix it." I grew up with the attitude that creativity and invention were part of life and every problem had a solution. I think Dad was the best part of my education. Dad, Dean, and I had some very choice times together. I remember well, cranking the handle on the blower for the old forge to heat some piece of metal so that Dad could mold it into special shape or to simply sharpen some plowshares.

Wanting to know Truth and to understand the world and the universe — how it all works and why — is core to my being. This desire is probably why physics became my quest, along with wanting to understand the Creator of it all.

Mapleton is predominantly a Mormon community near Provo, Utah, and I was raised with that ethic. However, my parents were not active participants in that faith. We had two Catholic families in town — Carnesecca and Bleggi. Growing up, I went to school with their children. They seemed every bit as good as the Mormon children. Both religions seemed good to me.

My thoughts were that if I were to know which church or religion were true — if any of them — I should investigate all of them to be fair and to do a comparison. Since I didn't know much about Mormonism, I started there, that

About the Author

being the most convenient. I became active in The Church of Jesus Christ of Latter-day Saints (Mormon Church) during my late teens as some of my good friends shared with me some of the teachings, and they resonated with my soul as being true.

I was particularly impressed with the doctrine that in God's plan everyone had an equal opportunity for eternal happiness. Subsequently, I have found no other religion that has this doctrine of equal opportunity for all. I was also touched by the message of the Book of Mormon, which is a religious history of ancient America, which gave The Church its "Mormon" nickname. It seemed right that God would have a religious and consistent record for the New World (America) as well as having the Bible for the Old World. In addition, I loved the Mormon philosophy that Truth is Truth no matter where it comes from. I got a book on comparative religions and studied it. I have found some Truths in every religion I have studied, and I have studied all of the major ones.

The atomic clock community is not a large one, so my travels have taken me around the world, and I have always taken the opportunity to learn about the religious beliefs in each country I have been privileged to visit. Over the years, I have found many hard questions answered as I have searched deeply in the scriptures, in the writings of history, and into other cultures and religions. Some of these will be an important part of the book.

So, am I a Mormon? I am a Christ-centered member of The Church of Jesus Christ of Latter-day Saints. I have learned to examine every tradition in the light of Truth, and in so doing I have discovered false traditions that have crept into essentially every religion as worldly influences creep in. I choose to not be of the world to the best of my ability — even though we live in the world. I can tell you much more about the future from the scriptures than from the media. I have seen estimates that half of Christian males suffer from pornography addiction with the media and internet contributing massively to this problem. Who do I follow? I follow Christ. I've found that by keeping my attention on the good things of God, that my heart is bolstered against the enticements of the flesh that surround us. As is well said, "But for the grace of God, there go I." We will offer some additional loving suggestions in reaching out to those with this addiction and other kinds of addictions as well in Chapter 8.

I went on a mission for The Church of Jesus Christ of Latter-day Saints to the Eastern United States. I fell in love with the scriptures and the Author of them. I loved sharing His message and met many wonderful people there. I was used to the Rocky Mountains and really missed them. The high relative humidity of the east was impressive to me as well. We used to joke that fall and spring were beautiful in the east — both weeks! We used to also joke how the "humility" gets to you.

I so enjoyed sharing the gospel that I considered changing my career path from physics to religion. I visited with a wise physics professor, Richard Hales. I then had my physics class from him at Brigham Young University. In summary he said: "God will bless you if you focus to serve in whatever career path you choose. If you choose a career that best uses your talents, then so much the better. Don't worry about money; you will be fulfilled and God will take care of your needs." So, I chose to stay in physics. His counsel has turned out to be prophetic.

The most important person I met while serving a mission in the eastern United States was my beautiful wife-to-be, Edna Love Ramsay. We met at the Hill Cumorah pageant in upstate New York. Under the direction of Dr. Harold Hansen, 150 or so young missionaries were joined with three busloads of young ladies coming out from Utah to put on the pageant so that they could have a balance of male and female as they shared the miraculous story of the coming forth of the Book of Mormon on stage. They had study groups while we were not practicing in preparation for the performance, and Edna and I happened to be in the same study group. She caught my eye, and I asked her to send me a picture, which she did. I had given her one of mine.

Toward the end of my mission, I wrote her a letter asking her for a date when I got home. In the middle of the letter, at the end of a sentence, I had a great big period — probably nearly a quarter inch in diameter. Then I said, "This is a thinking period!" When I returned home, I started dating this lovely lady, Edna Love Ramsay, from Snowflake, Arizona. I convinced her to add Allan on the back of Ramsay. We were married in the Salt Lake City temple and celebrated our 55th anniversary this year. Yet we have celebrated 57 anniversaries. How can that be?

In 1981 there was a seminar in New Delhi for 3rd-world nations to instruct

About the Author

them on atomic clocks, time and frequency metrology, and precise-time distribution systems. The United Nations paid my way to join with other experts in the field to teach at the seminar. I was able to obtain a $50 coupon that allowed me to get another ticket all the way around the world for my wife. After a week in India eating curry with every meal, we were looking for a McDonalds! Don't get me wrong. It is not that we don't like curry. We were just not prepared for this dynamic, cultural-dietary difference.

After the seminar, we had arranged to be in Guam for our wedding anniversary on the 20th of February. After celebrating our anniversary in Guam, we got on the plane that evening and flew to Hawaii. The next day was the 20th of February all over again — having crossed the International Date Line! So we got two for the price of one in more ways than one that year.

Then, more recently — the year of our 50th anniversary — we had two anniversaries. We had a private one for the two of us as we climbed Wolf Creek Pass to Treasure Falls on our snowshoes near the Continental Divide in colorful Colorado on the 20th of February. That summer we had another with the whole family, when all seven of our grown children and their children could come. Our children organized the whole thing, and I was very impressed with what they did for us. They held it at the old "White Church" in Mapleton, Utah, where we had had our wedding reception 50 years before. It was a grand celebration and a most delightful time having our family and friends together again.

After graduating in physics from BYU in 1960, I accepted a job at the National Bureau of Standards in Boulder, Colorado, who would pay for my graduate work at the University of Colorado (CU) if I worked full time. My intent was to get a doctorate so that I could go back to BYU to teach and do research — wanting to stay in the mountain west.

However, in 1965, while I was working on my master's thesis and teaching a physical-science class at CU in addition to taking one course, I was asked to make a life changing decision. In the congregations of The Church of Jesus Christ of Latter-day Saints there is no paid ministry. A person can be asked to serve in about any position. At that time, I was asked to serve in a bishopric in the Boulder Ward (congregation) for the Church. I felt I could not do justice to family, work, school, and church, and asked for some time to think and pray about it.

It was then that I entered the most intense fast and prayer time of my life. In the fourth day I received an answer that it was His call and that it was in preparation for me to serve as a bishop, which came as a total surprise. At the time, I shared this sacred experience with no one but my wife. The following year I was asked to serve as bishop of the newly formed Boulder 2nd Ward. I was humbled but not surprised at the call, having been told by the Spirit ahead of time.

I knew at the time I was blessed by the Lord with help in writing my master's thesis.[3] The inspiration from the Lord along with help from outstanding colleagues allowed me to finish my thesis in a timely manner. It was published in February of the next year in a special issue of the *IEEE Proceedings* entitled Frequency Stability. There were several other excellent papers published in that same issue. That thesis is the basis for what the world now knows as the "Allan variance." (If you Google "Allan variance," you will get about 50,000 hits.) At that time no one had a clue where this would go, but the Lord knew, and I fully believe that He blessed me with that work because I was willing to do things His way and accept His will in my life. Typically, a master's thesis will collect dust in the library! My thesis is the most referenced publication to ever come out of our department. I give the Lord the credit for the inspiration and help in that work.

I was surprised to receive the following e-mail from Patrizia Tavella — a friend and colleague from Italy — that had been sent out to the time and frequency community throughout the world:

Dear Colleague and Friend

In 2016 the Allan variance will be 50 years old!

The first papers appeared in 1966 [Proc IEEE, vol. 54, n.2, Feb 1966] and since then we have learned many things using the two sample variance and also we have enlarged and improved its capacity.

The fields of applications have also enlarged by the contribution of many different colleagues.

---

3) *http://www.allanstime.com/AllanVariance/inspiration.htm*

About the Author

We would like to celebrate this anniversary by a Special Issue of the IEEE Transactions on UFFC which should appear in early 2016, just in time to say Happy Birthday to the Allan Variance.

Please have a look to the web announcement on http://www.ieee-uffc.org/publications/tr/special-issue-variance-50th.asp and consider the possibility to submit a paper to this special issue.

The final deadline for submission has been fixed for April 2015, we would be pleased to receive a short abstract and title from you within Oct 2014 to be able to correctly size the special issue process.

Many thanks for your consideration and looking forward to receiving your papers.

Judah and Patrizia
Patrizia Tavella
Istituto Nazionale Ricerca Metrologica INRIM

I have been asked to write a couple of papers for this special issue. Judah Levine — a friend and colleague at NIST in Boulder has been asked to give a paper this year on the Allan variance. He has asked me to help. I thank the Lord for this methodology being as useful as it has been, and it continues to grow.

Many have asked me to describe the Allan variance. The equation representing it and the meaning of the mathematics behind it are explained in Chapter 20. A good friend and colleague from Sweden, Magnus Danielson, has written a detailed description of both the Allan variance and the modified Allan variance on Wikipedia.[4] In addition, I have a brief explanation on our web site.[5] In Chapter 25, I pull some fascinating facets together tying the Allan variance to time as it relates to what is time, man's time, God's time, etc., which never occurred to me until Professor Riddle gave me his definition of time. Here I give a simple word explanation of the Allan variance as relates to clocks.

The Allan variance is a measure of the change in clock rate over different

---

[4] http://en.wikipedia.org/wiki/Allan_variance and http://en.wikipedia.org/wiki/Modified_Allan_variance
[5] http://www.allanstime.com/AllanVariance/

averaging times. Since everything changes, the Allan variance can be applied — and often is — in a much broader context. For example, it is used in assessing the stability of gyros. Since everything changes in time, if those changes can be quantified, then you could measure the Allan variance of those changes.

By the way, I did not give it its name; Don Halford — my Section Chief at the time of its development — did that. When giving lectures, I call it the "two-sample variance," which is mathematically a better descriptor.

After that experience of receiving direct guidance, I knew that the Lord can and will bless us in all that we do as we seek for light and Truth and to do His will. Further, it doesn't matter whether it is in religious activities or school or family or work or wherever. After that experience, I sought Him at every opportunity and that has made a great difference in my life. Also, I learned that if we seek not for self-aggrandizement but to serve and use our talents, as my wise BYU physics professor counseled me that makes a big difference as well. I believe that the Truths of God will harmonize in all areas of life if we will but try to understand the Truth from Him, who is the author of all Truth, and live in harmony with those Truths to the best of our abilities once we receive them. I believe God is the Master Physicist along with being the Master of the Universe.

I have had several other similar and significant experiences like the one I just shared over the course of my life, once I learned that very valuable lesson about God's desire to help the scientists also. That is to say, briefly, a problem was encountered, a solution was sought in prayer, and an answer came in due process after exercising faith, which includes lots of "works" with an emphasis on the works!

As another example, the idea for GPS common-view came as I was praying. It is a very high-precision time transfer technique that allows clocks around the globe to be compared at the few nanosecond (billionth of a second) level. It became the primary way of communicating atomic clock data to the International Bureau of Weights and Measures (BIPM) for the generation of official world time (UTC, universal time coordinated) — for more than two decades. It is still being used today. This technique moved the precision of international-time comparisons forward by about a factor of 100. I will share some other interesting and similar experiences in the body of the book.

About the Author

An interesting irony occurred during the '70s and '80s. I was using the Loran-C navigation system to transfer the times of the atomic clocks in Boulder, Colorado, to the BIH (International Bureau of the Hour) in Paris, where official time was then being generated for the world. The measurement-noise of Loran-C was so much worse than the instabilities in the atomic clocks that it created a real challenge. I then made the comment, if I could find another way to transfer the timing data for our clocks, I would never look at the Loran-C system again. Thankfully, the time-transfer technique we found, as we developed the GPS common-view technique, solved that problem.

Ironically, about nine years ago Oak Ridge National Laboratories asked me to "look at Loran... again," and to my great surprise, by some tricks (and further inspiration) we were able to get timing and positioning with an accuracy of about five meters, which is nearly as good as GPS. The government, "in their great wisdom," rejected the results we obtained — feeling that Loran was an albatross needing to die and that GPS needed no back-up, an opinion which was shown to be seriously not true as GPS was jammed during the Iraq war.

We had hard data to show our results using the transmissions from six Loran-C stations to validate our hypothesis. Significant government misdirection is another reason I am writing this book — to help return to those fundamentals given us by the founders of this, the greatest nation in history.

I am an experimental physicist. I love theory, but without the data to support a theory, one cannot know of the validity of a theory. It can be empty philosophy, and we see way too much of that. The scientific method has served us well since it was introduced by Galileo, William Gilbert, Francis Bacon, William Harvey, Descartes, Robert Hooke, Newton, Leibniz and others — coming out of the Dark Ages. Essentially all of the technology that we have today is a result. Clearly, this method has been a great blessing to us, in both the good and bad of it, but for the most part it has been our servant and has helped bless mankind in a major way.

It is interesting, as was pointed out to me by my good friend, Chauncey C. Riddle, who is Professor Emeritus of Philosophy, Brigham Young University, that the Industrial Revolution is really about advances in technology. The advances in technology have affected us more than those in science. The scientists would like us to believe that science is more important than technology, but it is technology that brings science into practice. Also, advances in technology may have benefited science more than advances in science have benefited technology — an interesting perception of Professor Riddle's.

Regardless, one of those enormous blessings of these advancements is the United States of America becoming the greatest nation in history, as founded upon the principles of liberty and free enterprise. Trusting in God, and being in compliance with Biblical ethics for living our lives in righteousness, have been backbone to America's becoming the greatest nation in history. "In God we trust" has been more than just on our coins. Up until the last half century, Americans have — for the most part — lived it, taking us through two horrible world wars.

Toward the end of his career, Albert Einstein worked on the Unified Field Theory (UFT) and did not finish that work. It is still "a puzzlement" as Yul Brynner said in *The King and I*. Several others have worked on it as well. It still remains "a puzzlement" to the world.

A little over a decade ago I was impressed to try to understand the UFT, and I approached it with the intention to know how God does things by use of all six senses. The resulting research efforts opened to us a fifth-dimension beyond the four dimensions that Einstein uses in his special and general theories of relativity. We call this fifth dimension the "Eternity-Domain." In this book, you will see that this fifth dimension has enormous implications as we discuss God's Time, Man's Time, Transcendental Time, and where we are in God's Time. We have done seven experiments thus far validating this new UFT approach. These will be shared in detail in Chapter 21.

The last half-century has seen science without God, and we have seen the consequences with 93% of our major scientists in America being atheists or agnostics. Darwinian evolutionary theory is pervasive throughout our education system and secularism is increasing. Moral decay is rampant. Integrity in the work place and in government has greatly decreased. The percentage of divorces continues to increase and more and more couples are living together without the marriage commitment. Same-sex relationships are now commonplace. There are more abortions in Washington D. C. than live births. Pornography, child abuse, and drug abuse are on the increase, all indicators of the degeneration of society.

About the Author

This century, with the inclusion of God in science, we will see His glorious Truths revealed — in fulfillment of the Lord's promises and revealing His tender mercies in the midst of the storms ahead. This harmony between science and religion will come because we re-introduce and include God in our efforts to receive additional light and Truth. The fruits of that harmony will be most delicious to the souls of all who partake.

Many worry about where society is going and about what the future holds. I once heard a wise man say, and I wrote it in my journal, "We may not know what the future holds, but we do know who holds the future, and to be in step with Him is the greatest opportunity of our existence."

The message of this book is to move out of worry, fear, anxiety, etc. and into the power of love, hope, faith, peace, and joy as we follow Him, who is the giver of these gifts. I believe in the Truths of God and that His promises are sure. Living in harmony with them is like an anchor and gives great stability to the souls of all those embracing them in the midst of the storms of life. We will show that you can move out of fear and worry and into faith for I know God is really in charge. He is filled with infinite love for each of us as His children. We will demonstrate with data that we can trust in Him regardless of life's challenges.

Over the course of my life, I have the habit of when I want to learn something, I go to the best experts in the world or read their books on that topic. I have met some incredible people in this process and it has been very fruitful. Hence, my references and resources will be broad-based — gleaning Truth from wherever it comes. My acknowledgements are also broad-based — having received great help from many colleagues, friends, and family, and from my sweet and supportive wife, Edna. But most importantly, I thank the Lord for His help. As my good Boulder neighbor and bishop years ago, Myron Crawford, used to say, "Two people can do anything when one is the Lord!" Without God, I am nothing. He is my All in all.

The last thing I would like to mention is a request of you, the reader, that you might get the most out of this book; be open to the Truth and be not afraid to change your paradigm as new Truth comes to you. The Truth will be logical and will be validated in your heart. Within my mortal limitations, the Truths I share will be Truths of God. There will be some surprises and differing viewpoints from anything you have known. Jesus said, "If you continue in my word, then are ye my disciples indeed; And ye shall know the Truth and the Truth shall make you free."[6] I believe this great promise applies to us as individuals (free from the weight of sin) and as a society (free from bondage of ignorance).

One more thing. It is my observation that two of the biggest things that get in the way of our learning new Truths are: 1) lack of desire to know what the Truth is, and 2) bigotry. As we study history, we see that bigotry is the root cause of most wars and conflicts. Bigotry is the opposite of "loving your neighbor as yourself."

For example, if you think, "He is a Mormon; I'll not listen to him," then bigotry is majorly getting in your way. I am, fundamentally, a Truth seeker, and am often bothered by the false traditions that I see creeping into society. I am also fundamentally a Christ-centered scientist as you will see for good reasons throughout the book, using reliable data to validate Truths in science as well as in religion.

If you are not a scientist and think, "I am not a scientist; it will be over my head," If you feel it is over your head, I apologize, I have tried my best to write to all people for the Lord loves all as should we. It is my experience that a good science teacher makes all the difference as to whether we like science or not. Let us be careful that bigotry is not a problem in this direction as well. I have seen science be fun and exciting for almost anyone that I have been able to talk to about the beautiful Truths we have learned from Him, who is the author of all Truth.

If you think, "He can't be a good scientist; he is too religious." Then again bigotry is a problem. If you are not a Christian, and you feel not to want to learn from a Christian, which I am, then bigotry is still a problem, because in my search for Truth, I have learned that God has revealed His Truths in different degrees to all of His children (Christian and non-Christian), and I have developed a love for essentially all religions.

How can the world learn to love one another if it cannot listen to and learn from one another with sincere hearts? My hope is that you be open to the Truth. I have made my best effort in sharing the Truth as I believe it, in most cases from reliable data validating these Truths.

---

6) John 8:31-32

INTRODUCTION

This book is for the scientist and the non-scientist, for the Christian, and the non-Christian; in other words, it is for everyone. Scientists like data (evidence); you will see some very interesting, yet not well-known, data. Non-scientists are typically right-brained — loving harmony, beauty, things aesthetic, etc.; I resonate with these feelings as well, and have catered to them in the writing of this book. My wife, Edna, has helped me a lot in this regard; she sees things in nature and beauty that I find many people miss. To me, the true Christian is motivated by love, and that is one of the central messages and themes herein. In Chapter 4, using logic, I share from the Bible a new and very important definition of obedience as it relates to love that is not taught in any denomination of which I am aware, including my own, and I have studied all the major ones.

While working with the Atomic Clock in Boulder, Colorado, for 32 years, I had the opportunity to visit colleagues in other timing centers in many places around the world. I always took the opportunity while there to learn about other non-Christian religions. I found so much of good in them and many common teachings to Christianity. A typical problem the atheist has is all the terrible things that have been done historically in the name of Christianity. For me, those un-Christian like acts are a satanic distortion. Somehow the atheists seem to ignore the multitudinous millions killed under the atheists: Stalin, Mao, and Hitler, which, if you count, were a great number more deaths than all the so-called Christian wars. You will also find some uniqueness as I tie all of the chapters together into the title: *It's About Time.*

Time. What is time? Can you touch it, taste it, feel it, smell it, or hear it? Einstein is reported as saying, "If you ask me what time it is, I will tell you. If you ask me, what is time? I cannot." I cannot find this actual quote from Albert, but he should have said it! Did God give us time as part of our mortal-earthly existence? As I will show you scientifically, time as we have it in society is a mortal construct. Some cultures have no clocks, and God operates outside of time in the Eternity-Domain, as we will demonstrate experimentally in Chapter 21.

It is my belief that we live in the most exciting "time" in the history of the planet as we witness the "signs of the times" and the imminence of the coming of our Lord and Savior, Jesus Christ. You cannot extract time from the universe. It is an abstraction. Scientists like to think of it flowing throughout the universe. They often write their equations with time (t) as the independent variable. Does it flow so? Since you cannot touch, taste, feel, smell, or hear time, why do many say they feel time is going faster? Time as we use it is an artifact of man's devices. I should know; I helped develop the timekeeping system for the USA and for the world. Clock rate or frequency can be extracted from physical phenomena; i.e. a pendulum clock has its period determined by the length of the pendulum and the local gravitational field. With the application of some insights from a new Unified Field Theory, we can change gravity and hence change the period of a pendulum, which means gravity is not as we think it is. From these experiments, we were able to deduce a fifth dimension, the "Eternity-Domain." Knowing what we know now, Newton's apple could have been made to fall up! Time, gravity, and space (as we view it) are constraints of our mortality. God and angels are not constrained by them; neither need we be, as we open our minds to the Truth's of God, His Time and the Eternity-Domain.

Recently, some fascinating experiments have been done in Turin, Italy, showing how time emerges from entanglement at the quantum mechanical level. I have some interesting inferences from these experiments that I will touch on in Chapter 21. It will be interesting to see where these experiments lead.[1]

In this book, we will address some fun science and some real life issues while we discuss the time of our lives. We are excited to share some answers to life's greatest questions — including, why we are here, where we came from, where we are going, and where are we in time in the end-time scenario? I believe you will find the answers useful. My delight will be if this book helps you feel closer to God.

Victor Hugo proclaimed, "More powerful than all the armies of the world, is an idea whose time has come." It is time to understand time, and "it's about time" to come to that understanding. Indeed, I believe, that as we come to that understanding, it will be more powerful than "all the armies of the world," because it will be the Lord's way of doing His work — to bring to pass the immortality and eternal life of His precious sons and daughters. We will better

---

1 *https://medium.com/the-physics-arxiv-blog/d5d3dc850933*

Introduction

understand His work and His purposes for our mortal journey. We will see that as we use "our time" to align our thoughts and deeds with His "perfect plan of happiness," the peace, love, joy, and security we desire will come more and more in the midst of a world gone awry. This understanding will come as we build chapter upon chapter ending up with an understanding of God's Time in its grand majesty and His "Wheel of Time." Chapter 25 ties it all together in the title, *It's About Time* — what is it, really? We are immersed in God's Time, and His "wheel of time is motivated" by His love.

Specifically, Chapter 1 advances the scientific method by opening it up to our sixth sense based on some fascinating data to substantiate this approach. Then using this advanced scientific method, we will be assisted in answering some of life's most challenging questions: "Why is there so much of pain and suffering in the world? Where did we come from? Why are we here, and where are we going?" The information in Chapters 1 and 2 form a basis for us to examine several false traditions that have crept into modern society, which we address in the chapters in Section II. I believe most of these false traditions have crept in due to selective data collection supporting a certain thesis while ignoring contradictory facts. In Section II we share as best we can a variety of information and solutions useful in countering the devastating effects that have resulted from these false traditions. This information shared shows why we have so much disease now (of body, mind, and spirit), and how we can move out of disease into a state of health (of body, mind, and spirit.) In Section III, Chapter 20, we explain how man's time works in laymen's terms with the science behind it. Chapter 20 also explains how GPS works using precise time with some ideas on how it could be improved by as much as nearly a factor of 100. And you will learn how the four dimensions of Einstein's theories of relativity are tied into the workings of GPS. In Chapter 21 we will share our studies of a new Unified Field Theory (UFT), which include experimental evidence of a fifth dimension. We call it the Eternity-Domain as it ties to God's timing. While Chapter 20 explains physical time, and Chapter 21 God's Time. Chapter 22 describes spiritual time, which I call transcendental time. It ties directly to our prayer channel to God, and we discuss ideas on how to most effectively open that communication channel. Going on, Chapter 23 then proceeds to share where we are in God's timetable and shows that we live in the most exciting time in the history of the Earth. In Chapter 24 I share my thoughts on the great importance of God's own "Wheel of Time" and the unique concept of "One Eternal Round." It will be here where I describe why we are where we are in our galaxy and where we are going. This is the most exciting space trip ever. And Chapter 25 recaps what we have learned in the book and addresses the very difficult question, "What is time?" The implications and answers are far reaching and exciting.

Regarding a people without "time' as we typically think of it, science correspondent Richard Alleyne wrote on 20 May 2011 the following:

> The Amondawa people who live deep in the Amazonian rainforests of Brazil have no watches or calendars and live their lives to the patterns of day and night and the rainy and dry seasons.
>
> They also have no age, and mark the transition from childhood to adulthood to old age by changing their name. [I suppose the changing of their names could be called a clock.]
>
> The team of researchers, led by University of Portsmouth, said that it is the first time they have been able to prove time is not a deeply entrenched universal human concept, as previously thought.
>
> Professor Chris Sinha said: "We can now say without doubt that there is at least one language and culture which does not have a concept of time as something that can be measured, counted or talked about in the abstract." [Except that they changed their names with time! Their pendulum is just swinging at a much slower rate than our fast-paced civilization. Maybe these people are closer to the truth than we currently appreciate for God gives us a new name as we draw closer to Him.]
>
> And the Gentiles shall see thy righteousness, and all kings thy glory: and thou shalt be called by a **new name**, which the mouth of the Lord shall name.[2]

In this book, I will demonstrate with reliable data that there is a spirit in each of us that came from a pre-mortal sphere — outside of our time — and how that spirit is the "real you" housed in this mortal body. By

---

2 Isaiah 62:2

Introduction

extracting information from some of the 13 million documented near-death experiences (NDEs), we now have irrefutable data showing that our spirit continues on after death into a spirit world retaining the same character we have at the time of death. We will further see that that spirit couples to our mortal body and is able to couple to what we call the "Eternity-Domain," which is the sphere where God and His angels do their work, the work of encouraging us on the path to immortality and eternal life for a fullness of joy, regardless of our religious inclinations or affiliations. In a simple sense, we may call this coupling our conscience, as we move forward in time in our mortal journey.

What we perceive as "time" is limited by our mortal perspective. Man's time is our way of describing how time flows in our physical world; it is our limited physical perception. We may think of this as physical time.

At a more profound level, experimental evidence gives us direct evidence for what we have named "diallel-field lines," which couple us to God and all of His creations. These diallel-field lines couple us to Spiritual Time. Our emotions and the feelings of our hearts as well as our minds couple through diallel-field lines to God, to His angels, to each other, and to the universe, independent of proximity. I call this spiritual-time *transcendental-time*.

We can think of transcendental time as connecting us to God as He does His work of encouraging us to come back to Him, which is the path of joy in this life and a fullness of joy in the life to come. This transcendental time is the spiritual and emotional dimension of our beings. It allows us to couple into God's Time and His perfect plan of happiness, which is often hard for us to see in this very imperfect world. One of the main purposes of this book is to show how God's plan can bring about perfection and a fullness of joy out of all the imperfections we see in this world. It's About Time to do that!

This coupling between our physical sphere of existence and God's Time utilizes our intelligences, the feelings of our hearts and of our "guts." It is interesting that recently science has discovered brain cells' neurons in the heart and in the bowels.[3]

In the scriptures, it has been there all along. Perhaps you have heard the

---

3 *http://projectavalon.net/forum/showthread.php?t=1623 http://www.therealessentials.com/followyourheart.html*

touching phrases, "Listen to your heart," or "I have a gut feeling that this is not right!" Perhaps you have heard the following scriptural phrases: "…his bowels did yearn upon his brother …" "Let your bowels be full of charity toward all mankind…"; "The Lord looketh upon the heart…"; "Thoughts of thy heart…"; "Blessed are the pure in heart, for they shall see God." and "God is a discerner… of the thoughts and intents of the heart."

Pulling this all together in a book has been very rewarding for me. My overall goal in doing it is to reach out in love to help as many as possible to know of God's love for each of us and that we are His children of infinite worth with huge potential and destiny.

Let's take a trip through time, in time, around time, in which we will join time and eternity into a most meaningful relationship. It's about time, and I use the plural "we" and "us" because of the help I have received from family, friends, colleagues, and from on high. Thanks for joining us on this exciting journey.

For your information — and you will see why in the text — I use the King James edition of the Bible for most of the scriptures quoted.

DISCLAIMER AND PERSPECTIVE

I had to make the decision early on to make the contents of this book available as quick as I reasonably could or to wait until it was edited in significant detail. I chose the former because there are new truths highly relevant to our times, and I felt you would desire to have them as soon as possible. I am fundamentally a truth seeker, and I love to share as it seems appropriate to do so. There are new and exciting truths here that I believe need to be shared now. I know there are repetitions, but that often brings conviction. Going forward, I thank you in advance for your patience and forgiveness, and if you would like to share suggestions or comments, please go to our book's supporting web site and share thoughts and ideas; you can help it be a better book: *www.ItsAboutTimeBook.com* . I am excited to share, because the truths herein bring a harmony between science and religion that is so critically needed for our day, and It's About Time.

Section I

Expanded Scientific Method Brings Harmony with Religion and Answers Many of Life's Most Challenging Questions

Ghandi's life demonstrates how the simple power of Truth brings about a great good.

Chapter 1

It's About Time to Put God Back into the Scientific Method.

I Love the Truth, no matter where it comes from. And it is so exciting to live in a day when so many Truths are coming forth both in science and in religion. The beautiful dimension of it all is that Truth harmonizes with Truth, and in this century we are seeing science and religion coming into a grand harmony. It is as if God is opening the windows of heaven to understand His perfect and Divine Plan in the midst of all our challenges and imperfections in this world. In this process, we are seeing a great division as the righteous draw closer to God and as the wicked do more of wickedness. As Light and Truth are coming forth, we see a way of integrating God's ways into the scientific method, which is a major breakthrough, indeed, an exciting time.

I feel that another title for this book could be *Standing on the Shoulders of Giants* because we have learned so much from so many great people — either through their books or their talks, or in personal sharings as you can see from all the people included in the index. My sweetheart, who is also my wife, and I have gotten to know some awesome folks and feel so blessed to have learned so much from so many. Most especially we feel blessed from learning so much from on high, both from the scriptures and to our hearts and minds as we have asked God for His guidance.

This is not an ordinary book flowing in some elegant, enchanting, and heart grabbing sequence. Rather, It's About Time to capture the essence of the golden nuggets I have learned from so many wonderful and knowledgable people and especially from God. The 25 chapters build on each other while It's About Time to share a large number of topics in the spirit of Love. I believe these are very valuable in our lives, and we want to share them with you that you may benefit as well. I mentioned in my forward the reversal of the phrase: "Live and learn" into "Learn and Live" the abundant life God has promised,[1] and His promises are sure.

While Chapter 1 increases the richness of the scientific method, Chapter 2 utilizes that richness to deal with some of life's hardest questions: Why is there so much of evil and suffering in the world? Why are we here on this Earth? Where did we come from? And where are we going? We use our expanded scientific method to answer those difficult questions to a level that brings a measure of peace to the soul and real purpose to life. Section II then takes the basis provided by the first two chapters to cover a large number of false traditions that have crept into society and which have kept us from progressing as we could and should to higher levels of Light and Truth — both in science and religion and dealing with our desire to help you have a healthy and vital body, mind, and spirit filled with a knowledge of God's Love.

In the final section (Section III), I get to share in Chapter 20 exciting activities of my career in working with atomic clocks and GPS. You will see why atomic clocks have gotten a billion times better in my life time and how we may be able to make GPS even much better than it is. GPS includes the four dimensions of Einstein's special and general theories of relativity, and how that works is explained in one of the key references to Chapter 20.

In Chapter 21 we share data that gives evidence for a fifth dimension, and how the overtones of this are enormous. We call this fifth dimension the Eternity-Domain and share a unified-field theory around it. We have done seven experiments to date, all of which support this new UFT. The fascinating aspect of this new UFT is that it ties to the questions answered in Chapter 2 — regarding where we came from and where we are going and validating our relationship to God and what we can do to both enjoy His blessings and to please Him.

---

1) John 10:10

Chapter 1

We are told to experiment on His word. I use our extended scientific method and appropriate data to go along with this richness to share some remarkable results from this experimentation. A lot of the information shared herein will be new to the world at large and some has never been shared before publicly.

1.1 Science in the Past Century Has Been Without God and With Theories Contrary to God.

If I make a basic assumption that is wrong — no data to support it — and build a theory upon it, then that theory can potentially lead me down all kinds of strange paths. Those strange paths may well try to lead me away from God and a belief in His word. One very basic assumption of science is that the way things are now, is the way they were, but gradually evolving over millions and even billions of years. The Bible, in contrast, is a book of cataclysmic events contrary to this basic assumption of science and to the traditional theory of evolution. Hence, most scientists do not believe in the cataclysmic events of the Bible: in the cataclysmic creation events, in Noah's flood, "in [the] days [of Peleg] was the Earth divided," in Moses parting the Red Sea, in the cataclysmic destruction of Sodom and Gomorrah or of Babylon or Nineveh, in the cataclysmic events associated with the atonement of Christ, and in the events that are coming as the Earth is prepared for the glorious reign of Christ during the millennium, which will indeed be cataclysmic.

It is exciting that under the new science paradigm outlined below, we have direct evidence (data) that all these cataclysmic events of God are true. These data will enable us to show that the above basic assumption of science is wrong, and that indeed the theory associated with it has led millions of people down strange paths and away from God.

Because of the importance of the concepts in this book, I have sought out experts in the different fields to provide credibility to the new and refreshing paradigms shared herein, clearly bringing science and religion into harmony.

I learned the following very important concept from my good friend, Paul Rimmasch, author of the book, *The Lost Stones*. Paul is an expert forensic scientist and crime scene investigator. There are two different ways for a scientist to ascertain the Truth of something: deduction and induction.

Deduction is a top-down approach: one has an idea, a theory, or a hypothesis and then from experiments one observes whether the data fit or not. There is a danger with the deductive approach. If one has a pet theory, then there may be a tendency to pick the data so that they fit the theory. We see that integrity here is extremely important and to not be prone to cater to our pet ideas. If we know God's word is true and trust in it — and we will give you very good evidence that you can trust in His word — then this knowledge helps in the deduction approach. The deduction approach will be used where we know or strongly believe we can trust the theory or hypothesis.

The Savior is suggesting the deductive approach to us in John: "If anyone chooses to do God's will, he will find out... my teaching comes from God." "If ye continue in my word,... ye shall know the Truth, and the Truth shall make you free... If the Son therefore shall make you free, ye shall be free indeed," respectively.[2] This concept is true of individuals and of nations.

Induction, on the other hand, is a bottom-up approach. One observes all of the relevant data and then after patterns, trends, or reasonable models seem to fit all of the data, then an idea modeling the data, a hypothesis or theory that seems appropriate is developed that fits all of the data. We can then look at the implications of this theory or hypothesis. This process often allows us to refine the theory or hypothesis.

Induction is the approach the Savior suggests when he tells us how to know false prophets: "You will know them by their fruits... every good tree bears good fruit, but the bad tree bears bad fruit. A good tree cannot produce bad fruit, nor can a bad tree produce good fruit."[3] In this remarkable scripture we have both a necessary and a sufficient condition to know if a prophet and his teachings are true or false.

The induction approach is also very powerful in ascertaining Truth. It will be used throughout the book and at the end of this section to prove a fundamental and important point.

Deduction is the main approach in science today with the theory of evolution and the big-bang theory being very important cases in point. In forensics, they use induction to avoid the danger of unfounded biases coming from the high level of emotions involved with those associated with a crime scene.

---

2) John 7:17 (NIV) and 8:31-36 (KJB)
3) Matthew 7:15-20 (NAS)

Chapter 1

In the following important examples of theories contrary to God, consider that:

- The Darwinian theory of evolution based on "survival of the fittest" is based on the concept of a chance event causing the beginning of life. It has led many to atheism and has almost totally permeated the world's education system. "I'll be a monkey's uncle!" We will see some remarkable evidence to the contrary, showing direct evidence of God's existence and His Love for us. Evolution within species is well documented, and we see this as God's design for each species to adapt to its environment. We call this the law of evolution because it has solid data to support it. In contrast, there is no solid data to show evolution across species, which is the heart of the Darwinian Theory. It is an empty theory. In recent time, more and more data have come forth — as we will share — showing the emptiness of the Darwinian Theory. This is leading to the demise of the Darwinian Theory in this century and is greatly helping to bring science and religion into harmony.

- Einstein's theories and others have led to the " Big Bang theory," (some have called it the "Big Bank!"), "dark matter," and "dark energy," which have as an underlying philosophy to figure out how the universe came into being without God. Stephen Hawking said, "We don't need God." Hawking was the 17th holder of the famous " Newton's or Lucasian chair," which he held for 30 years. Sir Michael James Lighthill, the previous holder of the chair, said of Hawking's cosmology, "To me, it is not quite science, but more like creation myth." The " Electric Universe" community now has a large amount of data showing that both of these theories are not correct. You can go to their web site and listen to many scientific presentations along these lines. My youngest son, Nathan, and I have gone to their last three annual conferences, and their presentations have been exciting and cutting edge. Fortunately, they make all the talks available on-line. As a profound example, the talk last year (2013) by Rupert Sheldrake *Science Set Free* received a standing ovation. He has a book by the same title. He shared the 10 dogmas of modern science and showed flaws in every one of them.

As evidence of how science of the 20th century has had a great tendency to move people away from God, a 1998 survey taken by the National Academy of

Science (NAS), showed 93% of the "*greater*" scientists "expressing disbelief or doubt in the existence of God."

The people of the world, in their pride, think they are the enlightened and are part of a greatly civilized age. This is an egocentric paradigm. Those who know history know of civilizations that could do things and had technologies that we cannot replicate today and they had God in the center of their science and technology. My life has been enormously blessed because I have included God in my scientific endeavors. I know that without Him I can do nothing; He sustains my every breath. In Him we move, and live and have our being. Humility — the opposite of pride — is the doorway to God's Truths in science and religion. Gratitude for His uncountable blessings in our lives is a key to the door of revelation, wherein the mysteries of godliness are manifest. This attitude of gratitude is most helpful to the scientist as well.

Abraham is an inspiring example of a person, who knew God, and learned more of the heavens than we know today, because he was taught directly by God. In his book *The Blessings of Abraham*, E. Douglas Clark has some 2,000 citations from Islamic and Rabbinical traditions and from recently discovered apocryphal texts giving the detailed life of this remarkable prophet, and showing that his knowledge of the heavens was vast. His life was one of kindness, love, and devotion, and in humility he served those around him. He was promised by God that he would be the father of many nations. Douglas reports in his book that some current DNA studies indicate that some 95% of the people of the world are descendants of Abraham, showing that God's promises are sure; all nations have been blessed by the offspring of this remarkable prophet. We will see much more of those promises made to Abraham being fulfilled in the days ahead, and in those will be great blessings to the entire world.

In 1835, Joseph Smith translated the Book of Abraham from ancient Egyptian papyri. Using the deductive method for ascertaining the Truth of this record, if we assume he is a prophet of God and performed the translation with Divine help, then we look at the data found since authenticating the record — especially that found by Hugh Nibley and in Douglas Clark's book — then we have a confirmation of his prophetic stature. If we use the inductive method just using the information in the some 2,000 references in the bibliography of Clark's book, and then build a model based upon that information, we would find it consistent with Joseph Smith's translation, again validating Joseph Smith as a prophet of God. In other words, we could almost re-write the Book of Abraham from external sources as it was given to Joseph Smith.

1.2 A New Look at the Scientific Method

Let us look at the Scientific Method and build upon it:

- Oxford English Dictionary "a method or procedure that has characterized natural science since the 17th century, consisting in systematic observation, measurement, and experiment, and the formulation, testing, and modification of hypotheses." The great value of the scientific method as it came forth starting in the 1600s was that it was based on DATA (experimentation)! We saw great progress and then most of the scientists believed in God. In the last century, we have seen a reversal in that most major scientists do not believe in God, and there are blatant examples of where scientists are ignoring the data to preserve their pet theories. Darwinian Theory of evolution — natural selection and "survival of the fittest" — is a major case in point. We have excellent data showing evolution within a species, but no solid data showing evolution across species, which is absolutely necessary for Darwinian evolution to have validity. Agassiz, who was considered the greatest natural scientist at the time of Darwin, told him that he had extrapolated too far. Agassiz knew that Darwin did not have data to support his cross-species evolutionary concepts.

- Reasoning along with the five senses is used to ascertain scientific Truth with a lower case "t," ever evolving and changing over time as new Truths are discovered. In contrast, the Truths of God are absolute in that they don't change over time. We will demonstrate from data that He can be totally and completely trusted. This I know to be true. If we make sure that the Truths of science align with the Truths of God, then we are guaranteed harmony between science and religion. In concept, we could develop in that process a Bible of scientific Truth, as it were, which would be in perfect harmony with reality. This is an extremely important concept and principle that was utilized by the early scientists who got the scientific method rolling — then later on, especially with Isaac Newton and Louis Agassiz.

Again, let us remember that Newton is recognized as the greatest scientist who ever lived and Agassiz as the greatest natural scientist of his day. Both were totally devout believers; God and the Bible were their standard of reference. However, we have the significant challenge to know what the absolute Truths of God are. There are fallibilities in all the written texts of scriptures we have for a variety of reasons. The Apostle Peter gives us a clue on how to deal with this problem.[4] The scriptures were given by the Spirit, and the only way we can really know what God was saying to that prophet or prophetess is by the Spirit. Even then, we have a hard time appreciating the cultural and societal contexts for those words given by God. As we think about this dilemma, it points to the great importance for each of us having a personal relationship with God. He can then cater the interpretation He gives us — in an infinitely-loving way — to our needs and to our paradigm to best help us on our path back to Him. We will treat this important topic much more in Chapter 22.

The feeling that has grown over the last century that science and religion cannot be made to harmonize is exactly what Satan wants us to believe, and some say that a creationist cannot be a scientist. Those with that thinking seem to believe that they are greater scientists than Isaac Newton, who was also a devout creationist, and who is recognized, as we have said, to be the greatest scientist to ever live. We will share many examples showing that science and religion can harmonize.[5] Most do not know that Newton spent more time in the Bible than doing science as documented in his personal writings.

Louis Agassiz (1807-1873) said, "It is the job of prophets and scientists alike to proclaim the glories of God." He was greatly respected by his contemporary, Charles Darwin. In fact, Darwin shared his feelings with the poet, Henry Wadsworth Longfellow, "What a set of men you have at Harvard! Both our universities put together cannot furnish the like. Why, there is Agassiz — he counts for three."[6] All of Agassiz's peers knew he was a giant intellectually and spiritually, and he had proven Darwin's concept of natural selection invalid at the time based on fossil data, showing that there was direct evidence of

---

4) (2 Peter 1:19-21)
5) You may find the following link of interest regarding Sir Isaac Newton: [http://rsc.byu.edu/archived/converging-paths-Truth/brief-survey-sir-isaac-newtons-views-religion]
6) [Anderson, *The Other Eminent Men of Wilford Woodruff*]

intelligent design behind the "explosion" of the several different species coming forth, which could not have evolved. Agassiz's goal was to understand the mind of the Divine.

An atheist has no evidence that there is no God. One cannot prove the non-existence of anything. If there is anything, then it exists. If it does not exist, no one knows it, and therefore no proof can be had. So atheism is an empty theory with no data to support it. As a consequence we see atheists use tactics of name calling, of yelling louder, and of accusing theists of being intellectually dishonest, when, in fact, they are guilty of the same.

Morals have no meaning in a "survival of the fittest" mentality. So lying, cheating, and immoral conduct are teachings common to their lot. The resulting impact on the moral fiber of the masses is tragic. We see the world spiraling down as a result. Richard Dawkins has sold millions of his book, *The God Delusion*. It is dishonest for him and all the atheists to ignore the great work coming forth solidly documenting intelligent design in the DNA. The prophecies tell us that this downward spiral in morality will be turned around as a people look to the Truths of God and harmonize our lives with them. Science then will fall naturally into place.

The following gives interesting evidence and demonstrates that God is there for us and He can be trusted. For those who have eyes to see and hearts to feel, He is in the details of our lives.

The King James edition of the Bible (KJB) is admitted by experts — whether they believe in God or not — to be among the most beautiful literature on the planet. Handel's *Messiah* has been performed more than any other music and it is based almost totally on the King James edition. There are probably more copies of the Bible around the globe than any other book. Its translation into English from ancient languages is both inspiring and miraculous. Just the story of William Tyndale's dedication in bringing it into English for those who know that remarkable story of devotion and dedication touches one to the core. He was fluent in eight languages and totally focused his energies in making the Bible available to the English people and died at the stake at age 41 for his efforts. About 90 % of the King-James edition New Testament is Tyndale's translation.

Here is a sample example of the divinity of the Bible, and there are thousands of such samplings.

If you ask the question, what is the shortest chapter in the KJB, it is Psalm 117; it has two verses. If you now ask what is the longest chapter in the KJB, it is Psalm 119; it has eight verses for each of the 22 letters of the Hebrew alphabet (176 verses covering over six pages), and in the Hebrew Bible, each of the eight verses starts with that letter in the Hebrew alphabet — sequencing through all 22. Next if one asks, "What is the center chapter of the KJB?", it is the chapter between these two: Psalm 118.

Now we know the main message of the Bible is trust in the Lord. Lo and behold, Psalm 118:8 proclaims this message, "It is better to trust in the Lord than to put confidence in man." So we see that the center chapter of the Bible focuses remarkably on the center message of the Bible.

If that were not evidence enough, the Lord gives us one more, which makes it irrefutable. If you ask, "What is the center verse in the KJB?" It is the same chapter and the same verse (verse 8) as the center message of the Bible. So we have Psalm 118:8 being the center chapter and the center verse containing the center message of the KJB.

If that doesn't surprise you enough, if you divide 1188 by 2, you get 594. Guess what? There are 594 chapters before before Psalm 118 and 594 chapters after Psalm 118. Could this have happened by accident? Impossible! As we study and think about how this came to be, we, indeed, see the Lord inspiring the translators and compilers of the Bible, as well as being in the details of our lives, manifesting His ever caring Love for us and wanting us to have a witness of His message that will lead us safely home. I see Him in the details of my life every day.

As a devout Christian and a scientist, I fully agree with Agassiz that it is the responsibility of prophets and scientists alike to proclaim the glories of God, and I see the evidence of God's Love in every direction I turn. Agassiz knew the God of the Universe and had direct answers to prayer in his scientific investigations. One time he had the impression that there was a certain fossil that he was looking for in a rock he had found. He had a dream showing him the fossil inside. He chipped away the rock and found exactly what he had seen in his dream.[7]

---

7) Anderson, *The Other Eminent Men of Wilford Woodruff*

Another example of the harmony between science and religion can be found in the 4 ½-square-mile Earth works in Ohio, carbon-dated about 200 AD. It maps out the plan and purpose of our mortal journey. Archeologists can tell that it was a time of great peace because there were no battlements anywhere in proximity. This extremely-sophisticated series of Earth works also maps out the eight nodal phase points of the moon's rising and setting positions over its 18.6-year repeat period. It is considered as inspiring by experts as putting a man on the moon.[8]

1.3 Scientific Deduction and Induction and Its Relationship to History

Over the entrance of the Norlin Library at the University of Colorado is a famous quote by George Norlin, "Who knows only his own generation remains always a child." Those who do not know history's mistakes are doomed to repeat them, and when you stumble on your own mistakes it hurts, more than just your toes! In the Bible, most of the last chapters of Deuteronomy (28-33) are devoted to the blessings or cursings that would come to the children of Israel depending upon whether they lived the commandments of God or not. The history recorded in the Bible and documented from secular historic writings show that the words of Moses were literally fulfilled. The Assyrian captivity of Israel around 723-721 B.C., the Babylonian captivity of the Kingdom of Judah culminating in the destruction of the temple in Jerusalem around 586 B.C., and the destruction of Jerusalem (70 A.D.) are vivid fulfillments as they turned from God, and are a direct attestation of the validity of these prophecies of Moses as given by God. When they turned to Him and lived the commandments, they were blessed. Those who know history, and have eyes to see and hearts to feel, know this very important historic fact that as we keep God's commandments we are blessed; otherwise, we have no promise; and the consequences of disobedience are vividly apparent, as those who turn from God have not His promised blessings in their lives.

The Lord often uses ironies to show His hand in our lives. The King of kings was born in the most humble of circumstances and had no home to call his own. Jerusalem means the city of peace and is the most fought-over city in the world, but finally it will become what the Lord named it, as a witness to all

---

8) See Steve Smoot's DVD *Lost Civilizations of North America*, and http://www.allansTime.com/Spiritual/Hopewell_Civilization_Great_Octagon/

people and nations that His work will not be frustrated and His divine purposes will be fulfilled. We will see this yet in the years ahead.

Many say, "Look at the horrible atrocities committed, especially during the Dark Ages — in the name of Christianity!" I simply say, "That is a false Christianity." If you want a good look at the original Christian movement from a source outside the Bible, read the book, *Will the Real Heretics Please Stand Up* by David W. Bercot. He spent a good part of his life researching the writings of the early Christians that are not in the Bible over the first two to three hundred years of the original Christian Church. His book is very enlightening, and later I will use it as resource in addressing the question, "Are we a Christian Nation?"

Now, as a nation, we are turning from Him, and the nation is spiraling down, losing its God-given Liberty. It is critical that we learn this "Timeless" lesson from history, that as we do His will, we will be blessed — individually and as a nation. If we do not, we lose that promise. It is our choice; He will force no man to heaven. To do so would make it not heaven. In humility — coming to know our total dependence upon Him — and by prayer — as we ask, seek, and knock — we can find and know His will in our personal lives and for those we are privileged to serve.

In the inspired wisdom of the founding fathers of America, the Declaration of Independence and the Constitution avoid the establishment of a state church, even though the founders were essentially all Christian. These inspired documents were configured so that under the principles of liberty, any of America's citizens could find that God who would set them free from personal sins as well as to keep them free as a nation if we would but keep His commandments like He told Moses.

In the context of this extremely important historical lesson, with a desire to bring God into science, let us see what we can do with the scientific method that will bring harmony with God's science.

Let us use the deductive approach as we compare our information with the absolute Truths of God and the inductive approach as we observe data generally. We will see that this process gives us a broader breath and an additional richness to the scientific method that will bring harmony between science and religion.

Chapter 1

1.4 Absolute Truths of God

What are the absolute Truths that we know from God? As we have said earlier, the Truths of God are absolute and don't change with time. The Truths discovered by science commonly evolve and change over time, because we are not given the Truths of God's science in a science Bible. Even the Truths of the Bible and other sacred books are filtered through the particular paradigm of the prophet or prophetess who wrote them as well as the cultural setting at the time. In addition, we have the problem of accuracy of translation into languages other than the original.

The Holy Ghost, the Spirit of Truth, can reveal these absolute Truths to us. I feel the Holy Ghost has revealed the following glorious Truths to my heart, mind, and soul. God, lives and He Loves us. As it says in Genesis, He created man (male and female) in His image. We see the God of the Bible has a female part as well. So, perhaps, a more accurate way to say it is, "He/She Love us!" As the Apostle Paul shares, our spirits are their " offspring."[9] How exciting; our spirits are the offspring of Heavenly Parents, who Love us with an infinite Love. I use in the book the generic and traditional "He, Him, and His" meaning both male and female parts of God. Christ has taught us that God likes to be called our Father, as in the Lord's Prayer[10]. Ancient texts talk of a Mother in Heaven.[11] Heavenly Father shares little about Her; the sense is that He knew how His name would be taken in vain, and He wanted to protect Her from that abusive language. Each of us is precious to God. Their greatest desire is to have us come back and live with Them in Celestial-Heavenly realms. To this end, God gave us the Creation, the Fall of Adam and Eve, and the Atonement of Christ. " Adam fell that man might be, and men are that they might have joy."[12] Because of Father's Love for us, He gave His Only Begotten Son[13] to show us the way and to work out the perfect atonement. He is the way, the Truth and the life, and no one comes to Father but by Him.[14] The Savior worked out the infinite atonement in a seven-step process to bring perfection to Father's glorious plan of happiness, showing His infinite Love of the Father and of us.[15]

---

9) Acts 17:29; Hebrews 12:9
10) Mathew 6:9-13
11) *https://www.google.com/search?q= Mother+in+Heaven+ancient+texts*
12) 2 Nephi 2:25
13) John 3:16
14) John 14:6
15) *http://ItsAboutTimeBook.com/spiritual-science-seven-steps-in-atonement/*

Free choice is also essential to the plan, and everyone will have an opportunity to hear His glorious gospel message, either in this life or in the spirit world beyond the grave.[16] For this cause was the gospel preached also to them that are dead that they might be judged according to men in the flesh, but live according to God in the spirit.[17]

The Messiah came to redeem the children of men from the fall. And because they are redeemed from the fall they are free forever, knowing good from evil. . . Wherefore, men are free according to the flesh; and all things are given them which are expedient unto man. And they are free to choose liberty and eternal life, through the great Mediator of all men, or to choose captivity and death, according to the captivity and power of the devil. . .[18]

Opposition is part of the plan, and Earth's challenges have a Divine design, as will be explained in detail in the next chapter. Viewed through celestial eyes, all of these Earthly experiences can be for our good, and as we ask, we can learn important lessons from each. God's Time is designed so your "phone line" to Him is always open; it is never busy. He always has "Time" for each of us and enjoys communicating with each of His precious sons and daughters. He knows you by name. Isn't it exciting that you never get a busy signal when you call on the Lord. Heartfelt-meaningful prayers always get through, and He wants it to be two way. Let us always have our minds and hearts attuned to the " still small voice." He tells us that we can always trust in Him and to not trust in man; they are all imperfect.

Then we have the capstone promise of the Father. Since our spirits are His offspring, we are gods in embryo, and through the atonement we can become like Him and follow the instruction of our Savior to be perfect even as our Father-in- heaven is perfect.[19] We cannot perfect ourselves, but Christ can as we come unto Him.[20] Each person is unique, with their own set of strengths and weaknesses to overcome through Christ; and each of us has a unique mission to fulfill in helping to bring the kingdom of God on earth as it is in Heaven, as He asks us to do in the Lord's prayer. The final offering is fullness of joy and to receive all that the Father has as we are perfected through Christ's

---

16) 1 Peter 3:18-20
17) 1 Peter 4:6
18) 2 Nephi 2:27
19) Matthew 5:48
20) Moroni 10:32-33

Chapter 1

infinite atonement as we come unto Him and this because of His infinite Love for each of us. How could we imagine a better plan?

1.5 Application of this New Scientific Approach

Using the above fresh approach to the scientific method, I have gathered a significant amount of evidence to know that our spirits existed in a pre-mortal sphere as the Lord told Jeremiah in Chapter 1:

4 Then the word of the LORD came unto me, saying,

5 Before I formed thee in the belly I knew thee; and before thou camest forth out of the womb I sanctified thee, and I ordained thee a prophet unto the nations.

In her book, *Life Before Life*, our dear friend and author, Sarah Hinze, has investigated the question of our spirits existing before we were born and shares several cases affirming that this is true. Several others have investigated that question with positive affirmation as well. In the remarkable story by Roy Mills, recorded in his book, *Soul's Remembrance*, he recounts much of his pre-mortal life with some profound and very important insights as to our purpose in coming here. He writes:

[Life's] main purpose is to create greater spiritual Love for us all . . . Some people become so Earthbound in their thinking that they use their free will to ignore their mission after they are born . . . We must always remember that we are much, much more than our minds and bodies. . . Our choices during our most bitter trials. . . are part of our Earthly experience,. . . We can use them. . . to make positive choices. . . . . .how much joy we experience is up to us. When we realize that we, ourselves, chose many of the difficulties in our lives, and that all things happen for a reason, we begin to let go of Earthbound thinking. . . We begin to understand that life isn't cruel or unfair, but is a spiritual journey that leads us to meaning and joy and real happiness. Knowing the Truth about ourselves sets us free to be the happy, loving people God wants us to be. . . While we are in Heaven, we promise to help each other learn and grow through our Earthly relationships and experiences... Every year, every day, every minute of our time here is important, because our lives are connected to so many others... We should be thankful for each moment of our lives, even the ones that cause us pain, because while we are here, we are keeping promises that we made in Heaven: promises to friends, to family, to ourselves, and to God.[21]

We will learn more of Roy's remarkable and inspiring experience later.

In 1961, Gustaf Stromberg of the Franklin Institute in Philadelphia proposed a fifth dimension based on some experiential information. Einstein's theory gives us four dimensions. Stromberg suggested that this fifth dimension, he called the " Eternity-Domain," gives time elasticity while maintaining sequence. He further explains that this Eternity-Domain can "...be described as an Almighty, Wise and Living Person, the Creator of all things, physical, mental, and spiritual. In our mind we have an 'image' of this Person, and an idea of His existence and nature. We are created in His image,... We can also understand that this Person in His wisdom may select one or more souls to carry important messages and admonitions to other human souls,... Some of these messages we can also hear directly when we listen to the voice of the 'Cosmic Conscience.' They tell us unequivocally that the essence of Divine law can be expressed in the simple admonition: 'Love ye one another!'"

This article had a significant impact on me over 50 years ago. Stromberg used as his scientific data base the several near-death experiences (NDEs) that he was aware of at that time.

People were not so prone to talk about them then. Since then, there have been well over 13 million documented NDEs, and a large number of books have been written sharing their stories. Many of them are irrefutable in terms of documented witnesses to the validity of their stories. In other words, they saw things while dead that they could not have known at the time they shared their stories.

As a scientist, I have to accept these as data. To say these data don't exist would be dishonest. One cannot rationally explain away the large number of experiences, which using the inductive method of determining Truth gives us a beautiful and consistent model of our life-before-life.

---

21) Roy Mills, *Soul's Remembrance*

We can also use the deductive method of determining the Truth of life-before-life since we have scriptures that tell us of this previous sphere of our existence. Large amounts of data have been compiled by our friend, Sarah Hinze, validating with amazing inspirational stories our pre- mortal existence. She has published these in her books which can be obtained from *Amazon.com*, or you can read about her on her web site.[22]

In 1975, Dr. Raymond Moody added significantly to the body of knowledge that Stromberg was alluding to with numerous case histories carefully documented in his classic book, *Life After Life*. These several experiences augment this fifth-dimension, Eternity-Domain thesis. One very well documented experience has been recorded in a separate book, *Return from Tomorrow*, the story of the physician and psychiatrist, Dr. George Ritchie. In his NDE he saw a spiritual laboratory where objects yet to be built on Earth were being created spiritually. Years later, he saw these objects in reality. Several others have now also documented similar experiences as those of Dr. Ritchie's.

In summary, the NDE experiences validate some very important facts, taken as a whole, and that are also consistent with a new Unified Field Theory (UFT) that I will describe in Chapter 21. One of them is that there is a spirit that lives in each of us and the "real person" had a pre- mortal birth and continues on after death. This is now well documented in the literature and now scientifically for me using this expanded scientific method.

1.6 The Need for Faith in Science and Religion

In general, we cannot absolutely-publicly prove God in this mortal sphere; He designed it that way. God wants us to start by faith, and grow in faith, until we come to a sure knowledge through a personal manifestation of God himself, coming face to face. Many prophets have come to know God face to face, but most of us live by faith. However, the promise is made to all for Him to be a personal part of our lives.[23]

Ironically, science, which is about facts, has to have faith as well. In 1931 Kurt Gödel proved the famous inconsistency theorem, which in essence says that science cannot prove the consistency of a system; or in other words, they

---

22) www.sarahhinze.com
23) John 14:21-23; Doctrine and Covenants 93:1

need to have faith in some set of assumptions or axioms. Neither can anyone prove there is no God; yet all the wonders of nature — the heavens and the Earth — portray the glory of His creations.

Science has moved the world to doubt the authenticity of the Bible story; and this may be in large measure Christians' fault because of their belief in the infallibility of scripture. When science finds a contradiction with scripture, they say the scripture is wrong; they most likely don't understand the scripture. As we will discuss in detail in Chapter 3, the scriptures are of infinite depth and are designed to help anyone who reads them in their quest to know God, to find guidance in their lives, and to find that path leading them to happiness here and a fullness of joy in Heaven above. They are an incredible wealth of enlightened principles as well as important historic references. It is each person's choice to apply them in their lives, and we each have the responsibility to discern what is true or not, both in science and religion. The Savior is our center focus; the prophets lead us to Him. As he studied the life of Jesus, C. S. Lewis said there can be only one of two conclusions: 1) He is the greatest deceiver of all time, or 2) He is exactly what and who He said He is: the Son of God and the Redeemer of mankind. I fully agree with Lewis and take Jesus' word as Truth. For example, since Jesus told of Noah's flood, there was a flood just as the Bible tells it. Most scientists don't believe it. There is not enough water in the atmosphere to cover the highest mountain they say. They ignore the statement in the Bible that not only did it rain, but the fountains of the deep opened up. There is now scientific evidence documenting the flood. Recently, scientists "have discovered evidence for oceans of water deep beneath the U.S. [24] How much water is there in the Earth to make Noah's flood? Enough!

All major civilizations on Earth have in their history the story of the flood.

Several years ago archeologists and scientists discovered Noah's Ark on Mount Ararat, just as the Bible says. There are other geological findings that support this very important Bible story.[25]

One may ask, "Why would God destroy all those people?" One answer is that if He had left them alone, they would have only gone deeper into wickedness, leaving less chance to redeem them beyond the grave as outlined

---

24) http://www.ibTimes.com/scientists-find-evidence-oceans-water-far-below-Earths-crust-1601296

25) The following link shows pictures and gives 14 correlations with the Bible story. There is a lot of disinformation as well as good information on the web about Noah's Ark. *http://www.arkdiscovery.com/noah%27s_ark.htm.*

in the third and fourth chapters of 1 Peter. Everything God does is motivated by Love. As a father rebukes a wayward child, out of Love, so does God rebuke us. He knows we're capable of so much better. He knows what we can become, and if we knew, we would be so motivated.

As Jesus clearly stated, the same scenario will be fulfilled in the last days. When the cup of iniquity of the world is full, Babylon will fall, and the Earth will be purged again to prepare for the Lord's glorious millennial reign.

I personally feel the Lord's Love in my life and the Father's Love as well, and it is the most wonderful feeling. This feeling apparently was common among most early scientists for they not only believed in God but often received Divine help in their research.

Science and other conspiring forces presented Charles Darwin's theories in a way that brought a basic change in scientific thinking, proposing no need for God since everything happened by chance. The 1900s saw the theory of evolution being taught throughout our education systems to this day, and those who do not accept it or are not willing to teach it have often been expelled. Ben Stein, in hi eye-opening movie *EXPELLED, No Intelligence Allowed*, documents that this is our current situation in America. Ben interviews folks from both sides of the issue, a fascinating film.

Darwin's theory implicitly teaches that there was no fall of Adam and Eve and — bypassing the beautiful " Garden of Eden" with the idea that man is evolved from lower forms of life. If there were no fall, then there is no need for an atonement to overcome the effects of the fall. So implicitly, Darwin's teachings are anti- Christ and often lead to atheism. The way people have applied this theory has had a great negative impact on the morality of mankind. Accountability goes away if we are evolved animals. If that is true, then what we do doesn't matter, and good and evil and moral laws have no meaning in this theory.

This theory, along with some other philosophies, has led to our current social paradigm that science and religion are incompatible. They say that if an idea is based on religion, then it is not scientific. And vice-versa. This is in total contrast to the early devout scientists — coming out of the Reformation era — and is dishonest in its logic.

In this book, using the scientific method, I will reintroduce God back into science. I do this by introducing the sixth sense into the scientific method, beyond the five traditional senses. The data from experimentation are fundamental to whether a hypothesis or theory has validity. This century and the last have seen scientists relying more on the theory and less on the data, especially if it did not agree with the theory. I will show that the inclusion of this sixth sense — using reliable data — brings in God, thereby bringing a harmony between science and religion. This is extremely important and fundamental to what will be shared in this book based on real data for all six senses.

Current science misses the sixth sense which is now bringing forth a new and most refreshing science with God fully integrated. Read the books *Signature in the Cell* and *Darwin's Doubt* by Stephen C. Meyer, and the link we mentioned previously shows the great harmony that existed between science and religion as documented in the Ohio Earthworks. These Earthworks are part of an estimated 200,000 mounds built in North America.[26]

Now with this new and fresh approach to the scientific method, let us move forward in the remaining chapters and share some additional major insights and perspectives with data to support them.

In addition, it goes without saying that an honest heart never lies. Let us listen to our hearts along with the five senses and do so with integrity, as we follow the path of Truth and Light and Love.

Also, with this fresh approach in the next chapter, we will have the basis to answer some of life's most difficult questions.

---

[26] For more details about the above concepts: *http://www.allansTime.com/Spiritual/Truth_vs_Truth.htm*

Chapter 2

It's About Time We Tackle Some of Life's Most Difficult Questions.

2.1 Where Did We Come From and Why Are We Here?

God establishes His word in the mouth of two or three witnesses so that we may know of its surety.[1] As mentioned in the previous chapter, the Lord told Jeremiah, "Before I formed thee in the belly, I knew thee; and before thou camest forth out of the womb I sanctified thee, and I ordained thee a prophet unto the nations."[2]

We have a second witness in our good friend, Ralph V. Jensen, who is a humble sweet soul. He had two NDEs, and during his time across the veil, he was taught by Christ about our pre- mortal existence as well as about the Garden of Eden and the ministry of the Messiah. He has written his remarkable experience in a book entitled *Taught by Christ*. My wife and I along with some other friends have had the privilege of sitting down with him and having him share some of the details of this experience. It is a very inspiring story, and fortunately there is an interview with him on the internet.[3]

---

1) Matt. 18:16
2) Jer. 1:5
3) *http://www.youtube.com/watch?v=yjg0wBBTlMA*

We learned more from him than is in this interview or than is in the book and feel fortunate to have him as a friend. We feel his story to be true because of the integrity of the man and the consistency with the scriptures of all that he shares.

A third witness is another of our dear friends, Sarah Hinze. Her biography, *Memory Catcher*, is one of her books, and it will touch your soul to the core. It documents the life of this amazing lady.

Sarah and her husband, Brent Hinze, are also friends of Roy Mills, whose book also tells of his pre- mortal existence. They have written up his story as well, which has not yet been published, and they have additional details, some of which they have shared with my wife and me regarding his very unusual story. There is a lot of interesting humor in their relationship. It was fun to hear it and to feel the witness of the Spirit of the truth of their experiences with Roy. Brent shared with me a preprint of their write-up of Roy's story. It will be a significant contribution to the life-before-life literature and evidence bank.

Using the induction method for determining truth from the above data along with other data we have allows us to develop a fascinating model for our pre- Earth life, which is totally consistent with the scriptures. This model is fundamentally necessary in answering the above question and for knowing our purpose here on Earth.

All people on this Earth were born with spirit bodies in this pre- mortal sphere of existence from our Heavenly Father and Mother. There we knew and felt their love. Heavenly Father explained His grand plan of happiness to us. It was apparent to us that during our mortal journey and growth opportunity we would sin. The need for a Savior was apparent. To satisfy this need we all witnessed the foreordination of Jesus Christ who volunteered to be our Savior (Jehovah, as His name was known there) to be born in the meridian of time as the Only Begotten of the Father in the flesh and as the Redeemer of the world.

It was explained to us there that a veil of forgetfulness would be placed over our minds as we came to Earth and gained a physical body so that here we could grow spiritually by living by faith. All of us were born with a conscience (the Spirit of Christ) to know good from evil and to help us in our opportunities here in mortality. We knew this was a perfect plan and we

Chapter 2

rejoiced in the anticipated opportunity to receive a physical body, to live by faith, and to learn from life's experiences. We made covenants with Heavenly Father and with family and friends to help one another along the way. We knew that if we listened to the " still small voice," we could come to know our loving Heavenly Father and Mother, and His Beloved Son again, as we knew them there, and that in that knowing we could gain eternal life with a glorious resurrected body — thanks to the infinite atonement of our Savior. As it says in Job, we shouted for joy at the anticipated opportunity to be like our Heavenly Parents, and thus the heavenly choirs sang at the birth of Jesus. Life's challenges are but opportunities for growth and to exercise compassion and love for others as we help them along their way.

It is our grand opportunity here to come to know the Father and the Son. To know them is eternal life.[4] To feel their divine love and to be one with them here as we were with them there is what Jesus prayed for in His great and last High Priest's prayer.[5] We chose to follow Him there; let us also choose to follow Him here.

If the atheists' requests were granted that God be more evident, then that would destroy our opportunity to grow by faith in Him and deny us the joys that come as we come to know Him through the process of living by faith.

Our missions in mortality were designed in the economy of souls to be that which would best help us find our way back to our heavenly home. In the Apostle Paul's great sermon on Mars Hill, he said it well, "[God] hath made of one blood all nations of men for to dwell on the face of the Earth, and hath determined the times before appointed, and the bounds of their habitation."[6] In other words, when and where we are born is in our eternal best interest.

Roy Mills is an additional profound witness of our premortal existence. He was so impressed with the beauty and loving dimension of our pre- mortal sphere, he asked Heavenly Father for special permission to not have the veil placed over his memory that he might share it here in mortality. He was granted that with two stipulations: 1) that he would not share the incredible beauty of our pre- mortal sphere until he was given permission, and 2) that he would write a book about it. His book is called *Soul's Remembrance; Earth Is Not*

---
4) John 17:3
5) Chapter 17 of the Gospel of John
6) Acts 17:26

Our Home. I was so impressed with the book that I wrote a book report about it back in 2001, and it is on our web site. The following is taken from that report.

> Roy Mills' book was published in 1999. It has extremely interesting information about where we came from and why we are here on Earth. Roy was given heavenly assistance both in its writing and in its publication. He was not given permission to share this information until he was 45 years-old.[7]

We learn from his book that we each have a "life book," and that we chose, within the massively important influence of the infinite love and wisdom of our Heavenly Father, our life's mission. Roy was shown his life Book - showing some of the experiences that he would have during his life. He then tells of some of the training he received to prepare him for his life's mission. He describes that his angelic trainers took "great pleasure in making me laugh. . ., but their jokes were all harmlessly amusing. . .. But they took their role in preparing me for my birth very seriously. They taught me a lot, but the things they stressed the most were pure and perfect love (the original text reads "loving unconditionally," but Chauncey Riddle pointed out to me that God's love is not unconditional). It is pure and perfectly designed for each person. God does not love Jesus the same as Satan; so I made this word replacement], using free will wisely, praying from the heart, gaining knowledge and experience, and seeking truth. They said these were the building blocks of a good life on Earth."

In his pre- mortal state, Roy became very good friends with the three sons that he would have in mortality. He saw his mother and other family members and friends with whom he would spend time during his Earthly sojourn. He saw that they would mean a great deal to him. He tells of his incredible experience with the Father in helping him determine his life's mission, and of the marvelous mission instructions that Father gave to him. He learned the fundamental truth that we are literally the spirit offspring of God. He also tells of his marvelous experience with our Lord and Savior, Jesus Christ. He tells of the incredible light emanating from the Savior's Being, which goes out to all of Father's children, and which will guide and direct us in our paths if we will choose to listen to that inherent guidance toward goodness that is within each of us. Roy felt the Savior's "tremendous pure and perfect love." "Not only did

---

7) http://www.allanstime.com/Spiritual/BookReports/souls_remembrance.htm

Chapter 2

I feel His great love for me as one of God's spirit children, but I also felt His love as a close friend. Jesus is the best friend we will ever have in this life or the next, and I knew in that moment that He would always be there for me, no matter what happens. . . . . .nothing can separate us from the love of Christ."

Roy tells of his pre-birth experience, which all by itself makes the book worth reading, the description fascinating as he was being placed in his mother's womb. He describes, "My spirit made contact with the mind in my body, and a complete sharing of information took place. I was amazed that this mind could already function on its own, and that it was so absorbent, with the ability to comprehend what I was communicating to it. Once connected with my spirit, this wonderful mind worked with great speed, and it didn't take long before the information exchange was completed. Soon, my spirit and my mind merged into one consciousness; my spirit settled completely into my physical body, and the two began to feel and act as one."

He explains that even as a very young person, "Because my spirit had a heavenly knowledge, I often knew what people were going to say or do before it actually happened. When my mother took me to various places, I sometimes recognized them, even though I was seeing them for the first time in my Earthly body. And I recognized a lot of people the first time I saw them, because I had met them in Heaven before I was born. It wasn't their physical faces that I knew, but their spirits. I believe infants can recognize a person's spirit when they see it, because the child's ties with Heaven are still strong so soon after being born. But before long most children forget that they are mighty spirit beings who came from Heaven. That forgetfulness is part of God's plan, because if we could keep talking to angels and remembering Heaven every waking moment, we would limit the Earthly experiences we are meant to have by constantly choosing the right." We often learn and progress spiritually the most from the mistakes we make. When Roy met Sarah Hinze, he had tears in his eyes, and Sarah asked, "Why are you crying?" Roy told her that he recognized her from pre- mortal and they were to work together.

The importance of the ministering of angels is clearly portrayed. Roy had a very difficult childhood, and he was visited on numerous occasions by angels to help him through various crises. He says, "It may seem unusual that I've been helped so many times by angels, but the truth is that we are all helped by angels. Most of us are just not aware of it. They look out for us in all kinds of ways. If you have an uneasy feeling that won't go away, a thought or warning that suddenly pops into your head, an unexpected delay, or something that throws a monkey wrench into your plans, it may be that your guardian angel has just kept you from a mistake or a disaster. They. . . come to us. . . at moments of our greatest grief and sorrow and need. God sends angels to help us. We might not always see or hear or feel their presence, but they are there. And they will bring us God's love and help if we will let them and if we will be humble and listen. . . Angels are wonderful beings, who serve God by serving His children. They do many different things, but most of what they do is to help us in completing our missions. . ."

In speaking of the ministering Angels he says, "The Angels were especially good to me, and there are no Earthly words to describe the love that flowed between us. They performed many roles - guide, teacher, escort, friend, and at times they even attended to my needs almost like servants. They are glorious beings who act upon the knowledge that serving others is the best expression of love." One of the Angels he was privileged to meet was the mother of the Son of God. He says of their meeting, "There was so much love and strength and beauty in her presence! As I stared into her penetrating eyes, I was filled with awe and great love, and I told her I loved her. ' I love you, too,' she replied with a beautiful smile. . ."

"Some spirits choose a life of extreme sacrifice or nearly unbearable suffering. The light of God shines brightly around these people, and in heaven they are thought of as mighty spirits, because they come to Earth to suffer so that others can grow in understanding and love." He further says, "When we see a child who is confined to a wheelchair because of a problem with his or her muscles or nerves or brain, we feel compassion for the child. But the deeper truth is that this individual had compassion for us first, having chosen their condition to help us grow. . . These courageous souls choose their conditions in order to inspire love, compassion, caring, patience, or forgiveness in others. . . I was told that many people live with their personal problems so long that they sometimes just accept them. They say to themselves, ' That's just part of who I am. I can't change it.' But this is the result of Earth-bound thinking. While in heaven, we didn't choose these problems because we wanted to live with them. We chose them because we wanted to grow and to learn by overcoming them. God doesn't want us to merely endure hardship; He wants us to learn from it and find happiness." He further explains that God does not send us to Earth without the necessary tools to overcome life's challenges. By overcoming we grow closer to God.

The Angels ". . .explained that the more difficult a challenge was, the more knowledge and wisdom and love it could bring. I knew that it would be pleasing to God if I dealt with many experiences in a positive way, so I began selecting all kinds of challenges. . .the angels finally warned me that I was choosing too many. . .So when I selected my Earthly experiences, I had no concern for what they would do to my physical body. I knew that the more I suffered for others, or because of others, the more I would grow in spirit. And with that knowledge, I had a completely different perspective on adversity and the experiences I chose for my own life. . .always knowing God has a perfect plan." One of our greatest opportunities in this life is to be grateful for every experience and that in it is the potential for spiritual growth. God is the master in turning evil into good. He will help us do the same if we will but turn to Him.

Roy further explains that, "A mission can be very short. . . Some. . .are sent here to be spiritual teachers, or good stewards of the Earth, or care givers. Others are sent to be healers, or artists, or helpers. Many are sent to be good parents or friends. . .no two people have the exact same one. And they are all important.. Whether He sends us here to take care of one child or to change the world, each and every mission is noble and valuable.

"Some people become so Earthbound in their thinking that they use their freewill to ignore their mission. . . The angels taught me that when we feel dissatisfied or restless or empty, our spirit is trying to tell us that we have strayed from the course God laid out for us. Those are the times when we need God's guidance the most. If we ask for direction and stay open to the answer — no matter what it is — God's Spirit will lead us back where we belong. All we have to do is listen to our inner spirit, which recognizes and promotes feelings of peace when we are correctly directed. Feelings of joy and contentment are the fruits of doing God's will, and our spirit helps us feel and know our choices are right." The importance of what Roy says here cannot be overemphasized.

In regard to our missions, one of the pearls that Roy shares is that, ". . .how much joy we experience is up to us. When we realize that we, ourselves, chose many of the difficulties in our lives, and that all things happen for a reason, we begin to let go of Earthbound thinking... We begin to understand that life isn't cruel or unfair, but is a spiritual journey that leads us to meaning and joy and real happiness. Knowing the truth about ourselves sets us free to be the happy, loving people God wants us to be."

Roy learned that we are each given different gifts, and that one of our greatest opportunities is to use our gifts in serving others. As we seek, ask and strive in our quest for truth and light, we will be blessed with additional gifts, which will bless our lives as well as the lives of others. He learned "that all of God's gifts are special and marvelous in their own right." The gifts of God that we receive are just the right ones to help us fulfill our missions in mortality.

"While we are in Heaven, we promise to help each other learn and grow through our earthly relationships and experiences... Every year, every day, every minute of our time here is important, because our lives are connected to so many others... We should be thankful for each moment of our lives, even the ones that cause us pain, because while we are here, we are keeping promises that we made in Heaven: promises to friends, to family, to ourselves, and to God."

"Everything in heaven moves steadily and smoothly, like the finest clockwork. Its pace is nothing like the frenzy here on Earth," says Roy of his experience there. He also saw that all of us are important to God; there are no inferior children. "While I was in heaven, I never saw anyone who was depressed, maimed by an accident, or suffering from disease. Our spiritual bodies are always healthy and whole; they are perfect, and our spirits shine with a radiant beauty. Our physical bodies are only a temporary home for our spirits..."

"The angels told me that every soul is born equal. No one is more important than anyone else, no matter how it seems on Earth. From the lowest to the highest, from murderers and thieves to world leaders and saints, everyone is equal and is a child of God. But for most of mankind's history, we have fought over our differences — especially in race and religion. If people only knew what it is really like in Heaven, they would stop hating and fighting others who are different from them. They would start caring and loving instead......in Heaven, I never saw different races... only spirit beings who were filled with love for each other... God chooses what our earthly race will be... for our spiritual growth... The angels taught me that when we judge someone else, we are really judging ourselves... Insulting or judging another only makes all of us weaker...

Chapter 2

We should love everyone with a pure Christ-like love, no matter who they are or what they have done. . . You should always treat them the way you want to be treated. And you must try to help others without expecting anything in return. That is true love." Overcoming prejudices is one of the great challenges of life. When we do so, we bring about great spiritual growth in our lives. When you think about it, all the major wars have been caused by prejudices.

Roy also talks about the review of our lives after our time on Earth. "During our life-reviews, we re-live how we treated other people during our time on Earth. . . For all the times we treat people with love or kindness, we get to experience the positive emotions we made them feel. And for the times we treat people wrong or abuse them in any way, we experience the pain and hurt we caused them."

Roy says that, "Of all my Heavenly memories, receiving my mission from God is the one I cherish most. No matter what I write about this extraordinary event, it is impossible for me to do it justice. No words can ever come close to conveying the incredible love, grace, wisdom, power, and majesty I felt in the presence of God. . . I am fortunate to remember heaven and to have had the company of Angels in my life, but these things aren't necessary for us to remember to talk to God and have His healing influence in our lives. We can pray for assistance and feel confident in knowing that our guardian Angels are watching over us, . . .we have family and friends that agreed to come into this life to help and support us. We can rely upon them and build our relationships with them, just as we can with God. Our greatest friend is Jesus, who loves us so completely and with a pure love. He has done so much for us already, and continues to do all He can for us."

Roy learned that agency — free choice — is fundamental to Heavenly Father's plan. We received and concurred with our missions before we came here. Here our choices determine whether we will fulfill our missions or not. Our avenue to find out the will of the Father is prayer. We should not be surprised, therefore, that the Lord commands us to pray always. It is the most oft' repeated commandment in the scriptures — for our own best good.

He relates how prayer is our connection to God and that all sincere prayers are heard. In the Lord's infinite love he appropriately responds to them in a way that will be for our eternal best good. If we sincerely pray by the Spirit according to His will, then will follow answers to those prayers in our eternal best interest and in God's infinite wisdom of the proper timing for those answers to come.

In conclusion Roy says, "The missions God gives us are often difficult. But we did have a say in choosing them. God's love and guidance are always with us, and we are never alone as we try to remember who we are. Our loved ones on the other side, the Angels, and God, Himself, are always near, ready to help, and are doing everything possible to uphold us with strength, understanding and, most of all, with love."

As I mentioned before, Brent and Sarah Hinze have also written his story as well. One insight from their book is that Roy was led with unique angelic assistance to Sarah's book, *Life Before Life*, and told that she was the best expert on life-before-life experiences. Hence, he made contact, and Brent and Sarah became very good friends with Roy and his wife, Phyllis. As I mentioned before, when they met, Roy had tears in his eyes. Sarah asked him why he was crying. With joy in his heart, he said, "I remember you from before we were born. We agreed to work together!"

I feel that Roy is one of those special messengers God has sent alluded to by Dr. Gustaf Stromberg of the Franklin Institute in Philadelphia back in 1961. Roy's information was meant by God to come forth in our day.

As we take the information we have shared before and apply the induction method of reasoning, the consistency with the fullness of the gospel as restored through the Prophet Joseph Smith, and from that restoration we learn not only that we had a premortal existence[8] with our loving Heavenly parents, we are literally their spirit offspring — making us gods in embryo — and we are each of infinite worth![9] We sat in premortal councils where Father's plan of happiness was outlined for all of us and each of us. We rejoiced — "shouted for joy"[10] — as this infinitely loving opportunity was extended to us to make this mortal journey in our path to godhood, to perfection like our Heavenly parents are perfect.[11] As we looked at them, we knew that any price was worth paying to be like them. We saw that any trial or challenge would be worth it; in fact, we saw that life's challenges were actually extended to bring growth in

---

8) Abraham 3:22-28
9) Acts 17:28; Heb. 12:9; Doctrine and Covenants 76:24
10) Job 38:7
11) Matt. 5:48

Chapter 2

our lives and were necessary because of free choice. Further, we saw that we would sin. The need for a Savior was clearly manifest, and Father's firstborn[12] offered to be our Redeemer, foreordained[13] to bring perfection to Father's plan because of His infinite love for us, as well.

In our premortal existence, Lucifer offered a different plan that would save all mankind, but would take away free choice.[14] Father knew that to take away our agency would destroy the plan. Satan uses force; Father will force no man to heaven. As mentioned before, it would not be heaven were that the case. Lucifer rebelling against the Father's plan brought about a war in heaven, and a third part of the hosts of heaven followed Lucifer, and they became the devil and his angels and were cast down to the Earth.[15] The Lord could have cast them into outer darkness, but that would have denied the opportunity for opposition for us here on Earth, which we needed for our growth and development. We learn further, through the restoration, "that there must needs be opposition in all things," otherwise, there would be no existence.[16] We knew that through the atonement, we could overcome all things, and that this was the path to perfection. We saw that sin, suffering, pain, and challenges would help us better appreciate the sacrifice of our Lord and Savior, Jesus Christ, and these would bring us closer to Him only when properly appreciated.

Isaiah asks the question, "How art thou fallen from Heaven, O Lucifer, son of the morning!"[17] And then he gives the answer in the verses following as Satan sought to "exalt" himself "above the stars [angels] of God." His pride had led him to believe he had a better plan to save all by taking away the agency (free choice) of man. We honor God because he is the author of our agency, and perfectly personifies love, patience, humility (opposite of pride), long-suffering, gentleness, meekness, and all the other wonderful attributes of godliness and goodness. As we come to know God, we come to honor and emulate Him because of these marvelous and perfect characteristics. We exercise true faith in Him as we focus our bodies, minds, and spirits on 1) the truth that God exists, on 2) modeling our lives after His character and attributes, and on 3) asking the question, "Are we doing everything we can to

---

12) Romans 8:28-29; Colossians 1:14-15
13) 1 Peter 1:19-20
14) Moses 4:1-4
15) Rev. 12:1-10
16) 2 Nephi 2:11-29
17) Isa. 14:12

fulfill our own personal missions in mortality to honor Him.?" In this faith we will find His love, joy and peace, as we treasure up His word. In so doing, we are promised that we will not be deceived.[18] The great deceiver, who sought to destroy the agency of man, will not have power over us. What a remarkable promise and assurance from our infinitely loving Heavenly Father and His Beloved Son.

In the Book of Job is a great lesson regarding pain and suffering. He was foreordained and agreed to perform an incredibly difficult mission, as a type and shadow of the Savior's mission. Job teaches us to trust always in the Lord in spite of any and all pain and suffering we may endure; the Lord will more than make it up to us, and in the end we will say, "Thank you, Lord!" with deep feelings of gratitude. This he did for Job.[19]

When we succor the suffering, reach out to the poor, the widow, the fatherless, etc. the Lord is very pleased when we do it with His pure love in our hearts, and especially when we share His glorious gospel message. There is no more important message. That is why the gospel means the good news! It is THE BEST NEWS! Through the atonement, Christ has overcome all things, and as we come unto Him, we know that He will help us overcome all obstacles and sins in our lives on our path to eternal life, which is the greatest of all the gifts of God.

2.2 Scientific Validation of Our Ultimate Destination

Here we use the scientific method of inductive reasoning taking from the very large data base of well documented NDE experiences (there are now more than13 million recorded). To ignore this would be dishonest. We need to be as Agassiz and seek for the creative mind of the Divine in the facts at hand — being honest with ourselves and to all of our senses. In many documented cases, the person having the NDE tells of things that would have been impossible for them to know in their state of unconsciousness. These things they know by their spirit and intelligence that had left their bodies.

The counsel given by Polonius to his son, Laertes in Shakespeare's Hamlet

---

18) JS Matt. 37
19) Job 42:12-16

seems 'apropos,' "This above all: To thine own self be true. And it must follow as the night the day, Thou canst not then be false to any man." Indeed, the importance of listening to our hearts is implicit in Polonius' counsel, but as I have studied the several NDE experiences, we need to take this counsel to the next step, "To God be true." Then everything will work together toward a good end without fail. We have validated as we examine a large number of these NDE experiences a consistency with the new UFT to be described in Chapter 21. In addition, these NDE experiences validate that there is a spirit that lives in each of us and the "real person" continues on after death with the same character traits that it has in this life. This is now well documented.

The inductive evidence tells us that where we go when we die depends on how much we love God and our neighbor — the first two and greatest commandments. We will discuss a lot more about where we go and our purpose in life in the final chapters of this book. It is wonderful to live in a day when these truths are being made known. Let us thank the Lord for His goodness.

2.3 Why Does God, Who Is Good, Create Evil?

I form the light, and create darkness:
I make peace, and create evil:
I the Lord do all these things.

— Isaiah 45:7

And how and why does the Lord create evil or allow it? Why is there so much of pain and suffering in the world? The great thinkers of the ages have never come up with the answers to these questions. C.S. Lewis, after going through his own apostasy from Christianity, may have come the closest — becoming one of Christianity's most avid advocates. During his teenage and apostate years, he felt that the strongest force for atheism was expressed in a poem of Lucretius:

Had God designed the world, it would not be
A world so frail and faulty as we see.

After J. R. R. Tolkien, and others of his friends, persuaded Lewis to deeply consider Christianity, it turned his life around. As he considered the Christ, he came to the conclusion that "only two views of this man are possible. Either he was a raving lunatic of an unusually abominable type, or else He was, and is, precisely what He said. There is no middle way. If the records make the first hypothesis unacceptable, you must submit to the second. And if you do that, all else that is claimed by Christians becomes credible — that this Man, having been killed, was yet alive, and that His death, in some manner incomprehensible to human thought, has effected a real change. . . in our favour."[20] After gaining his deep appreciation for Christianity, in regard to evil, pain, and the suffering that come with life, Lewis poignantly states that if you "Try to exclude the possibility of suffering which the order of nature and the existence of free wills involve, . . .you find that you have excluded life itself."[21]

Albert Schweitzer — considered by some to be the greatest humanitarian and one of the greatest thinkers of the last century — was a devout student of the teachings of Jesus. He mastered philosophy, music, theology, and medicine, and became the world's authority on Bach's timeless organ music, and he also could build an organ. His great intellect and all his learning led him to "Reverence for Life," all life. He built a hospital in Africa to relieve the suffering there and had great success in doing so. He dealt with both world wars and wrote much. In 1945-1950 he was considered the greatest man in the world. He won the Nobel peace prize. He had and is still having a great influence on the world's thinking.

In all of his 80 years of learning, Dr. Schweitzer could not answer the "great questions: What is the meaning of evil in the world? ...In what relation do the spiritual life and the material life stand to one another? And in what way is our existence transitory and yet eternal?"[22] He left these questions unanswered, but knowing of God's love he focused his life on loving by relieving the sufferings of others.

Herman Melville had similar questions, "How can a God that is good create evil?" He wrote the classic book, Moby Dick, trying to deal with this question. In the Rubáiyát of Omar Khayyám — the great Persian thinker — Omar struggled with the same questions, and after listening to the world's philosophers discuss these questions, he said, "I went out by the same door wherein I went."[23] The questions remained unanswered!

---

20) C.S. Lewis, *The Problem of Pain*, pp 13-14
21) C.S. Lewis, *The Problem of Pain*, pp 13-14
22) Albert Schweitzer, *Out of My Life and Thought*
23) *The Rubáiyát of Omar Khayyám*

Chapter 2

One of the great validations of the truths of the restoration of the fullness of the gospel of Jesus Christ through the instrumentality of the Prophet Joseph Smith is that it answers not only the above questions but also several other challenging questions: Who are we? Where did we come from? What is our purpose in life? Where do we go after death? None of the great thinkers, including Lewis, give in-depth answers to these questions, which are gloriously now available to us through the latter-day restoration in this last dispensation of the gospel of Jesus Christ, the dispensation of the fullness of times. We will find that why a loving Heavenly Father allows evil is agency (free choice). Our right to choose is at the heart of His perfect plan of happiness for His children. One of the grand ironies of the plan is that He brings perfection out of imperfection, a fullness of joy out of the deepest of sufferings, pain, and despair and all through His Son's infinite atonement. We can become totally free only when we voluntarily give our will to Him. "Thy will be done;" represented the perfect life of the Savior. We each have a mission that is designed for our eternal best good. He knows it and we don't, but by asking in faith, we can find that direction in our lives that will be most fruitful and enjoyable here and in eternity.

Said Jesus, "And ye shall know the truth, and the truth shall make you free."[24] This He does in the midst of trials and adversity, and amongst many in bondage, as we come unto Him. For He is ". . . the way, the truth, and the life. . ."[25] He means us to be free individually from sin and as a heavenly society, as part of His ultimate plan of happiness filled with unspeakable joy.

The Apostle Paul so aptly said, ". . .the natural man receiveth not the things of the Spirit of God: for they are foolishness unto him: neither can he know them, because they are spiritually discerned."[26] The evil in the world has led many to doubt God's existence; it is one of those doctrines not understood or appreciated by the "natural man." Even though this doctrine has been brought into light and understanding through the restoration of the fullness of the gospel, much of it is taught in the Bible, as we will see.

Properly viewed then, every challenge and trial is an opportunity for growth. Like C. S. Lewis, we will be *Surprised by Joy* (the title of his book), as we come to understand Father's plan. " Adam fell that men might be, and men are

---
24) John 8:32
25) John 14:6
26) 1 Cor. 2:14

that they might have joy."[27] Rejoicing in our trials as well as our blessings will lift us out of the mire and darkness of life, in which so many find themselves. ". . .all things work together for good to them that love God."[28]

Neal A. Maxwell, one of our modern Apostles, said that there are three reasons for life's challenges: 1) a natural consequence of mortality; 2) the sins or mistakes we make; and 3) those given us by a loving Lord for our growth and benefit. "For whom the Lord loveth he chasteneth,"[29] and he loves us all. Let us thank the Lord for our chastisements and challenges as well as our uncountable blessings; they all have a divine purpose. When they come, and they will, let us ask the Lord what lessons He wants us to learn from them. Viewed from His perspective, we know that they will always be for our eternal good. He loves us with an infinite love and can and will turn every bad thing into good as we listen to the Spirit, asking how we may best benefit from these challenging opportunities.

In addition to these three, we learn that in our pre-mortal realm of existence we were a partner with our Heavenly Father in choosing our life's challenges. We could see that they would best help us in our eternal journey to perfection. Often, that is so hard for us to see in this mortal sphere of existence. It is clear from the scriptures that the Lord gives challenges for our eternal benefit:

"Thou shalt also consider in thine heart, that, as a man chasteneth his son, so the Lord thy God chasteneth thee."[30] "Blessed is the man whom thou chastenest, O Lord, and teachest him out of thy law."[31] "He that spareth his rod hateth his son: but he that loveth him chasteneth him betimes."[32] "For whom the Lord loveth he chasteneth, and scourgeth every son whom he receiveth."[33] "As many as I love, I rebuke and chasten: be zealous therefore, and repent."[34] "Thou shalt thank the Lord thy God in all things."[35] "And in nothing doth man offend God, or against none is his wrath kindled, save those who confess

---

27) 2 Nephi 2:25
28) Romans 8:28
29) Heb. 12:6
30) Deut. 8:5
31) Ps. 94:12
32) Prov. 13:24
33) Heb. 12:6
34) Rev. 3:19
35) Doctrine and Covenants 59:7

not his hand in all things, and obey not his commandments."[36] "Verily, thus saith the Lord unto you whom I love, and whom I love I also chasten that their sins may be forgiven, for with the chastisement I prepare a way for their deliverance in all things out of temptation, and I have loved you."[37] "I, the Lord, have suffered the affliction to come upon them, wherewith they have been afflicted, in consequence of their transgressions; Yet I will own them, and they shall be mine in that day when I shall come to make up my jewels. Therefore, they must needs be chastened and tried, even as Abraham, who was commanded to offer up his only son. For all those who will not endure chastening, but deny me, cannot be sanctified."[38] In other words, His challenges will come where they will benefit us the most.

Thus we see agency (free will) led to the Fall and thence to sin, pain, suffering, and life's challenges. Ironically, the Fall is a major and necessary step in the Father's perfect plan of happiness. After the Fall, Eve was glad, and in rejoicing responded, "Were it not for our transgression we never should have had seed, and never should have known good and evil, and the joy of our redemption, and the eternal life which God giveth unto all the obedient."[39] We could say, " Adam and Eve fell UP!" It is interesting that east is the direction of the Lord's coming,[40] and the Lord planted a garden eastward in Eden.[41] When Adam and Eve fell, they exited the garden to the east.[42] In other words, they fell in the direction the Lord wanted them to go, but they needed to be responsible for the decision that brought about their fallen condition. Their fall opened the door for Father's spirit offspring (you and me) to gain a physical body and gain life's experiences.[43]

Another of the grand ironies of the universe — perfectly exemplified by our Savior — is that as we surrender our will to the will of the Father, and then we can be totally free. As our Redeemer so succinctly said:

> "And ye shall know the truth, and the truth shall make you free. . . If the Son therefore shall make you free, ye shall be free

---

36) Doctrine and Covenants 59:21
37) Doctrine and Covenants 95:1
38) Doctrine and Covenants 101:2-5
39) Moses 5:11
40) Matthew 24:27
41) Genesis 2:8
42) Genesis 3:23-24
43) 2 Nephi chapter 2

indeed."[44] The Holy Spirit will guide us to all truth, if we choose to listen and to hearken to "the still small voice."[45] His angels are there for each of us and all of us and speak the words of Christ by the power of the Holy Ghost to bring about the Father's great work of bringing to pass the immortality and eternal life of His precious children, as they may choose to hear and hearken.[46] If we choose otherwise, we will find ourselves following the ways of the world and in opposition to our eternal happiness. We are promised the Holy Ghost as our constant companion, which is most desirable, as we love the Lord, exercise charity toward all mankind and to the household of faith, and let virtue garnish our thoughts unceasingly.[47] Wherefore, how great the importance to make these things known unto the inhabitants of the Earth, that they may know that there is no flesh that can dwell in the presence of God, save it be through the merits, and mercy, and grace of the Holy Messiah, who layeth down his life according to the flesh, and taketh it again by the power of the Spirit, that he may bring to pass the resurrection of the dead, being the first that should rise...And the Messiah cometh in the fulness of time, that he may redeem the children of men from the fall. And because that they are redeemed from the fall they have become free forever, knowing good from evil; to act for themselves and not to be acted upon, save it be by the punishment of the law at the great and last day, according to the commandments which God hath given. Wherefore, men are free according to the flesh; and all things are given them which are expedient unto man. And they are free to choose liberty and eternal life, through the great Mediator of all men, or to choose captivity and death, according to the captivity and power of the devil; for he seeketh that all men might be miserable like unto him.[48]

Therefore, let us use our agency in the midst of all the challenges of this life to be grateful for them and to follow Him, who will lead us through them all back to the Father and a fullness of joy with our loved ones; for we can overcome all things through Christ.[49]

---

44) John 8:32, 36
45) 1 Kings 19:12; 1 Nephi 17:45
46) 2 Nephi 32:2-5; Moroni 7:27-32; Moses 1:39
47) Doctrine and Covenants 121:45-46
48) 2 Nephi 2:8, 26-27
49) Matt. 11:28-30

Chapter 2

2.4 Why Does God Not Make Himself Better Known and More Apparent if He Wants Us to Believe in Him?

This is a chief question of the atheist and is often used to justify his lack of belief. As mentioned in the previous chapter, we now have scientific data giving evidence that we all lived in a pre- mortal sphere of existence with our loving Heavenly Father. There we knew Him and felt His love and understood, as it was explained to us, His grand plan of happiness with free choice being the center of it.

It was explained to us there that for our maximum growth and benefit as we came to Earth we would need to live by faith. Faith has been explained as the principle of action in all intelligent beings. Our spiritual growth comes as we exercise faith in God: that He exists and to learn His character and attributes that we might better emulate them in our progress toward perfection — and that we may learn through prayer (by asking, seeking, and knocking) that our path is aligned with God's will. He promises that whosoever asks will receive, and whosoever seeks will find, and whosoever knocks, it will be opened, and His promises are sure. Those answers will come in His time for He knows what is best for us.

Following this path of faith in this life will lead to peace, love, joy, and the greatest happiness in this life and a fullness of joy in the life to come. This we knew before and rejoiced as we understood that by exercising faith here, we would have the greatest opportunity for spiritual growth.

Faith leads to repentance so that we are sufficiently humble and free from sin to hear the " still small voice," then the truths of God and the marvelous mysteries of His magnificent ways will flow into our beings. Often when we sin, we think God will not hear our prayers. God is there for us all and we are all sinners.

If we put anything ahead of God (idolatry), then that makes it harder for us to hear Him. He is ever there to commune with us if we are sufficiently repentant and humble. This is typically the problem with the atheists; they are caught up with the things of this world and typically occupied with their own selfish desires and often tend to be prideful, the opposite of being humble.

2.5 God Spoke Anciently to Prophets and Prophetesses. Can He and Does He Speak to Them Today?

The answer is obviously, yes. God has not lost His power to speak to us today. We will address the very important question in detail later in Chapter 22, and answer the key question to the issue, "How can we know He is speaking?"

I find C. S. Lewis' thinking to be right on. "Those who believe that Christ is just a great moral teacher," Lewis points out, "this is absurd. Either Christ is what he said he was, the Son of God, or he is a liar. If a liar, then He cannot be a great moral teacher. If He is the Son of God, then we best get on with true Christianity — loving God and our fellow man and looking forward to being like Him."[50] This was Christ's great promise, as Lewis summarized Christ's teachings.

I find it fascinating that in a large percentage of the near-death experiences across the globe amongst all religious beliefs, many return having seen the Christ and have felt of His pure and perfect love, and they are changed forever. Using the inductive scientific method of reasoning given this large data set, one concludes the reality of His glorious resurrection. And we know from the scriptures that the resurrection was capstone to finishing the perfect atonement, which broke the bands of death and opened the door for every sinner to repent, have faith in Christ and be forgiven. Being clean before God makes you a candidate for Heaven.

Regarding prophets, they are mortal and thus make mistakes. We read of true and false prophets in the Bible. The Savior warns, "Beware of false prophets."[51] If there were to be no prophets in the future, He would have said that. Instead, as we have said before, He gave us a sure recipe to know. In the same chapter in the gospel of Matthew, He said, "By their fruits ye shall know them." He said, "A good tree cannot bring forth evil fruit, neither can a corrupt tree bring forth good fruit."[52] As one analyzes this statement logically, it is both a necessary and a sufficient condition of proof. A true prophet or prophetess' words will come to pass and the truth of them can be validated. Their words will always bring us

---

50 Michael Wilcox, *Lions, Dragons, and Turkish Delight*
51) Matthew 7:15
52) Matt. 7:16-18

to Christ and His teachings. They will never put anything between us and God, for that is idolatry to worship anything but God. A true prophet will never get between you and the Lord. He will lead you to Christ and ever encourage you to follow Him.

I have weighed Joseph Smith, Jr., the American prophet, on the Lord's judgment scale and found him to be a prophet of God. As any mortal, he made mistakes, but the fruits of prophecies and labors stand him among the giants of the prophets who have walked this earth. The Father and the Son appeared to him in a most profound epiphany and told him they had a work for him to do, and he did it. Later, God sent an angel (Moroni) to him and told him of a record of the religious dealings of God with the ancient inhabitants of America, and that he would be given the means to translate it, and he did it in a miraculously-timely way. He also translated from some ancient papyri the writings of Abraham, and since that time the accuracy of his translation has been validated. (Read *The Blessings of Abraham* by E. Douglas Clark). He received many additional revelations and they have come to pass. Two of them of particular interest are: 1) He said, "The world will prove Joseph Smith a true prophet by circumstantial evidence." And that is happening in this book as well as elsewhere; I can give you great detail on that one. 2) He prophesied of his martyrdom, and it happened as he said. He gave his life as a testimony of the truth of the work God had given him to do.

His brother, Hyrum, was martyred with him. Hyrum was five years older and stood by Joseph all of his life. All who knew Hyrum (both Mormon and non- Mormon) knew him as a man of integrity and with a great and charitable heart. His witness of Joseph's divine calling by itself is proof as a man of undeviating integrity before the world. But my most important witness is that of the Holy Ghost, whose mission it is to testify of the Father and the Son and of their work.

Harold Bloom in his book, *The American Religion*, called Joseph Smith "an authentic religious genius unique in our national history," with "insight [that] could have come only from a remarkably apt reading of the Bible… So strong was this act of reading that it broke through all the orthodoxies — Protestant, Catholic, Judaic — and found its way back to elements that Smith rightly intuited had been censored out of the stories of the archaic Jewish religion." Regarding Joseph Smith, I have probably weighed most of the arguments against him being a prophet. As I weigh all the reliable evidence and use the scientific method of reasoning , Joseph comes up a Prophet of God in almost all that he did. The amount he accomplished in his short 38 ½ years is most inspiring. John Henry Evans wrote in his book: *Joseph Smith The American Prophet*

> Here is a man who was born in the stark hills of Vermont; who was reared in the backwoods of New York; who never looked inside a college or high school; who lived in six States, no one of which would own him during his lifetime; who spent months in the vile prisons of the period; who, even when he had his freedom, was hounded like a fugitive; who was covered once with a coat of tar and feathers, and left for dead; who, with his following, was driven by irate neighbors from New York to Ohio, from Ohio to Missouri, and from Missouri to Illinois; and who, at the unripe age of thirty-eight, was shot to death by a mob with painted faces.
>
> Yet this man became mayor of the biggest town in Illinois and the state's most prominent citizen, the commander of the largest body of trained soldiers in the nation outside the Federal army, the founder of cities and of a university, and aspired to become President of the United States.
>
> He wrote a book which has baffled the literary critics for a hundred years and which is today more widely read than any other volume save the Bible [probably today the Quran is read more than the Book of Mormon]. On the threshold of an organizing age he established the most nearly perfect social mechanism in the modern world, and developed a religious philosophy that challenges anything of the kind in history, for completeness and cohesion. And he set up the machinery for an economic system that would take the brood of Fears out of the heart of man—the fear of want through sickness, old age, unemployment, and poverty.
>
> In thirty nations are men and women who look upon him as a greater leader than Moses and a greater prophet than Isaiah; his disciples now number close to a million; and already a granite shaft pierces the sky over the place where he was born, and another over the place where he is credited with having received the inspiration for his Book.[53]

---

53) 1933

Chapter 2

When one realizes Joseph had only a third-grade education, studying his life shows direct evidence that his instruction came from on high. Emma Smith, Joseph's wife, who helped him with the translation of the Book of Mormon, was another close and reliable witness of the truth of this work. She knew Joseph was a prophet as sure as the eleven other witnesses who were privileged to see the plates from which the record was translated. For archeological and other evidence of where the Book of Mormon took place see the link below[54].

You will see the profound significance of the quotes I use throughout the book that have come forth in modern times authenticating that God has again been speaking to us. One of the most beautiful dimensions of God's love is that He will speak to all who will listen to "the still small voice." We cannot overemphasize the importance of personal revelation. All who desire may receive. He wants to communicate with you in a most loving way.

Now that we have some answers to some of life's most challenging questions along with our new-expanded, scientific method, we are ready to focus on several, life-altering, false traditions that have crept into society. In addition to uncovering them, we will give solutions and new perspectives in many cases from the best qualified experts in the world. If these solutions and perspectives were internalized by folks across the globe, I believe the world would be a much better place to live.

---

54) *http://ItsAboutTimeBook.com/book-of-mormon-geography/*

It's About Time

Section II

Applying Our New Basis for Scientific Thinking to Several False Traditions

Chapter 3

What Is God's Word?

If we are asked, "What is God's word?" most would say, "It is the Bible!" When we think about it, we come up with some interesting and very relevant questions. God's word is His word coming from His mouth; it is pure Truth, because He cannot speak otherwise. In fact, He doesn't even need to speak; He can communicate mind to mind in purity. Adam and Eve walked and talked with God. He has talked directly with many of the prophets and prophetesses. We have learned that He talked with each of us in our premortal sphere and gave us instructions that would best help us in our mortal journey, and those instructions are still resident in our intelligence that was downloaded into our brain at birth. We do not remember His Words to us because of the veil that was also placed so that we could live here by faith and have the opportunity to grow in so doing. In other words, we have the exciting Truth that His Word is resident in each of our intelligences which we can access through faith and obedience to His commandments, and by asking Him to reveal our missions to our hearts and minds and which are unique for each of us. The Savior did this perfectly. The best thing we can pray for in humility is, "Thy will be done." As we choose to follow His will, He can and will lead us on that path that is best for us in His infinite wisdom. We can totally trust in Him.

If we say the Bible is God's word, then we may ask, "Which translation is the most correct?" In spite of differences, the accuracy of the current translations we have is remarkable when compared with ancient texts — a validation that the translators were assisted by the Holy Spirit. Logically, we want to get as close to the source of His word as possible, and His word was not given to the ancient prophets and prophetesses in English! We have only a few of the original ancient Biblical texts. Do we have anyone who can give us a perfect translation of those few texts into English? The others have come to us through a series of translations and of copies of copies in many instances. How accurate are they?

If we ask the question, "Who has the greatest impact on the teachings of modern Christianity?" David W. Bercot in his excellent book, *Will the Real Heretics Please Stand Up?*, documents that the writings and teachings of St. Augustine and Martin Luther have had more influence than those of Jesus. In other words, the translations we have — as remarkably accurate as they are — have been biased by the paradigms of the translators. Can we expect otherwise? They bend the translations to their beliefs, for they believe that what they believe is true.

Many scientists believe the Bible is a myth. Should we trust science more than the Bible? The world, in general, seems to believe more in science, and the world is spiraling down as they remove themselves from the God of the Bible. Observing this dangerous trend should be a great big wakeup call to repentance. Christianity used to be the world's predominant religion; now it is Islam. Is there an important message in that trend? Ironically, Islam became a larger percentage of the world population about the time Barak Obama became president, 2008 — and the year of the financial crises.

If Christ is who He says He is, then why is He not the most influential person in history? The world (with Satan's enormous influence) is doing everything it can to put down the teachings of Christ, for Satan knows their value. This also is a witness to us of the importance of Christ's teachings and atonement as we see Satan fight so hard against them.

Then we may ask, "Has God talked to others outside of the Bible prophets and prophetesses?" "Yes!" is the exciting answer. Ancient writings have now been found documenting God's words to Enoch, to Zenos, to Jonah at

Chapter 3

Nineveh, and to several others. We have in Joseph Smith, Jr. a documented modern prophet as shown from logic in the previous chapter. By their fruits ye shall know them.

We will now show through inductive reasoning that Joseph Smith comes up with high probability as a prophet of God. You will see some beautiful and very important Truths coming out of this study. It was a major miracle how Joseph was able to translate and bring forth the Book of Mormon, which is a unique record of an extremely important piece of religious history here in ancient America. In his deep study of this book, our oldest son, Sterling, has written a book, *The Vision of All; Our Past, Present, and Future as Foretold in Book of Mormon History*.[1] The phrase, "The Vision of All," is from the 29th chapter of Isaiah, which is a prophecy of the coming forth of this record.[2]

Today, I believe there are thousands who are able to hear the " still small voice" of God, and they are serving humanity in many wonderful ways. It is exciting to live in a time when so much light and Truth are coming forth, and that in the midst of a world going awry. Fascinating.

At the personal level, about a year ago I received inspiration regarding a scriptural phrase with far reaching implications, as you will see below. Based on the information I was able to gather in the spirit of inductive reasoning, I was able to come up with a hypothesis that fit the scriptural information (the data). Then I was able to come up with an experimental test of that hypothesis. The test came out in 100% agreement with the hypothesis. Then about a month ago, another scriptural phrase was given to me that capstoned the previous one, and this one had even more relevance to our day and to the future. It is one of those cases told of in the scriptures where out of small things, the Lord will work great things. To develop this experience, I will share the following story.

Our study starts in the 1970s when the Lord had prepared a person to translate the Book of Mormon from English into Afrikaans in South Africa. There was a critical need for this translation at that time. Professor Felix Mijnhardt of Pretoria University and a member of the Dutch Reformed Church shares the following incredible story in 1972 at a conference for the Church as reported by a missionary, John Pontius, who was present. We got to meet John

---

1) It is available on his web site: http://www.greaterthings.com/Books/Vision/
2) Isaiah 29:11

before he passed. He died just four years ago, and his sweet wife Terri has been most helpful with additional material.

I attended the Stake Conference in Johannesburg on May 14, 1972, when the new translation of the Book of Mormon into Afrikaans (Die Boek van Mormon) was presented. It was an electric moment. People wept. Some had waited all of their lifetimes to read the Book of Mormon in Afrikaans. Many people had learned English for the sole purpose of reading this scripture. The Spirit was strong among us as we rejoiced. . .

Professor Mijnhardt was invited to come to the stand and speak about his experience in translating the Book of Mormon. He recounted how he had been given a gift of languages from God from his youth. He said that he was fluent in many languages, including English, Afrikaans, Hebrew and Egyptian, as well as many others. He was presently employed at Pretoria University as a language professor. He said he had been praying that the Lord would give him some task, some divinely important task, that would justify his having this gift of language from God.

He said. . . that he had visited a group of Mormon leaders, including a Bishop Brummer, Mission President Harlan Clark and others, who sought to commission him to translate the Book of Mormon from English into Afrikaans. He said that he knew of the Book of Mormon from his religions studies, and his initial reaction was that he did not want to be involved in translating it.

However, that evening, as he prayed upon his knees, as was his habit, he said the Spirit of the Lord convicted him. The message was something on the order of, "You asked me for a great, divinely inspired task of translation, I sent it to you in the form of translating the Book of Mormon, and you declined." Professor Mijnhardt said he could not sleep through the night because he knew that translating the Book of Mormon would get him into trouble with his university, which was owned and operated by the Dutch Reformed Church. When morning came he telephoned Elder Clark to inform him that he would begin the translation immediately.

Chapter 3

He stood at the pulpit and described the experience. He said, "I never begin translating a book at the beginning. Writing style usually changes through a book, and becomes more consistent toward the middle. Accordingly, I opened to a random place in the middle of the Book of Mormon, and began translating." He said, "I was startled by the obvious fact that the Book of Mormon was not authored in English. He said, "It became immediately apparent that what I was reading was a translation into English from some other language. The sentence structure was wrong for native English. The word choices were wrong, as were many phrases." He said, "How many times has an Englishman said or written, ' And it came to pass?'" We all laughed, and knew he was right, of course.

He continued, "When I realized this, I knew that I had to find the original language, . . .and then proceed to translate it into Afrikaans. He listed a half-dozen languages he tried, all of which did not accommodate the strange sentence structure found in the Book of Mormon. He said, "I finally tried Egyptian, and to my complete surprise, I found that the Book of Mormon translated flawlessly into Egyptian, not modern, but ancient Egyptian. I found that some nouns were missing from Egyptian, so I added Hebrew nouns where Egyptian did not provide the word or phrase. I chose Hebrew because both languages existed in the same place anciently."

"I had no idea at that time why the Book of Mormon was once written in Egyptian, but I can tell you without any doubt, that this book was at one point written entirely in Egyptian." I heard him say this over and over. Then, he said, "Imagine my utter astonishment when I turned to chapter one, verse one and began my actual translation and came to verse two, where Nephi describes that he was writing in the language of the Egyptians, with the learning of the Jews!"

He said, "I knew by the second verse, that this was no ordinary book that it was not the writings of Joseph Smith, but that it was of ancient origin and was in fact scripture. I could have saved myself months of work if I had just begun at the beginning. Nobody but God, working through a prophet of God, in this case Nephi, would have included a statement of the language he was writing in. Consider, how many documents written in English, include the phrase, "we are writing in English!" It is unthinkable and absolute proof of the inspired origins of this book.

He paused, then noted, "I am one of the few people in the world that is fluent in ancient Egyptian. I am perhaps the only person fluent in ancient Egyptian who is also fluent in Afrikaans and English. And I know for a fact, that I am the only person alive who could have translated this book . . . into Egyptian, and then into Afrikaans. . .

Professor Mijnhardt spoke of many other things regarding the translation of this book, and then said, "I do not know what Joseph Smith was before he translated this book, and I do not know what he was afterward, but while he translated this book, he was a prophet of God! I know he was a prophet! I testify to you that he was a prophet while he brought forth this book! He could have been nothing else! No person in 1827 could have done what he did. The science did not exist. The knowledge of ancient Egyptian did not exist. The knowledge of these ancient times and ancient peoples did not exist. The Book of Mormon is scripture. I hope you realize this. . ."

"I have taken this book of scripture, this Book of Mormon, and presented it to my Board of Regents, and urged them to embrace it as scripture. They declined, of course. I took it to the head of our Dutch Reformed Church and demonstrated why the Book of Mormon is scripture, and urged them to at least study it, even if they did not canonize it or even share it with the people of the church. I urged them to just think what having a new and profound book of scripture could mean to the church, to my church, the Dutch Reformed Church. I pointed out that they need not become Mormons, in the same way that they did not need to become Jews to embrace the Old Testament. They considered my presentation for a very few seconds and then rejected it. . . I am deeply disappointed, but I am not deterred. I will keep promoting this book as scripture for the remainder of my life — simply because it is scripture, and I know it."

He paused then added, "I am not a member of your church, and do not expect to become one. I have been asked why I have not joined your church many times, and my answer is because God has not directed me to join you. If He had, I would be standing here as a fellow Mormon. Perhaps my mission in life is better served outside of your church. I haven't studied your doctrine or your history since Joseph Smith. The only thing I know about you is that you have authentic, ancient scripture in the Book of Mormon, that your church was begun by a living and true prophet of God, and that all of the world should embrace the Book of Mormon as scripture. It simply can't be denied. I believe every religion could embrace the Book of Mormon without becoming a Mormon. . .

It feels so good that the Lord would provide scriptures coming from both the Eastern and Western hemispheres. In 2012, after learning of this remarkable story regarding Professor Mijnhardt, our bishop, Gene Peckham, of the Fountain Green First Ward asked us to read the Book of Mormon in a hundred days, knowing that in the doing we would have a significant spiritual uplift. I did, and it was as Bishop Peckham promised. But in addition to a spiritual uplift, knowing of the insights of Professor Mijnhardt about the ancient language nuances, as I read, it was almost as if several phrases and passages jumped out at me that were recognized by Professor Mijnhardt as being translated from ancient Egyptian. It was a great experience, and I filled up several pages in my journal as the Lord taught me beautiful new Truths. About a year ago, after having this experience, I felt to study the occurrences of one of these phrases: " And it came to pass."

The theory of evolution implies that civilizations and languages evolve over time. The scriptures teach just the opposite: the most sophisticated and advanced language was the language of Adam and Eve. The confusion of tongues at the tower of Babel brought about a degeneration of the Adamic language. As will be validated below, to my great delight, I found direct literary evidence that the phrase " And it came to pass" translates from the ancient Adamic language as a marking of a significant event in the Lord's doings. I believe we will yet learn of a special Adamic character like an asterisk or flag signifying that the Lord wanted a deeper focus on the message being shared following this marking. Apparently, such a character existed in at least the ancient Hebrew and Egyptian languages, for as we will see it carried on for a few millennia after Adam. That fingerprint of the Lord or marking was lost to usage after the writings of Matthew, Mark, and Luke and in modern revelation. All the Book of Mormon prophets used it except for Moroni, who was the last prophet in the Book of Mormon, and he used it profusely only when translating the Jaredite records, but not in any of his personal writings except in a very interesting place as he closed the Jaredite record. The enlightenment on this important place is explained in the article at the link below.[3] While preparing this article, I found a third witness of the importance and additional significance of the phrase "And it came to pass." As explained in this article, there are fifty-two paired occurrences of this phrase around seven clustered events. These double occurrences also seem to map back to the Adamic language as a third independent witness. His father, Mormon, used it at a significant level.

    Here is the grand key to this understanding. As recorded in the Book of Mormon, the Jaredites did not have their language confused at the Tower of Babel.[4] As Moroni brought forth a translatoin of their record and made an abridgement of their history as recorded in the book of Ether in the Book of Mormon, 100% of the chapters describing the Lord's doings among the Jaredites during their 1,600-year history contain this phrase, "And it came to pass." There are 160 occurrences of this phrase in the Book of Ether, by far the highest density in all of scripture. On average, every other verse uses this phrase. Fascinatingly, in the three chapters in the Book of Ether where Moroni makes commentary there are zero occurrences of this phrase except as cited above. In the Book of Genesis, 61% of the chapters contain this phrase, a high preponderance in the chronological beginning, and 63% of the Book of Mormon chapters contain this phrase, whereas only 22 % of the Old Testament taken as a whole and 13% of the New Testament chapters contain this phrase. Specifically, Matthew, Mark, and Luke — the synoptic gospel writers — use this phrase. These three writers were documenting the life of the Savior as proof to their constituency that Jesus was the promised Messiah and tying their writings to the Old Testament as part of the proof. In contrast, there are zero occurrences of this phrase in the writings of Peter, James, John, Paul, and Jude. These other

---

3) *http://ItsAboutTimeBook.com/science-and-religion-it-came-to-pass/*
4) See chapter one of the Book of Ether in the Book of Mormon

Chapter 3

five New Testament writers were addressing the members of the Church in the meridian of time; they had already accepted Jesus as the Christ. Their epistles were more along the lines of living the gospel they had received. Similarly, there are zero occurrences in the personal revelations given to Joseph Smith. On the other hand, in the Pearl of Great Price there are 57 occurrences of this phrase, and they occur only in the books of Moses and Abraham, where Joseph Smith was the prophet to bring forth these ancient revelations as found in these two books. There are zero occurrences of this phrase in the other books, which are modern revelations, in the Pearl of Great Price. The Pearl of Great Price along with the Doctrine and Covenants and the Book of Mormon combine with the Bible to make up the canon of scripture for The Church of Jesus Christ of Latter-day Saints. Interestingly, the Doctrine and Covenants contains zero occurrences of this phrase except in three places where Bible and Book of Mormon scriptures are quoted in the context of the revelation given.

More recently, I had a second epiphany when I learned of an exciting new dimension to the above: our past and future are tied to the two phrases " And it came to pass" and " And it *shall* come to pass." Another friend of ours, Lee Nelson, told me a month ago of a fascinating book beautifully augmenting what we have learned about the phrase " And it came to pass." In addition, this book brought in another phrase, " And it shall come to pass," which added a totally new dimension to the above and further testified from an independent source the authenticity of the scriptures and the inspiration that has come in translations.

The book is entitled *The Secret Doctrine of the Kabbalah; Recovering the Key to Hebraic Sacred Science* by the late Dr. Leonora Leet. She obtained her Ph.D. from Yale and was a Professor of English at St. John's University. She spent 20 years in a massive re-envisioning of the Kabbalah. It ties to the tree of life, and we read from Wikipedia:

> Kabbalah seeks to define the nature of the universe and the human being, the nature and purpose of existence,.. it [has] its religious origin as an integral part of Judaism, to its later Christian, New Age, and Occultist syncretic adaptations.

Kabbalah means "to receive" and is felt to reveal how the universe and life work, and is the study of how to receive fulfillment in our lives. Dr. Leonora Leet's work is monumental in this regard, and this book is the second of a four-volume set on the Kabbalah — studying this ancient Hebrew tradition bringing added depth to their beliefs and history.

On page 12 she shares a fascinating perspective.

> The form of the verb chayah, meaning "to be," appearing in the word vehayah, normally translated "and it shall come to pass," has the same Gematria number as the Tetragrammaton, the holiest of divine Names, since it simply transposes its letters from YHVH to VHYH, the Gematria value of these letters being 26 (Yod = 10, Hey = 5, Vav = 6, and Hey = 5). But its cognate term veyehi, normally translated "and it came to pass," has a Gematria value of 31 (Vav = 6, Yod = 10, Hey = 5, and Yod = 10). What is interesting about this number is that it is the same as the word for God, El (Aleph = 1 and Lamed = 30). The significance that can be drawn from these Gematria equations is that the word referring to a future product of causality is associated with the personal name of God, the Tetragrammaton, while that referring to a past causality is associated with the impersonal Name, or rather word, for God, El.

She goes on to say that "the personal realm with the future can be taken to support the main cosmological thrust ... that the course of cosmic evolution is toward the development of divine personality..." In other words, the phrase "And it came to pass" has a numerically equivalence to EL (God) and ties to important events of the past. While the words in the phrase "And it shall come to pass" have a numerical equivalence to Jehovah and tie to events in the future – moving us toward "divine personality."

When I learned about the numeric significance of these two phrases, I was excited. I felt to use deductive reasoning taking Dr. Leet's hypothesis that the phrase, " And it shall come to pass," numerically equates to Jehovah and ties to the future. I was able to find 157 occurrences of the phrase " And it shall come to pass" in the canon of scripture used by The Church of Jesus Christ of latter-day Saints, and almost without exception they all look to the future, just as she said, and in particular to the Second Coming of Christ and the events leading up to this world changing epoch. Essentially all of them tie to Jehovah, consistent with the Kabbalistic numeric equivalent as she explains in her book. I was so excited with this discovery and thanked the Lord for the inspiration

Chapter 3

to look in this detail at the significance of that phrase.

Looking to the past significant markers, the following are some very interesting examples of the use of the phrase " And it came to pass": (The bracketed [ ] notes after are my comments on that particular scripture — showing its great importance as indicated by the Lord with this marking.) They seem to tie in beautifully with this Kabbalistic view shared by Dr. Leet.

In the article at the link below, I share several examples of scriptures including the phrase "And it came to pass."[5]

These will give you a good feel for why the phrase is an important marker that EL, God, is using to help us understand His word. I share here two notable examples.

> And it came to pass after seven days, that the waters of the flood were upon the Earth. In the six-hundredth year of Noah's life, in the second month, the seventeenth day of the month, the same day were all the fountains of the great deep broken up, and the windows of heaven were opened.[6]

> And it came to pass after these things, that God did tempt Abraham, and said unto him, Abraham: and he said, Behold, here I am. And he said, Take now thy son, thine only son Isaac, whom thou lovest, and get thee into the land of Moriah; and offer him there for a burnt offering upon one of the mountains which I will tell thee of.[7] [This is the first time "love" is mentioned in the Bible as Jehovah (Jesus Christ) asks Abraham to go to the same mountain (a three-day journey), where Christ knew He would be crucified at the same time of the year — some 19 hundred years later — as type and shadow of the atonement, where the Savior would spend "three days and three nights in the heart of the Earth."[8]]

As a humorous aside, as Mark Twain reported in his book *Roughing It* during his encounter with Brigham Young and the Book of Mormon, he called it "chloroform in print," and said that if you removed all the " And it came

---

5) *http://ItsAboutTimeBook.com/science-and-religion-it-came-to-pass/*
6) Genesis 7:10
7) Genesis 22:1-2
8) Matt. 12:39-40

to pass" phrases, you would reduce it to a pamphlet. Outside of his being humorous, I wonder how he really felt. Little did he know of the great literary and religious significance of this phrase.

As Dr. Leet points out in her Kabbalistic view, the phrase, " And it shall come to pass" ties numerically to Jehovah and the future. This phrase has a very different sense, and almost always the prophets who use this phrase are a very different set of folks. When you look at who they are, their words align beautifully with Dr. Leet's view, as this phrase ties to our future.

I believe that in these two phrases we have a surprise for Mark Twain and for the world. Alma says to his son, Helaman: "…behold I say unto you, that by small and simple things are great things brought to pass; and small means in many instances doth confound the wise."[9] We see in the simple phrase " And it came to pass" a marker that El (God) is telling us something very important and this phrase gives us direct-extrapolated evidence for the existence of the Adamic language. In addition, from the data, we see that that language with its important markers filtered down in diminishing use into all the scriptures — both in the old world and in the new. In the second phrase " And it shall come to pass," which points us to the future according to Dr. Leet, and which as she shows is tied numerically to Jehovah, we see it pointing to the scriptures telling us of events associated with the glorious Second Coming of our Lord and helping us better prepare for that. How much more "gold" is hidden in the scriptures as we feast upon their infinite depth and beauty, each nugget designed to bring us ever closer to Him the Author of them.

To see direct evidence of the differences in text usage of these two phrases and to simplify, let us use P for past to represent " And it came to pass," and F for future to represent " And it shall come to pass." Whereas Isaiah only uses P three times, he uses F 27 times as he prophesies of our day leading up to the Second Coming of the Great Jehovah. I counted 59 occurrences of P in the book of Genesis and only 10 of F, which again aligns with Leet's view of past and future usages. Of the 135 occurrences of F in the Bible, they occur in those books prophesying of the latter days: Isaiah, Jeremiah, Ezekiel, Daniel, Amos, Micah (which Jesus quotes from extensively and which He shares with the people in America when He visits them after His

---

9) Alma 37:6

resurrection, telling of the latter days), Hosea, Joel (which Moroni quotes from when he first visits Joseph Smith — telling him of his work to translate the Book of Mormon in preparation for the Second Coming of our Lord), and Zechariah.

There are 63 occurrences of F in the Book of Mormon, which, remarkably, are all prophecies of the latter days given mainly by Nephi, Abinadi, and Jesus — telling of future events as the Book of Mormon goes forth and joins with the Bible. These scriptures tell of the role of the Gentiles in helping in the prophesied gathering of Israel in the latter days. Also, these scriptures share the Lord's gathering of the righteous from the "four-corners" of the earth as the world is purged. All of these great latter-day events are so relevant to us in preparation for the Great Day of the Lord as the Father brings about the fulfilling of His covenants and promises to us.

While Joseph Smith uses P zero times in the revelations given him, he uses F 37 times in revelations given him in the Doctrine and Covenants, and these 37 all relate to prophecies of latter day events and the coming of the Lord. All of this clearly shows the finger-print of the Lord and also this information lines up profoundly with Dr. Leet's Kabbalistic view of the past and future significance of the phrases P and F, respectively. The probability of all of this happening by chance is essentially zero. Hence, using inductive reasoning for the phrase, "And it came to pass," we have found direct evidence of the existence of the Adamic language and as being the most sophisticated. This is just the opposite of evolutionary theory. This phrase also validates a beautiful harmony between ancient and modern scriptures for both the eastern and western hemispheres. When this phrase occurs, God is telling you of some past important event.

In addition, using the deductive reasoning based on the model of Professor Leonora Leet, the phrase, "And it shall come to pass" from a totally independent data set perfectly validates her model. Her model is based on ancient Hebrew texts, traditions, etc. Our data set uses all available scriptures (both ancient and modern) and shows a perfect harmony with the model, providing direct evidence of the inspiration Joseph Smith received in bringing forth these records. In the occurrences of this phrase, Jehovah is telling you of our day and of important events leading up to His Coming. To see additional information on this very important phrase, "And it shall come to pass," see the article at the link below.[10]

---

10) *http://ItsAboutTimeBook.com/and-it-came-to-pass-jehovah-science-and-religion/*

A Second Witness

There have been some who have doubted the authenticity of the remarkable story around Professor Felix Mijnhardt's translation of the Book of Mormon into Afrikaans. On John Pontius' wife's, Terri's, blog site, "Unblogmysoul" you may find John's journal entry regarding this important historic event.[11]

As another witness of this transcendent event, I share the following: I gave a colloquium at Snow College on 14 March 2013, the science faculty and President of the College took us out to dinner after. I shared Professor Mijnhardt's story with them. Dr. Brian Newbold (Professor in their Engineering Department) came up to me and shared that he had served as a missionary in South Africa as well. Here was another witness of the authenticity of this remarkable story. While there, he met Brother Swanepoel, whom the Church had appointed to work with Professor Mijnhardt to assure the translation was doctrinally correct. Brother Swanepoel told Dr. Newbold that for the most part his translation was "spot on." I asked Dr. Newbold if he could give me more details. Fortunately, Newbold kept a journal, and he subsequently retrieved it for me to share details of the experience. I summarize those here. He had recorded, "Bro. Swanepoel's Afrikaans was excellent to my novice ear, and I'm sure he had an excellent education in his native tongue." He showed Newbold an autographed copy of the first edition off the press of this Afrikaans translation signed by Professor Mijnhardt, Bishop Brunner, and Brother Swanepoel. Newbold wrote in his journal:

> I remember Bro. Swanepoel describing the brilliance of the professor. He also explained how the professor felt "called" or "prepared" for this work. Bro. Swanepoel emphatically considered the translation a miracle. He openly stated that he rarely questioned any doctrinal concerns with the translation. He said it was the most spiritual experience of his entire life. At this moment as I write, my memory of that evening is clear, and I feel again the Spirit I felt that night testifying of God's hand in the translation. I remember Bro. Swanepoel testifying in clear language that he knew that God was in charge of the translation process.

One can read this in the Church's *Ensign* magazine, March 1973 issue:

---

11) http://unblogmysoul.wordpress.com/2014/06/24/die-boek-van-mormon/

Chapter 3

Bishop Johannes P. Brummer of the Johannesburg Second Ward, one of those who shepherded this valuable and important translation, told of the divine guidance that made its publication possible. He had translated about a third of the Book of Mormon into Afrikaans, but it had been a long, tedious effort, and it was imperative that the translation be completed without further delay so that the building up of the Church in South Africa could progress with greater speed and with every possible advantage. But where could a person be found with the necessary academic excellence and sufficient spirituality to complete such a task?

One day an acquaintance of Bishop Brummer brought his friend, Felix Mijnhardt, a language teacher from Pretoria, to meet him. The man not only had a consuming interest in everything related to the scriptures, but he also had been raised in a home with a spiritual atmosphere. His father, the Reverend C. F. Mijnhardt, compiled the first concordance of the Afrikaans Bible.

We have in the Bible and the Book of Mormon much of the word of God, but there is much more that we don't have. We have the following inspiring scripture in the Book of Mormon:

> For behold, the Lord doth grant unto all nations, of their own nation and tongue, to teach his word, yea, in wisdom, all that he seeth fit that they should have; therefore we see that the Lord doth counsel in wisdom, according to that which is just and true. [12]

Over the centuries, the Lord has raised up special people to help preserve and bring forth His word in many different languages. William Tyndale (1494-1536) is a classic example. He both devoted his life and gave his life as a martyr to bring forth the Bible in English. At one point he said to a fellow priest, "If God spare my life, ere many years I will cause a boy that driveth the plough shall know more of the scripture than thou doest." Tyndale was conversant in eight languages, and 90% of the King James edition of the Bible's New Testament is his translation. We have documented and demonstrated the accuracy of his translation above. His work was the inspiration for the Geneva Bible, which inspired the Pilgrims as they came to America.

---

12) Alma 29:8

I have eight different translations of the Bible in English. Upon checking all eight, only the King James Edition and the Geneva Bible retains the integrity of these two phrases, " And it came to pass," and "It shall come to pass." This gives direct indication of the inspiration given by the Lord to William Tyndale. How could he have known of their significance and placement — tying to the past and to the future of our Lord's coming? And the scriptures coming forth later through the instrumentality of the Prophet Joseph Smith were not even available at the time of Tyndale's incredible work of translation.

As we dig deeper into what is the Word of God, we have the following extremely important scripture from the first chapter of the Gospel of John:

> In the beginning was the Word, and the Word was with God, and the Word was God. The same was in the beginning with God. All things were made by him; and without him was not any thing made that was made. In him was life; and the life was the light of men... That was the true Light, which lighteth every man that cometh into the world. He was in the world, and the world was made by him, and the world knew him not. He came unto his own, and his own received him not.

Here we see that Christ is the Word because it is His life, His teachings, and His atonement that bring perfection to Father's perfect plan of happiness. Our hearts should be filled with gratitude for the sacrifices and miraculous efforts of all those who have brought forth His word to us that we may understand who is the WORD. As we prepare our hearts, the Lord will give us more.

I believe God's word is Truth in capitals, and can be totally trusted; it is the same yesterday, today, and forever. But, how can we know it is God's word? The Apostle Peter gives us the answer. We think of the scriptures as Truth, but are limited by the language in which they were written, by the prophet or prophetess who received them, by the translation errors bringing them to the language in which we read them, and by the mortal limitations of the reader. This is why Peter emphasizes that they were given by the Holy Ghost, and they can be correctly understood only by that same power.[13]

It is exciting to know that we can personally receive God's word. The

---

13) 2 Peter 1:19-21

Chapter 3

Apostle James made the following very important promise: "If any of you lack wisdom, let him ask of God, that giveth to all men liberally, and upbraideth not; and it shall be given him. But let him ask in faith..."[14] So His promise is that as we pray and ask, the Lord is there, and by the power of the Holy Ghost He will reveal to our minds and our hearts answers as we humbly and sincerely ask, seek, and knock.[15] The challenge most of us have is that answers will come in His time and in His way for He knows us better than we know ourselves, and He will answer in that way which is best for us from an eternal perspective. It is wonderful to know that He always answers a humble, heartfelt prayer.

As He reveals His Truth in our minds and in our hearts, it will be accompanied by the sweet affirmation of the Spirit — bringing love, joy and peace to our souls. It will always be enlightening. It is good to write these experiences and impressions received in a journal — to be shared, as guided by the Spirit — with our families and those we are privileged to serve. I have had this experience many times as have several of my family and friends. In His infinite capacity to love, He wants to commune with each of us, but He will not force Himself on us. Let us open the door of our hearts and let Him in.

It is also exciting that many ministers are now using the Book of Mormon. I have the following interesting letter from a Southern-Baptist, ordained-minister friend of ours. He gave me permission to share this letter he wrote to his friend, Mark:

Dear Mark,

What a delightful surprise your email was!

Thank you for your very kind words.

Yes, I'm an ordained Southern Baptist minister who for these past 22 years has been preaching out of both the Bible and the precious Book of Mormon. You might be happy to know there is a wonderful move of the Holy Spirit among us Protestants. More and more of us are embracing the wonderful message of the Book of Mormon. I have Lutheran ministers, Baptist ministers, Pentecostal ministers, Disciples of Christ ministers, who

---
14) James 1:5-6
15) Matt. 7:7-8

preach out of the Book of Mormon and who have a testimony of the book. I even have a Jewish Rabbi friend who teaches out of the Book of Mormon.

These are wonderful days!

Book of Mormon Christians and Bible-believing Christians, believe it or not, are coming together. I have a saying we have far more in common than all our differences. Are there differences? Of course. But it's time we began celebrating our commonalities rather than continue magnifying our differences.

You might be interested in this bit of history.

A couple years ago a group of LDS brothers and sisters in Salt Lake City, Utah, somehow discovered that there was a Baptist preacher living in Independence, MO who was preaching out of the Book of Mormon. They got my phone number and called me, asking if I would come out to Salt Lake and share my testimony. I went in January of 05.

While preparing my heart regarding what I was to speak on, I asked the Lord, Lord, what do you want me to say to these people? He spoke to my heart, "Ask their forgiveness."

Mark, that caught me off guard. But I had recently been delving into my family history and I ran across some information that startled me. On my mother's side of the family, one of my great, great, great uncles was a member of the Missouri mob that ran the Saints out of Missouri.

As a Baptist minister, that really ate on me knowing that I had relatives that were part of the Missouri mob. Anyway, here's what I did. I obeyed the Lord and prepared a formal written Statement of Repentance to read to my precious LDS brothers and sisters in Utah — got on the plane and flew out, and spoke to about 200 Saints.

Chapter 3

What can I say?! It was awesome. Very moving. No Baptist preacher had ever flown out to Salt Lake to ask forgiveness before…

I said all that to say this. Somehow the statement of repentance got all the way up to the First Presidency, including to President Hinckley. It wasn't long afterwards that I received a phone call from President Hinckley, asking if when the next time I came out to Utah, if I would not be his guest at the Lions House for a luncheon. Of course, I flew back out and had a wonderful luncheon sponsored by the First Presidency of your church.

For the past two and half years or so I have been flying back and forth from Independence to the Utah valley sponsoring what we call Building Bridges Conferences. The purpose of the weekend conferences is to get Book of Mormon believers and Bible-believing Protestants to attend, so that the two of us can see how much in common we have. And Mark, I don't mind saying the Lord has really blessed our efforts.

We hold our conferences all the way up and down the valley as far south as St. George all the way up to SLC. And we hold them in Protestant churches. We have hundreds of LDS brothers and sisters who attend, as well as Protestants. Pray for us as we continue building bridges.

Anyway, I thank you for your kind words. I'm glad you enjoyed the booklet *The Baptist Version of the Book of Mormon*. I have attached another booklet that seems to have caught on. It's titled *What Do You Think of Joseph Smith? 100 Evangelicals Interviewed*…

His Blessings as we labor together in His vineyard,

Lynn Ridenhour
Southern Baptist Minister
Independence, MO

When Lynn Ridenhour researched his ancestry, he was devastated to find that some of them were participants in the horrible atrocities committed against the Mormons as they were driven out of Missouri in the 1800s. Many were killed; there properties were destroyed, and Governor Boggs passed a law in 1838 to exterminate, or murder, all Mormons who would not leave the state. Lynn came to Utah with a "Statement of Repentance," asking forgiveness for him, and for his ancestors, and for all who participated in these horrible and un-Christian acts against the Mormons.

I was in the meeting where he shared these feelings of regret and repentance and all were in tears. He also shared it with the Leadership of the Church.[16] When President Gordon B. Hinckley, President of The Church of Jesus Christ of Latter-day Saints read Lynn's statement, he was deeply touched and invited Lynn to Salt Lake City for a special meeting to augment the building of bridges between our different faiths.

Moving on to Isaac Newton's testimony of the Bible, "We account the Scriptures of God to be the most sublime philosophy. I find more sure marks of authenticity in the Bible than in any profane history whatever." And, "No sciences are better attested than the religion of the Bible," and the world seems to agree with him.[17]

In their pride, modern scientists think to say, "Since the Bible doesn't agree with us, the Bible must be a myth." Newton would have said it just the opposite: "If something does not align with the Bible, then it must be false." So our procedure in this book is to take God's word as absolute and then show how the Truths of science line up with His word. We will show scientific data that beautifully line up with the Creation, with the fall of Adam and Eve, with Noah's flood, and with the life and infinite atonement of our Lord and Savior for all Father-in-Heaven's children and for the Earth.

God's word is Truth given specifically to bring to pass the immortality and eternal life of His precious sons and daughters. As we anticipate the glorious return of His Beloved Son, we will see much more of His word to bring about this the most important work on the face of the Earth in fulfillment of the Apostle Peter's prophecy of the "restitution of all things" that the Lord will bring forth before His Coming.[18]

---

16) *http://www.greaterthings.com/EzekielConference/press_releases/January24_2005/*
17) *http://itsabouttimebook.com/the-book-that-influenced-the-world-religion/*
18) Acts 3:19-21

Chapter 3

A Third Witness of the Grand Harmony of the Scriptures:

We have in the two phrases, "And it came to pass" and "And it shall come to pass," two witnesses of the continuity and validity of all the scriptures. These two phrases illustrate a continuity of scripture between the Eastern and Western hemispheres, as they exist in the Bible and modern scriptures. The first phrase I have tied back to the Adamic language with both inductive and deductive evidence when joined with the work of Professor Leonora Leet in her book, *The Secret Doctrine of the Kabbalah, Recovering the Key to Hebraic Sacred Science*. The following plot illustrates the evidence for the phrase "And it came to pass" having its origin in the Adamic language:

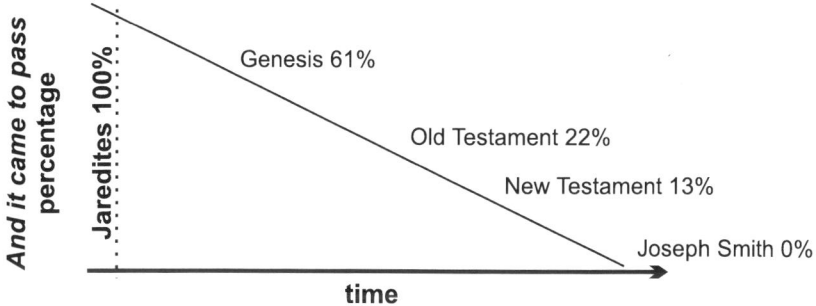

We have in the phrase, "And now it came to pass," a third witness.

As noted before, Professor Leet shares that the first phrase, "And it came to pass...," has a gematria equivalent to "EL," which refers to God, and ties to past events. She also shares that the phrase "And it shall come to pass" has a gematria equivalent to "Jehovah" and ties to future events.

In September of 2015, while reading the Book of Mormon, I discovered the significance of this "third witness" phrase, "And now it came to pass," as it ties to the Lord's definition of truth in conjunction with the other two phrases:

And truth is knowledge of things as they are, and as they were, and as they are to come;[19]

---
19) *Doctrine and Covenants* 93:24

We see the following correspondence:

And now it came to pass corresponds with *things as they are;*

And it came to pass corresponds with *things as they were;* and

And it shall come to pass corresponds with *things as they are to come.*

There are 96 occurrences of the phrase "And now it came to pass," in the Book of Mormon, and this phrase is exclusively in this book and in no other scriptures. As you study the usage of this phrase, it corresponds exactly with "things as they are." In the same vein, when you study the phrase "And it came to pass" and the phrase "And it shall come to pass," they correspond in usage to "things as they were" and "things as they are to come," respectively. These three phrases give profound evidence for Divine design and harmony amongst all the scriptures.[20]

In addition, in 40 of the 96 verses – nearly half – where the phrase "And now it came to pass" occurs in the Book of Mormon, the phrase starts the chapter. It is as if the Lord is saying, this is an important truth "now" that I want you to understand, and then the contents of the chapter are shared. Similarly, in the broadened search for this phrase cited below, in 13 of the 21 uses found in the Bible – over half – the phrase also starts the chapter. This again shows a grand harmony between ancient scripture and modern. Both in the Bible and the Book of Mormon, the usage of this phrase stretches across several books in both BC and AD time periods and both the Eastern and Western hemispheres. It is not unique to a particular prophet, but illustrates the fingerprint of the Lord. That Joseph Smith could have known the information shared in my book and in the book's appendices about these three phrases and introduced them as they are has a probability of zero. Divine design and the inspiration behind all of the scriptures is manifestly evident. It seems that these three phrases have their origin in the Adamic and tie directly to the Lord's definition of truth.

The Book of Mormon preserves the phrase perfectly as translated by the gift and power of God. Who uses it, as well as how it is used, is both inspiring and fascinating. Nephi, son of Lehi and the first historian, uses it four times; his brother Jacob uses it two times; Jacob's son, Enos, uses it once, Mormon,

---

20) *http://ItsAboutTimeBook.com/religious-science-translation-of-it-came-to-pass/*

the main historian, uses it 80 times; and his son, Moroni, uses it once as follows:

> Behold I, Moroni, do finish the record of my father, Mormon. Behold, I have but few things to write, which things I have been commanded by my father.
>
> And now it came to pass that after the great and tremendous battle at Cumorah, behold, the Nephites who had escaped into the country southward were hunted by the Lamanites, until they were all destroyed.[21]

It is as if he notes his father's historic style and starts his record out the same way – telling of his current lonely situation after the great destruction of his people. Mormon quotes letters written by Helaman to Captain Moroni, and in those letters, Helaman uses the phrase eight times in the three chapters quoted by Mormon. Helaman is a very spiritual, military leader.

As one examines the distribution, context, and who and why these three phrases are used, it gives both inductive and deductive evidence of the validity of the Bible, as Joseph in Egypt prophesied would happen in the day the Bible would be believed to be not true. Most of the leading scientists in America believe the Bible to be a myth. Now, if they have integrity of heart, they can see otherwise and be brought back to faith in God. The Lord told Joseph in Egypt the following:

> But a seer will I raise up out of the fruit of thy loins (Joseph Smith, Jr.); and unto him will I give power to bring forth my word (The Book of Mormon) unto the seed of thy loins—and not to the bringing forth my word only, saith the Lord, but to the convincing them of my word, which shall have already gone forth among them (The Bible).
>
> Wherefore, the fruit of thy loins shall write; and the fruit of the loins of Judah shall write; and that which shall be written by the fruit of thy loins, and also that which shall be written by the fruit of the loins of Judah, shall grow together, unto the confounding of false doctrines and laying down of contentions, and estab-

---

21) Mormon 8:1,2

lishing peace among the fruit of thy loins, and bringing them to the knowledge of their fathers in the latter days, and also to the knowledge of my covenants, saith the Lord.[22]

In addition, these three phrases, as three witnesses combined into one, giving a grand harmony to God's definition of truth – tying modern scripture to the ancient Adamic language, which is apparently more pure than any other.

If we broaden our search to not require the exact phrasing, "And now it came to pass," to include "now" and "it came to pass," then the translations in the King James Version of the Bible gives us 21 such scriptures. All of them fit the model of an important truth "now" – "things as they are." Most of the occurrences, both in the Bible and the Book of Mormon, as you will see, start with verse 1 of the chapter in which it occurs. With this broadened search, we pick up 43 more scriptures in the Book of Mormon, which fit the same model. In the following link, I share these 160 scriptures, 139 of which occur in the Book of Mormon.[23] In addition, the two links following this one help us gain greater insights into the scriptures and their significance for us today.

---

22) 2 Nephi 3:11-12
23) http://ItsAboutTimeBook.com/a-third-witness-spirituality-science/
http://ItsAboutTimeBook.com/religious-science-translation-of-it-came-to-pass/
http://ItsAboutTimeBook.com/shall-is-sure-spiritual-science/

Chapter 4

The True Nature of Love

There is so much of false love in the world. In this chapter are three articles to counter the false traditions that have crept into the world. The first one, 4.1, *All About Love*, is available with some variation on our web site.[1] The family is the most important unit in time and in eternity, and love is the primary mover to bring about the maximum happiness, joy and security in the family and in the home. I wrote this first article some years ago for our bishop to help him in counseling people in order to help strengthen the families in our ward (congregation) in Fountain Green, Utah. I was called to serve as a bishop in 1966, and have access to some excellent materials since that time to help in this regard and felt impressed of the Spirit to write this.

The second article, 4.2, is to help define from the scriptures what love is in God's eyes as best we can ascertain His divine view, which is the most important perspective we can have. In this article we show that love is the first law of heaven with an identity relationship to obedience. I have not seen this perspective taught in any church. As you will see, the implications of this are profound — given many of the false paradigms that have crept into societies over the years. I believe the principles taught here are timeless and

---

1) *http://www.allanstime.com/Spiritual/all_about_love.htm*

fundamental to our maximum joy, happiness, peace and development of love in our mortal journey.

The last article, 4.3, teaches us how to love as God would have us love. These teachings are from His word and from some excellent resources to bring the "how to" into a balanced perspective that we may best incorporate these teachings and the above principles mentioned into our lives for maximum benefit in time and in eternity — helping Father-in-Heaven to do His work of bringing to pass the immortality and eternal life of families.[2]

There is some duplication in these three articles, but repetition brings conviction. Besides, learning to love is the most important message we can learn. It is the solution to all the world's problems. Heavenly Father's work is motivated by His infinite love. For God so loved the world that He gave His Only Begotten Son...[3] to bring the opportunity of perfection out of all the world's imperfections. How great is that love? We can taste now, and have the promise that we can come to know His love as we truly come unto Him.

4.1 All About Love:
Finding, keeping, understanding, and enjoying
true love in your life

True love is pure and is the path to the greatest happiness in time and eternity. Additionally, it leads to the solution of all problems. In contrast, distorted love leads to most of the problems we have in society.

I. Perspective

This is written with the hope of helping in a fundamental area of societal needs of what love is and how to love and enjoy love as the Lord designed love to be. Love is elusive for many, because they are often chasing after love in the wrong places. True or pure love — as designed by a loving Heavenly Father and Mother with all of its enormous and most significant blessings — is all that the heart can desire. It is not elusive, but is constantly ever there for us to partake of, if we will turn to Him (meaning both the male and female energy of God) to find it. It is when we turn to the world's ways of loving that distortions often enter in, and pain, misery, and unhappiness follow.

---

2) Moses 1:39
3) John 3:16

Chapter 4

Here, we share how to enjoy love, how to partake of love, how to give love using the guidelines from inspired writings we have received from the Lord, and how to solve problems using love in the Lord's way. The Lord's love gives every hope, and fills the soul with joy. In opposition, distorted love is empty and devastating to the soul. Distorted love is currently destroying the fundamental unit of society, the family.

As a serious sign of the devolution of true or pure love into a distorted form, consider the results of a ChristiaNet poll: 50% of all Christian men and 20% of all Christian women are addicted to pornography.[4] They conclude: "Pornography is corrupting the church and destroying our nation; without action the problem will only get worse." And it has gotten worse.[5] With easy Internet access, pornography is now epidemic and rapidly growing its tentacles of seductive and titillating enticements. Also, it is now the trend in America for couples to live together out of wedlock, and the commitment to marriage and family are on the decline. Now, only 5% of marriages reach their 50th anniversary. Physical and mental abuse and sexual molestation in the family setting are increasingly devastating families, which is one of the byproducts of pornographic indulgence.

Divine love lifts our hearts, heals our spiritual and emotional ills, and wholesomely satisfies our inherent need for love. With sufficient faith, the Lord can and will heal wounds of every kind, of every person, and of every nation who will turn to Him. The Lord's love, in His infinite capacity to heal and to change us (patterned after His Divine nature), is ever there for us. Without Him, it is hopeless. With Him, all things are possible, and we know that in Him we can overcome these enormous problems that face us. His "arms are stretched out still," and his invitation is, "Come unto me, all ye that labour and are heavy laden, and I will give you rest. Take my yoke upon you, and learn of me; for I am meek and lowly in heart: and ye shall find rest unto your souls. For my yoke is easy, and my burden is light."[6] And, "If my people, which are called by my name, shall humble themselves, and pray, and seek my face, and turn from their wicked ways; then will I hear from heaven, and will forgive their sin, and will heal their land."[7] If we turn from Him, we are doomed. If we turn to Him, He will show us the way to true and pure love, even His love.

---

4) reported by Marketwire.com (August 7, 2006)
5 http://blazinggrace.org/pornstatistics.htm
6) Matthew 11:28-30
7) 2 Chronicles 7:14

As explained in more detail below, His love also helps us overcome addictions in addition to finding pure or true love, herein is true freedom given only by Christ.

In the Garden of Eden Satan taught "partake." "Ye shall not surely die." We see that same deceitful spirit continuing. Now, it is the enticements of the world, and Satan says, "partake; ye shall not surely die (spiritually), but have pleasure therein." This is one of the greatest deceits of the "great deceiver." Most of his enticements lead to addictions: drugs, alcohol, tobacco, pornography, sexual sins, which lead to abusive behavior and broken families. Unfortunately, many of us imbibe to one degree or another in some of these deceitful strategies that take us away from true or pure love, the Lord's love, and such behavior makes us miserable, discouraged, or depressed as Satan would have us be, the opposite of joy and peace.

How are we led into these distorted forms of love? Satan uses the strong desires and needs with which we are born, and deceives us into believing he has a more desirable way than the Lord's. Let us look at the main tool of deceit the Adversary uses; as defined in Noah Webster's 1828 Dictionary:

DECEIT, noun [Norm. deceut, contracted from Latin deceptio.]

1. Literally, a catching or ensnaring. Hence, the misleading of a person; the leading of another person to believe what is false, or not to believe what is true, and thus to ensnare him; fraud; fallacy; cheat; any declaration, artifice or practice, which misleads another, or causes him to believe what is false. My lips shall not speak wickedness, nor my tongue utter deceit.[8]

2. Stratagem; artifice; device intended to mislead. They imagine deceits all the day long.[9]

3. In scripture, that which is obtained by guile, fraud or oppression. Their houses are full of deceit.[10]

4. In law, any trick, device, craft, collusion, shift, covin, or underhanded practice, used to defraud another.

The Lord gives a sure way to not be deceived:

---

8) Job 27
9) Psalm 38
10) Jeremiah 5, Zephaniah 1

And whoso treasureth up my word, shall not be deceived, for the Son of Man shall come, and he shall send his angels before him with the great sound of a trumpet, and they shall gather together the remainder of his elect from the four winds, from one end of heaven to the other.[11]

Our key then is to "treasure up my word." Jesus is the Word,[12] and his words give us the sure path, and His Spirit the sure witness that we are on the right path. For He said most succinctly: "I am the way, the truth, and the life: no man cometh unto the Father, but by me."[13] As we listen and hearken to His voice, He has promised, and His promises are sure:

And at that day, when I shall come in my glory, shall the parable be fulfilled which I spake concerning the ten virgins. For they that are wise and have received the truth, and have taken the Holy Spirit for their guide, and have not been deceived—verily I say unto you, they shall not be hewn down and cast into the fire, but shall abide the day. And the Earth shall be given unto them for an inheritance; and they shall multiply and wax strong, and their children shall grow up without sin unto salvation. For the Lord shall be in their midst, and his glory shall be upon them, and he will be their king and their lawgiver.[14]

We are preparing for the great day of the Lord. If we think not, we are deceived, and Satan will lead us down distorted paths of love, which will leave us as the five unwise virgins (unprepared and empty). The Lord's counsel is: "Wherefore, be faithful, praying always, having your lamps trimmed and burning, and oil with you, that you may be ready at the coming of the Bridegroom."[15] We will discuss how we do that. First, let us consider, "What is love?"

II. What Is Love?

Probably, more has been written about love — letters, books, plays, movies, dramas, poems, and more has been sung, spoken, thought, taught about love, and more emotions have been stirred about love, more souls have been touched and changed by love, — than about any other topic, word, verb, or activity in human existence. Rather than try to capture the essence of all of the above, we

---

11) JS-Matt.37
12) John 1:1
13) John 14:6
14) D&C 45:56-59
15) D&C 33:17

will go to the heart of the question (What is love?) on a scriptural basis for in the Lord we know we can trust. Satan can and does lurk in avenues of love — subtly and deceitfully — to destroy true or pure love. Often, in the philosophies of men he seeks to entice us to fulfill our appetites in distorted ways.

It is wonderful to think that love is the principal power by which Heavenly Father governs the universe. He loves each of His children with an infinite love and everything else, which was the pure motivation for sending us His Son to work out the infinite and eternal atonement. Through His Son He has opened the incredible door, through which if we choose to enter, we may receive a fullness from the Father, of eternal joy, and be even as He is.

As we go to the scriptures, it seems significant that the first mention of love in the Bible is when the Lord commanded Abraham to take his son to Mt. Moriah to be sacrificed: "Take now thy son, thine only son Isaac, whom thou lovest, and get thee into the land of Moriah; and offer him there for a burnt offering upon one of the mountains which I will tell thee of."[16] This act is symbolic to Arab, Jew, and Gentile of the Father offering His Beloved Son on that same mountain some 1,900 years later as the greatest act of love in the history of mankind — as an offering for all our sins — breaking the bands of death and hell. "For God so loved the world that he gave his only begotten Son, that whosoever believeth in him should not perish, but have everlasting life."[17] We see that true love is purely unselfish and considers only the best interests of those who are the recipients of that act of love.

The love that the Lord designed for us to enjoy, we call true or pure love; any other kind we call distorted love. In contrast to true or pure love, distorted love is selfish and does not consider the best interests of those so influenced. The traditions regarding love that have come down to us are often of the distorted form of love. Consider, as an example, the rapidly increasing use of pornography, which science has shown acts like a drug when the body chemistry is considered. More and more are imbibing in this activity to satisfy their longing for love, but it leaves them empty after working its destructive course. Being addictive, the pornographic path leads to a downward spiral.

From the scriptures we learn that the first and the greatest commandment is to love God with all of our heart, might, mind, and strength. In Noah

---

16) Genesis 22:2
17) John 3:16

Chapter 4

Webster's definition of love, he succinctly says: ". . .and if our hearts are right, we love God above all things, as the sum of all excellence and all the attributes which can communicate happiness to intelligent beings."

The scriptures tell us that our next great opportunity then is to "love our neighbor as ourselves," patterning our love after His divine attributes. It has been said that, "We may not know what the future holds, but we do know who holds the future, and to be in step with him is the greatest opportunity of our existence." Moving forward with Him, we move with confidence and pure love and without fear, for " perfect love casteth out all fear."[18] And, "There is no fear in love; but perfect love casteth out fear: because fear hath torment. He that feareth is not made perfect in love."[19] We know that ". . . charity is the pure love of Christ, and it endureth forever; and whoso is found possessed of it at the last day, it shall be well with him."[20]

In Henry Drummond's monumental book, *The Greatest Thing in the World*, which is one of the most poignant discourses ever given on charity from 1 Corinthians 13, he says that "The Spectrum of Love has nine ingredients:

- Patience: 'Love suffereth long.'

- Kindness: 'And is kind.'

- Generosity: 'Love envieth not.'

- Humility: 'Love vaunteth not itself, is not puffed up.'

- Courtesy: 'Doth not behave itself unseemly.'

- Unselfishness: 'Seeketh not her own.'

- Good Temper: 'Is not easily provoked.'

- Guilelessness: 'Thinketh no evil.'

- Sincerity: 'Rejoiceth not in iniquity, but rejoiceth in the truth.'"

And from the life of the Savior, we know that love is unselfish service as He gave His all that we may inherit all things of God through Him. Drummond goes

---

18) Moroni 8:16
19) 1 John 4:18
20) Moroni 7:47

on to say, "Where love is, God is."

"Beloved, let us love one another, for love is (springs) from God, and he who loves [his fellowmen] is begotten (born) of God and is coming [progressively] to know and understand God [to perceive and recognize and get a better and clearer knowledge of Him].[21] "He that loveth not (with a pure or true love), knoweth not God, for God is love."[22]

After quoting this scripture from 1 John 4:8, Drummond proclaims, "Therefore love. Without distinction, without calculation, without procrastination, love. Lavish it upon the poor, where it is very easy; especially upon the rich, who often need it most; most of all upon our equals, where it is very difficult, and for whom perhaps we each do least of all. . . Give pleasure. . . For that is the ceaseless and anonymous triumph of a truly loving spirit, 'I shall pass through this world but once. Any good thing therefore that I can do, or any kindness that I can show to any human being, let me do it now. Let me not defer it or neglect it, for I shall not pass this way again.'

> ". . .there is no greatness in things. Things cannot be great. The only greatness is in unselfish love. . . The most obvious lesson in Christ's teaching is that there is no happiness in having and getting anything, but only in giving. . . And half the world is on the wrong scent in pursuit of happiness. They think it consists in having and getting, and in being served by others. It consists in giving and in serving others."

Drummond has a particularly powerful discussion on the importance of Good Temper as a critical characteristic of pure love. He says, "We are inclined to look upon bad temper as a very harmless weakness. . . the Bible again and again returns to condemn it as one of the most destructive elements in human nature." He calls it, ". . .the vice of the virtuous." He refers to the story of the Prodigal Son and points out how society brands the Prodigal as much worse than his brother, but then he asks the question, "But are we right?" ". . .there are two great classes of sins — sins of the Body, and sins of the Disposition. The Prodigal Son may be taken as a type of the first, the Elder Brother of the second. . . the Elder Brother. . . 'was angry,' we read, 'and would not go in.'

---

21) 1 John 4:7, *Amplified Bible*
22) 1 John 4:8, KJ, parenthetical is added

Look at the effect upon the father, upon the servants, upon the happiness of the guests. Judge of the effect upon the Prodigal — and how many prodigals are kept out of the Kingdom of God by the unlovely character of those who profess to be inside? Analyze, as a study in Temper, the thunder-cloud as it gathers upon the Elder-Brother's brow. What is it made of? Jealousy, anger, pride, uncharity, cruelty, self-righteousness, touchiness, doggedness, sullenness, — these are the ingredients of this dark and loveless soul."

True love is gentle, is forgiving, is kind, "thinketh no evil," brings perfect peace, casteth out all fear, lifts the hands that hang down, gladdens the heavy heart; strengthens the feeble knees, helps us to draw closer to the Father and the Son and to know them, helps bring out the best in each of us and we are better able to help bring out the best in others, helps us to know that we are loved and that each of us is a child of God — known personally by Him. True love exposes not the weaknesses of others, finds no faults, encourages to the fullest degree, gives us a new heart and brings to us the "being born of God" experience. As Drummond goes on to say, "To love abundantly is to live abundantly. . . No worse fate can befall a man in this world than to live and grow old alone, unloving and unloved." Eternal life, the greatest of all the gifts of God, is to know God, and "God is love." Hence, to know true love is to know God. So as you learn to love, you will learn to know God, and thus be on the road to the greatest of happiness in this life and to Eternal Life in the world to come, the greatest of all the gifts of God.

As mentioned above, love is the power by which God governs the universe. The universe is organized to bring about His Eternal work. Love is His work and His glory "to bring to pass the immortality and the eternal life" of His children. Love is not ethereal; it is a real force and of real substance. God's emanation of love is manifest in all directions as we see love manifest in the harmony and marvels of the heavens. As well, we see His love manifest in the intricacies and the intimacies of life down to the minutest level. From astronomic to microscopic we feel and see His love manifest. For those who have eyes to see, His love is manifest in the challenges and oppositions that life brings, giving us growth opportunities and ever wafting us closer to Him if we will but come unto Him — who is willing and wanting to bear our burdens. As we work with Him, he is able to turn every evil into good, and to overcome every sin and addiction.

As we learn to love as He loves, it becomes a real force in our lives, and those around us can feel that love. It reflects in our countenances. It moves us ever closer to Him, so that He becomes more and more a part of our lives. The Savior says, "If ye love me, keep my commandments."[23] The Apostle John says, ". . .if we keep his commandments. . . hereby we know that we know Him."[24] And Jesus promises, "He that hath my commandments, and keepeth them, he it is that loveth me: and he that loveth me shall be loved of my Father, and I will love him, and will manifest myself to him. . . If a man love me, he will keep my words: and my Father will love him, and we will come unto him, and make our abode with him."[25] In other words, truly loving Him allows the Father and the Son to be an intimate part of our lives. What an incredible promise, and, again, His promises are sure.

III. Practically, How Do We Open the Doors to True Love?

We may best address this vital question by first understanding the basic needs that a loving Heavenly Father placed in each of us for our mortal journey that we may enjoy true love and intimacy. Guided by true or pure love, we can fully satisfy these basic needs, which lead to a fullness of joy and eternal family relationships. Guided by distorted love in trying to satisfy these basic needs, we are led to momentary highs, but in the long-term distorted love leads to pain, emptiness, loneliness, and destructive human relationships. The most harmful dimension of distorted love is that it can lead to the destruction of the family. Because the family is the most important unit in time and eternity, the Adversary is working hardest to destroy the family. Unfortunately, he is having great success. Currently, one of his most effective tools is to get more and more people hooked on pornography, because, most often, it eventually leads to the destruction of the family.

Mark Kastleman in his book *Healing Hearts & Mending Minds*, poignantly states that we crave intimacy. "Whether we realize it or want to admit it, we crave human intimacy. We have an innate need to love and to be loved. We need to be close and connected to others, especially those in our immediate families.

"Many of the teenagers and adults I have interviewed who got involved

---

23) John 14:15
24) 1 John 2:3
25) John 14:21, 23

with Internet porn, cybersex chat rooms and/or illicit sexual encounters, reported that they were 'lonely,' that they felt 'disconnected,' that they lacked real ' intimacy' in their lives.

"You don't have to have sex to be 'intimate.' In fact, most human intimacy has nothing to do with sexual relations. Rather, it's about communication, understanding, appreciation, affection, mutual respect, friendship, quality time, sharing, and many more non-sexual actions and factors. One of the great preventions and protections against pornography addiction is true human intimacy, the quality and quantity of time you spend together as husband and wife, parent and child. This is what matters most." Another excellent resource in this regard is the book by Willard F. Harley Jr., *His Needs, Her Needs: Building an Affair-Proof Marriage.*

Kastleman goes on to make the powerful point that our homes need to be a refuge from the storms of life. We need to create a spirit of Zion in our homes (the pure in heart, where there are no poor [unloved or lonely] among us, where we are of one mind and of one heart), where Satan or his minions are held at bay, "because of the righteousness" of those abiding therein.[26]

Dr. Victor L. Brown, Jr., in his book *Human Intimacy — Illusion & Reality* says that he has learned the "bedrock reality that at every stage of our life we seek intimacy as urgently as we seek food and drink. We seek our parents' love. We seek friendship. We seek emotional unity in marriage along with physical fulfillment. Out of the love awakened by our children we find ourselves seeking their love even as we give love.

"This universal human need is so powerful that we are vulnerable to deception. Loneliness brings a desperation that makes us willing to see almost anyone as desirable, almost any situation as endurable, if it holds out the promise of intimacy. . ."

Kastleman, in discussing the proper, wholesome, and heavenly designed sexual intimacy between committed husband and wife, says, "The Experience is More Than Just Physical: As husband and wife move. . . together, there is more to the experience than just chemicals released in the physical body — much more. In addition to physical feelings, the emotions, the mind

---

26) 1 Nephi 22:26

and the spirit are all joined together in the experience. As natural chemicals are released, feelings of love, closeness and appreciation are brought to the forefront. The joining or fitting together of physical bodies becomes a symbol of the joining together of minds and spirits. God meant for this experience between husband and wife to be a union of their whole selves — body, mind and spirit. He designed it that way from the beginning. Such a union is possible only when there is total commitment to each other, when the couple shares all things, when they are bonded, welded, joined — married.

"Sexual intercourse is a sacred covenant between husband and wife, symbolizing oneness, fidelity and unity in all things. Sexual climax is a culmination of everything they have sacrificed and shared together, a celebration of their complete and total commitment, a crowning symbol of their marriage. . . This. . . experience is not possible outside of marriage."

As declared by God in the beginning, ". . .and they shall be one flesh."[27]

Kastleman goes on to say that, "The key to raising sexually healthy children and teens is to engender an environment where physical intimacy is regarded as healthy, positive, beautiful and even sacred, something to anticipate and look forward to at the appropriate time and under the right circumstances. Yes, we should teach our children about the dangers and darkness of pornography, premarital sex, self-indulgence, etc. But of greater importance is to demonstrate appropriate love within the home, coupled with gentle teachings on the wonderful joys of intimacy." The sacredness in the husband-wife relationship is better appreciated when we realize that in this act we have the magnificent opportunity to be co-creators with God, as we provide bodies for His spirit children. What an enormous responsibility and opportunity.

This is the most divine attribute of men and women as explained by Dallin H. Oaks. He says, "The expression of our procreative powers is pleasing to God, but he has commanded that this be confined within the relationship of marriage. . . Outside the bonds of marriage, all uses of the procreative power are to one degree or another a sinful degrading and perversion of the most divine attribute of men and women."[28]

---

27) Genesis 2:24
28) Dallin H. Oaks, *The Great Plan of Happiness* p. 74.

In total contrast, Satan would take away all of the sacredness as he does with pornography and its associated sexual sins and activities (lusting, lewdness, licentiousness, fornication, adultery, petting, masturbating, etc.). In addition, as Kastleman points out, pornography behaves exactly like an addictive drug on the body and is in fact addictive. Coming in through the mind, the chemicals released within the body are like that which happens with a drug addict. He says that parents need to teach their children "'why' pornography is harmful to their brain" — creating distorted unhealthful images that have to be rooted out before fully wholesome and healthy loving relationships can be enjoyed. For the addicted person, repentance, confession, and total faith in the Lord's atonement are critical for his recovery. There are other extremely useful steps that a person can take, if addicted, as will be discussed later.

There are many studies that have shown that the most important time in a person's life for learning proper sexual and intimate relationships is in the walls of a healthy, nurturing family setting, and surprisingly, the first three years of life. Single-parent families, or when both parents are employed, create significant challenges in bringing about these healthy-nurturing relationships during the first three years of a child's life, that are so critical in providing a foundation and a basis for all human relationships throughout his or her life.

Dr. Brown quotes from Burton White of Harvard University,

> I have devoted my whole professional career to pursuing the question of how competent people get that way. On the basis of years of research, I am totally convinced that the first priority with respect to helping each child to reach his maximum level of competence is to do the best possible job in structuring his experience and opportunities during the first three years of life. . . Therefore I do not think that any (other) job is more important in humanistic terms than the [proper role of parents].

Then Dr. Brown goes on to say,

> "In short, the ability to love in whole-hearted intimacy as an adult begins in the experience of having been loved as a child. In an environment of warm and familiar predictability, surrounded by people who celebrate his individuality without suppressing their

own, a child begins to develop a strong sense of identity. Thus fortified, he can move on to the next major prerequisite to intimacy — acquiring roles through which he can express his identity."

These roles are based on healthy role-models learned as an infant in a fully-functional, healthy home, where love and trust abound.

Regardless of where we are in the spectrum of our loving, intimate relationships with others, the Savior can and will help us move forward, and help us overcome every obstacle and every addiction, if we will but "Come unto Him." In her inspired book, *Come Away My Beloved*, Frances J. Roberts pens the profound invitation under the heading *The Call of Love*:

> O My beloved, abide under the shelter of the lattice for I have betrothed you to Myself, and though you are sometimes indifferent toward Me, My love for you is at all times as a flame of fire.
>
> My ardor never cools. My longing for your love and affection is deep and constant.
>
> Tarry not for an opportunity to have more time to be alone with Me. Take it, though you leave the tasks at hand. Nothing will suffer. Things are of less importance than you think. Our time together is like a garden full of flowers, whereas the time you give to things is as a field full of stubble.
>
> I love you, and if you can always, as it were, feel My pulse beat you will receive insight that will give you sustaining strength. I bore your sins and I wish to carry your burdens. You may take the gift of a light and merry heart, for My love dispels all fear and is a cure for every ill. Lay your head upon My breast and lose yourself in Me. You will experience resurrection, life and peace: the joy of the Lord will become your strength; and wells of salvation will be opened within you.

Ms. Roberts then goes on to share our dependence on God:

> My people, heed My words; yes, do not walk carelessly, nor lay

out your own paths on which to travel. You cannot know what lies in the distance, nor what adversity you may encounter tomorrow. So walk closely with Me, that you may be able to draw quickly upon My aid. You need Me, and no matter how well developed your faith is or how mature is your growth in grace, never think for a moment that you need My support any less. The truth is that you need it even more. For I shelter the newborn from many of the trials and tests I permit to confront those who are growing up in spiritual stature. You cannot grow unless I bring into your lives these proving and testing experiences.

So hold more firmly to My hand as you journey on in your Christian walk. Trust not in your own increasing strength, for truly, it is not your strength but rather My strength within you that you feel. You are as vulnerable to the treachery of the enemy and as frail as ever; but your knowledge of Me has deepened, and because of this your trust in Me should come easier.

Move forward with courage and confidence; but always allow Me to walk ahead, and choose the right path.

Let us recall that the purpose of life is to gain a body, to partake of life's experiences, and to learn to "do all things whatsoever the Lord [our] God shall command" — the greatest of which is to love God and to love His children (all of them). As we learn to love as God loves, we become like Him and we come to know Him. Knowing Him and His Beloved Son brings to us the greatest of all the gifts of God — Eternal Life.

In the face of life's challenges, trials, temptations, tests, and offenses, which are intrinsic to life's experiences, we often find it difficult to love. People (even loved ones) offend or hurt us. Often, our enemies would harm us greatly. The more we can view life from God's perspective, the more it helps to learn to love all and to forgive all that we may be forgiven of all — that the Lord may present us spotless to the Father, being perfected in Him.[29]

There are those that teach that life is a test. One may say, that this teaching is substantiated, because the Lord said, "We will prove them herewith. . ."

---

29) Moroni 10:32-34

But in so quoting, we best read the rest of the scripture, ". . .to see if they will do all things whatsoever the Lord their God shall command them." Keeping the commandments is too often viewed as a drudgery, a duty, "you have to do it, or else!" attitude. This attitude is a false tradition. We fail to realize that God said His way offers a fullness of joy and the greatest peace and happiness in this life. The greatest commandment is to love, which brings about the greatest fulfillment. [30]

Is Our Greatest Opportunity in Life to Learn to Love as He Loves?

Indeed, Life is an opportunity to learn how to Love as He loves, and to glorify Him who has given us the greatest of all the gifts of love — the gift of His Beloved Son. How grateful we should continually be for this most important gift. Through faith in Him, we learn how to love as He loves as we come to know Him. In so doing, we see the manifestations of His love in every direction in the marvels of His creations. The gift of His Only Begotten Son opened the door of faith and repentance, and the path whereby we may overcome all because He overcame all. As we internalize His love, we are changed by Him into "new creatures" in Christ, and feel to sing the songs of Redeeming Love. We receive a new birth in Him, and being born of God, we are brought to exclaim as the people of King Benjamin:

> And they all cried with one voice, saying: Yea, we believe all the words which thou hast spoken unto us; and also, we know of their surety and truth, because of the Spirit of the Lord Omnipotent, which has wrought a mighty change in us, or in our hearts, that we have no more disposition to do evil, but to do good continually. And we, ourselves, also, through the infinite goodness of God, and the manifestations of his Spirit, have great views of that which is to come; and were it expedient, we could prophesy of all things. And it is the faith which we have had on the things which our king has spoken unto us that has brought us to this great knowledge, whereby we do rejoice with such exceedingly great joy.[31]

The Atonement of our Savior gives every person a reason for hope. He

---

30) Matt. 22:36-40
31) Mosiah 5:2-4

will personally help us overcome every problem as we come unto Him. He has made that promise, and His promises are sure. Respecting our agency, He waits for our invitation to come into our lives, for our sincere faith in Him, and for our total, heart-felt repentance. In every thought turn to Him. Let your prayers be fully drawn out to the Father through Him continually. The peace and the joy that come through their cleansing power and the grace extended by the atonement place us on that path to overcoming the world and to receive eternal life in the world to come, which is the greatest of all the gifts of God (a fullness of joy). The Apostle Paul summarized it well, "My little children, of whom I travail in birth again until Christ be formed in you."[32]

IV. How to Build Loving Family Relationships:

A Christ-centered home is a home filled with His love, and Satan may come, but he has no power there. This home comes into being with Christ-centered parents, who teach their children to understand (by word and deed) that having Christ in their lives makes all the difference in the amount of true or pure love that is in the home. The home becomes a true refuge from the storms of the world. Here hearts are welded together in one, and the joys and eternal bonds of love that build in such a loving environment are the best that can be had in this mortal journey. This environment brings the greatest of joys not only in this life, but also in the life to come.

This may seem too idealistic, but the Lord has asked us to do it, and He never asks that which we cannot do. As we set our goals with the Lord's help we will move ever closer to Him and to this ideal. It is worth every effort. The degree to which we partake of this ideal is the degree to which we apply the principles that the Lord has given to us.

Let us review the basics for loving the Lord:

We truly love the Lord and desire to show Him our love by having our hearts in the right place — turning to Him in every thought and desiring to keep His commandments that we may reflect His love into the lives of others. Jesus prayed that we may be one as He and the Father are one and that His Kingdom would come on Earth as it is in heaven. It is about harmony in the home, with Him, with each other, and with the heavens. The attributes of those who truly love the Lord are:

---

32) Galatians 4:19

- Prayer is continually our guard, our guide, and our stay.[33]

- The inspired writings of the prophets are the road map for our lives; we daily feast upon these sacred writings individually and as a family.[34]

- Our hearts are filled with gratitude for all His loving kindness in all the bounteous blessings of life, and we are most grateful for the gift of His Beloved Son, as well as each other.

- We are grateful for the trials, challenges, and the vicissitudes of life as well — knowing that they give us growth opportunities.

- We love to serve Him, His children, and to help build His kingdom. We pray, as Jesus did, for His Kingdom to come on Earth as it is in heaven and for the reestablishment of Zion (a society made of pure people) to prepare for the glorious return of His Son;

- We keep the commandments as opportunities for exaltation and not as constraints that keep us from doing our thing;

- We keep the commandments because we love Him;

- We serve in the Church and the community as we have opportunity — not as a duty, but as is our sacred privilege to let the love of the Savior flow through us to others;

- We truly love our neighbor — especially those of our own household.

- In pure love, we seek to know and to unselfishly fulfill the needs of our spouse, that they may feel and bask in our pure and unfeigned love; male and female needs are very different by heavenly design. Learning those differences, and then learning to be sensitive to them is the process of becoming, as the Lord designed, "one flesh."

- We seek to serve the other members of our family — praying to know how we can best help them enjoy the Savior's love in their lives — helping them to feel part of the loving family team, and that their part is enormously important. We seek to see that their needs are fulfilled in harmony with the Father's Plan of Happiness.

---

33 D&C 10:5
34 2 Nephi 31:20-21

- We love, from deep within our hearts, all of Father's children, and seek for ways to help them to come to know the Lord's love and to help them feel ever closer to Him, that they may partake of the full measure of the redemptive powers of the atonement freely offered to them.

IV-A. In Marriage

As a loving husband seeks to fulfill the needs of his wife, he needs to know those needs. Similarly, as a loving wife seeks to fulfill the needs of her husband, she needs to know those needs as well. Because of fundamental differences between the male and female — designed by a loving Father-in- heaven — we often have misunderstandings about those needs. But properly understood and fulfilled bring the greatest of joys in the husband-wife relationship.

Pure love seeks to serve without the thought of receiving in return. A husband — filled with pure love — will not get upset with his wife if all of his needs are not met, and, again, similarly, a wife — filled with pure love — will not get upset with her husband if all of her needs are not met. They are both forgiving and understanding of each other because of the pure, Christ-like Love, in their hearts for each other.

The gospel of Jesus Christ brings to marriage a three-way relationship: between husband and wife and between each of them and the Lord. Each of them being " Born of God" is the most important part of their preparation for marriage as the Lord designed it. This "Born Again" experience can, of course, occur after marriage as well and will open doors to marital oneness not possible without the Lord.

Entering into covenants with each other and with the Lord for time and eternity is (can be, should be, ought to be) the most exciting, most joyous, and provides for the most fulfilling of all human relationships and experiences during our mortal journey, as well as in eternity. Being deeply in love with each other, they gladly, willingly, and with great joy in their hearts enter into these sacred covenants. She willingly and gladly gives herself to him — knowing that she can totally trust him — having had a witness that he is the right one for her companion in this most important union. Similarly for him, he willingly and gladly receives her to himself and promises to provide for her needs, for her protection, and to live up to that trust she has placed in him. As Paul so profoundly states, "Husbands, love your wives, even as Christ also loved the church, and gave himself for it;" [35] In other words, a husband should cherish and love his wife so much that he is willing to do all things necessary to provide for her needs, and he would be willing to lay down his life for her to protect her if need be.

Lloyd Newell shares: Those who have been happily married rejoice in the depth of that spiritual identity they share. One man, who lived in the nineteenth century, wrote this prayer for his wife: "Oh God. . .wilt Thou bless her with peace and with a long life; and when Thou shalt see fit to take her, let [me] go with her; and dwell with each other throughout eternity; that no power shall ever separate us from each other; for Thou. . .knowest we love each other with pure hearts. . .Now,. . . hear Thy servant, and let us have the desires of our hearts; for we want to live together, and die, and be buried, and rise and reign together in Thy kingdom with our dear children."[36]

When we hear this depth of feeling, we almost wonder if such emotion is possible in our world today. All around us we see examples of love that are fleeting—commitment that is temporary. We see emotions that are stirred and then die again, vanished like yesterday's styles. Perhaps it is because we have forgotten what marriage is. We become distracted by the first flurries of romantic attachment—and call that love. We find someone physically attractive—and call that love.

The love that becomes a joyful marriage has rich spiritual dimensions and forgiveness of one another. It says, "If it matters to you, then it matters to me." It says, "I can count on you under all circumstances." It says, "I am safe in your love. When I stumble on my weaknesses, you are not critical. I am a soul in the process of unfolding, and you accept me where I am. Because you love me, I am free to grow."

Is this kind of love really possible today? Yes, it is. But we must remember that it is the Lord who makes it possible. "Beloved," said John, "let us love one another: for love is of God; and every one that loveth is born of God, and knoweth God."[37] The kind of love that transforms two into one is a gift from

---

35) Ephesians 5:25
36) Orson F. Whitney, *Life of Heber C. Kimball* (Salt Lake City: Bookcraft, 1945), p. 335.
37) 1 John 4:7

God; and, if we want it enough, He is the one who can show us the way. We have to be willing to be transformed by Him into people far more capable of love and of loving than we could ever do on our own. We have to learn the fine lessons of forgiveness, patience, courage in the face of trials. As our hearts are changed through His love, we can, in turn, learn to love. Then, and only then, can we put our days of loneliness behind us to move into days of joy and being "one flesh" as He designed us to be.[38]

IV-B. In Parent-Child Relationships

There is a book, *Arming Your Children with the Gospel: Creating Opportunities for Spiritual Experiences* by R. Wayne Boss and Leslee S. Boss that is an excellent resource in this regard.[39] It is a how-to book with inspiring stories. I know the authors well, and they and their children have been tried by fire and have weathered the storms. All of them are examples of true disciples of Christ filled with love.

IV-C. The Value of "Touching" In Healthy Human Interactions

A book that every parent should read is the classic: *Touching: The Human Significance of the Skin* by Ashley Montagu. Montagu taught and lectured at Harvard, Princeton (where he chaired the Department of Anthropology), University of California, and New York University. He has written over 60 books.

The following is a reader's comment on this book taken from Amazon books:

> An amazing book on parenting and the touching humans need! If you only get one book on parenting and how to raise a child get this one. This is a thick book with lots of research to support the ideas presented by Dr. Montagu. However scholastic this book might be it is very readable. . . This is one amazing book full of data that instructs us to hold our children close as long as possible and why this is so critical to their development in every aspect. Read it!

From a *Reader's Digest* article on improving marriage, we read:

---

38) Lloyd D. Newell, *May Peace Be with You* [Salt Lake City: Deseret Book Co., 1994], 126
39) Available from Deseret Book: *http://deseretbook.com/store/product?sku=4634502*

> Human touch aids the release of feel-good endorphins, for giver and receiver. So link arms as you walk into the grocery store. Brush her cheek with your fingertips when you smooch good morning. Revive the ways you touched in the early days — a kiss on the back of the ear, a hand through her hair. Touch is a complex language. It pays to improve your vocabulary.[40]

Your touch, as motivated by pure or true love, always enhances human relations (with husband and wife, in the family, and in general) as long as the recipient knows where you are coming from. Touch is a very important coupling technique as loving energy flows from giver to receiver. It can come in many forms: a warm hand shake, a touch on the arm or cheek, a hug, pat on the shoulder, etc.

As Montagu discusses in his book, touching of a mother with her infant is critical in its development in all aspect of its being. Without touch, the infant will literally and most likely die.

Touch, as motivated by distorted love, leads down paths that are soul destroying and have the opposite effect as that motivated by pure or true love. Satan, knowing the importance of healthy human interactions, has turned the "touch" thing upside down with all kinds of distortions from what could be and should be appropriate in healthy human interactions both sexually and nonsexually. Satan must be happy that now our society is litigious about it. A teacher who touches a student may get sued!

The Lord gave us these beautiful tactile sensitivities. Properly used and appreciated, they bring so much of that which is beautiful, lovely, and very enjoyable in human relationships. Interestingly, touching is also so much a beautiful part of our lives in all relationships with animals as well. They know, feel, and enjoy the pure and tender touch coming from a loving human being.

Though not tactile, loving thoughts are like spiritual touch. Humans, animals, and plants respond to these in a beautiful way as well. As we think on Christ, His love and spirit will reflect from our countenances and bless the lives of all that interact with us. Keeping the second great commandment includes not only our neighbors, but all of His marvelous creations. Being one with God includes being one with them.

---

40) *Reader's Digest*, February 2007 p.170

Chapter 4

V. Have You Been Born of God So That You May Enjoy Pure Love?

Alma asks this most poignant of questions: "And now behold, I ask of you, my brethren of the church, have ye spiritually been born of God? Have ye received his image in your countenances? Have ye experienced this mighty change in your hearts?"[41] If we have not, we cannot enjoy the full blessings of true or pure love in our lives.

V-A. How Are We Born of God?

King Lamoni's father asked this question of Aaron in the Book of Mormon:

> And it came to pass that after Aaron had expounded these things unto him, the king said: What shall I do that I may have this eternal life of which thou hast spoken? Yea, what shall I do that I may be born of God, having this wicked spirit rooted out of my breast, and receive his Spirit, that I may be filled with joy, that I may not be cast off at the last day? Behold, said he, I will give up all that I possess, yea, I will forsake my kingdom, that I may receive this great joy. But Aaron said unto him: If thou desirest this thing, if thou wilt bow down before God, yea, if thou wilt repent of all thy sins, and will bow down before God, and call on his name in faith, believing that ye shall receive, then shalt thou receive the hope which thou desirest.[42]

Upon the King's following Aaron's counsel, he received this most important desire and was born of God.

We often associate being born again with being baptized of water and receiving the Gift of the Holy Ghost, and indeed, these are essential ordinances, but are of no avail if our hearts are not right before God so that we can receive the baptism of fire, which is administered by the Father, Son, and Holy Ghost, and which can and does give us complete freedom from sin. And as the Savior said, "If the Son therefore shall make you free, ye shall be

---

41) Alma 5:14
42) Alma 22:15-16

free indeed."[43] We must be born again of water, fire, and the Spirit to see the Kingdom of Heaven.[44]

The importance of being born of God and its relationship with the pure love of God is beautifully taught by John:

> Whosoever believeth that Jesus is the Christ is born of God: and every one that loveth him that begat [the Father] loveth him also that is begotten of him. By this we know that we love the children of God, when we love God, and keep his commandments. For this is the love of God, that we keep his commandments:... For whatsoever is born of God overcometh the world: and this is the victory that overcometh the world, even our faith.[45]

Beloved, let us love one another: for love is of God; and every one that loveth is born of God, and knoweth God. He that loveth not knoweth not God; for God is love. In this was manifested the love of God toward us, because that God sent his only begotten Son into the world, that we might live through him. Herein is love, not that we loved God, but that he loved us, and sent his Son to be the propitiation for our sins. Beloved, if God so loved us, we ought also to love one another.[46]

Whosoever is born of God doth not continue in sin; for the Spirit of God remaineth in him; and he cannot continue in sin, because he is born of God, having received that Holy Spirit of promise.[47]

Alma tells his profound and powerful conversion story to his son Helaman in Alma 36. The finger-print of the Lord is evidenced as this whole chapter forms an inspired chiasmus with the focal point being his coming to Christ and feeling the depths of the atonement. Even more depth can be found as one reads A with A' and B with B', etc. — arriving at the focal point — forgiveness through the atonement of Christ, which should be the same for each of us.

---

43) John 8:36
44) John 3:3-5; 2 Nephi 31:
45) 1 John 5:1-4
46) 1 John 4:7-11
47) JST 1 John 3:9

Chapter 4

A. My son give ear to my words (v 1)
  B. Keep the commandments and ye shall prosper in the land (v 1)
    C. Captivity of our fathers—bondage (v 2)
        [symbolic of our captivity when we sin]
      D. He surely did deliver them (v 2) [as he will deliver us]
        E. Trust in God (v 3)
          F. Support in trials, troubles and afflictions (v 3)
            G. I know this not of myself but of God (v 4)
              H. Born of God (v 5)
                I. Limbs paralyzed (v 10)
                  J. Angel spake: If though wilt be destroyed,
                  seek not to destroy the church of God (v 11)
                    K. I was racked with eternal torment (v 12)
                      L. Racked with my sins. . . tormented with
                        the pains of hell. . .pains of a damned soul (v 12-16)
                        M. Harrowed by memory of my many sins. . .then
                          remembered coming of Jesus Christ (v 17)
                            Alma's conversion, v 18-19: Now, as my
                            mind caught hold upon this thought, I cried
                            within my heart: O Jesus, thou Son of
                            God, have mercy on me, who am in the gall
                            of bitterness, and am encircled about
                            by the everlasting chains of death. And now,
                            behold, when I thought this, I could remember my
                            pains no more; yea, I was
                            harrowed up by the memory of my sins no more.
                        M' Oh what joy, my soul filled
                          with joy as exceeding as was my pain (v 20)
                      L' nothing so exquisite and sweet as was my joy (v 21)
                  K' Nothing so exquisite and so bitter as were my pains (v 21)
                J' Angels in the attitude of singing and praising their God (v 22)
              I' Use of limbs returns (v 23)
            H' Born of God (v 26)
          G' Therefore my knowledge is of God (v 26)
        F' Supported under trials and troubles and afflictions (v 27)
      E' Trust in him (v 27)
    D' He will deliver me (v 27)
  C' Egypt—captivity (v 28-29)
  B' Keep the commandments and ye shall prosper in the land (v 30)
A' This according to his word (v 30)[48]

---

[48] Adapted from *Chiasmus in the Book of Mormon* by John W. Welch, BYU Studies, vol. 10 No. 1 - Autumn 1969

Robert L. Millet says in his book, *Alive in Christ: The Miracle of Spiritual Rebirth*:

> The good news is that through the atoning blood of Christ, and by the power of the Holy Ghost, we can be born again, born from above, "born of God, changed from [our] carnal and fallen state, to a state of righteousness, being redeemed of God, becoming his sons and daughters." What a joyous message: We may become "new creatures" in Christ.[49] As President David O. McKay explained: "What the sun in the heavenly blue is to the Earth struggling to get free from winter's grip, so the gospel is to sorrowing souls yearning for something higher and better than mankind has yet found."[50]
>
> That something is a rebirth, a renewal. The Fall brought death, and Christ brought life, not only in the form of resurrection but also in the form of spiritual regeneration. Those who are born again have come alive to God and to godliness, have broken forth from the tomb of spiritual death, from complacency and inertia. They no longer walk "in darkness at noon-day."[51] The dawning of a new day dispels darkness and disbelief. President Ezra Taft Benson spoke of an experience that President David O. McKay had in the 1920s. After falling asleep, President McKay "'beheld in vision something infinitely sublime.' He saw a beautiful city, a great concourse of people dressed in white, and the Savior. 'The city, I understood, was his. It was the City Eternal; and the people following him were to abide there in peace and eternal happiness. But who were they? As if the Savior read my thoughts, he answered by pointing to a semicircle that then appeared above them, and on which were written in gold the words: These Are They Who Have Overcome the World — Who Have Truly Been Born Again! When I awoke, it was breaking day.'" President Benson added: "When we awake and are born of God, a new day will break and Zion will be redeemed."[52]

Professor Millet's book is an excellent resource for helping us to appreciate the power of the atonement to cleanse us.[53]

---

49) Mosiah 27:25-26
50) Gospel Ideals, 3
51) D&C 95:6
52) A Witness and A Warning, 65-66
53) Robert L. Millet, *Alive in Christ: The Miracle of Spiritual Rebirth* [Deseret Book Co., 1997], 15 - 16.

Chapter 4

V-B. How Being Born Of God Helps In Relationships

Alma offers insight into integrity of values and behavior in testifying of his own change:

> For, said he, I have repented of my sins, and have been redeemed of the Lord; behold I am born of the Spirit.
>
> And the Lord said unto me: Marvel not that all mankind, yea, men and women, all nations, kindreds, tongues and people, must be born again; yea, born of God, changed from their carnal and fallen state, to a state of righteousness, being redeemed of God, becoming his sons and daughters;
>
> And thus they become new creatures; and unless they do this, they can in nowise inherit the kingdom of God.[54]

Henry B. Erying asks this question, "Have you thought very much about that great, mighty change that the scriptures talk about?" Then he amplifies the answer:

> Alma the Younger described it this way: "The Lord said unto me: Marvel not that all mankind, yea, men and women, all nations, kindreds, tongues and people, must be born again; yea, changed from their carnal and fallen state, to a state of righteousness, being redeemed of God, becoming his sons and daughters; and thus they become new creatures; and unless they do this, they can in nowise inherit the kingdom of God."[55]
>
> Have you ever thought, "Wouldn't it be nice to have that experience?" Well, the scriptures don't suggest that it would be nice; they say that it's necessary in order for you and me to have what it is we want, which is eternal life. Now, so that you don't get too discouraged, consider what President Ezra Taft Benson said: "When we have undergone this mighty change, which is brought

---

54) Mosiah 27:24-26
55) Mosiah 27:25-26

about only through faith in Jesus Christ and through the operation of the Spirit upon us, it is as though we have become a new person. Thus, the change is likened to a new birth. Thousands of you have experienced this change. You have forsaken lives of sin, sometimes deep and offensive sin, and through applying the blood of Christ in your lives have become clean. You have no more disposition to return to your old ways. You are in reality a new person. This is what is meant by a change of heart."[56]

I want you to know that what we're talking about is real. This is not just "Wouldn't it be nice?" but something that can be done. Let me tell you how. The key is found in one word, and the word is surrender. Now, some people are used to being tough and strong, and they feel that surrendering is not something you should do very often. But we're not talking about surrendering to a human being. President David O. McKay said that " human nature can be changed, here and now." He then quoted this statement from Beverly Nichols: "You do change human nature, your own human nature, if you surrender it to Christ. Human nature has been changed in the past. Human nature must be changed on an enormous scale in the future, unless the world is to be drowned in its own blood. And only Christ can change it."[57, 58]

    The Savior, after His resurrection, offered a most poignant prayer for those who would surrender themselves to Him. These are they whom the Father gives to the Son because of their faith: " Father, I pray not for the world, but for those whom thou hast given me out of the world, because of their faith, that they may be purified in me, that I may be in them as thou, Father, art in me, that we may be one, that I may be glorified in them."[59] He prayed similarly just before His passion: "I pray not for the world, but for them which thou hast given me. . . which shall believe on me through their [the Apostles'] words; That they all may be one; as thou, Father, art in me, and I in thee, that they also may be one in us: that the world may believe that thou hast sent me."[60]

---

56) "A Mighty Change of Heart," *Ensign*, October 1989, p. 4
57) David O. McKay, *Stepping Stones to an Abundant Life* [Salt Lake City: Deseret Book Co., 1971], p. 23; as quoted in Ezra Taft Benson, " Born of God," *Ensign*, July 1989, p. 4
58) Henry B. Eyring, *To Draw Closer to God: A Collection of Discourses* [Deseret Book Co., 1997], 105
59) 3 Nephi 19:29
60) John 17:9, 20, 21

He is the one who can change us as we surrender to Him. As Paul said, "My little children, of whom I travail in birth again until Christ be formed in you."[61] As we surrender and invite Him into our lives, He will come in and perfect us — helping us to overcome all things. We cannot do it without Him. King Benjamin taught this principle profoundly:

For the natural man is an enemy to God, and has been from the fall of Adam, and will be, forever and ever, unless he yields to the enticings of the Holy Spirit, and putteth off the natural man and becometh a saint through the atonement of Christ the Lord, and becometh as a child, submissive, meek, humble, patient, full of love, willing to submit to all things which the Lord seeth fit to inflict upon him, even as a child doth submit to his father.[62]

Let us therefore surrender to Christ, so that we may truly be " Born of God" and be filled with His perfect love. As we do so we will then love our companions and our offspring with that same perfect love. By this example and feeling the door is opened for them to respond in kind more readily than otherwise. So let us be wise.

VI. Overcoming Addictions That May Destroy a Marriage or Family

Addictions may come in a variety of forms: alcohol, tobacco, drugs, food (overweight issues, e.g.), pornography, etc. We are so blessed to have the Savior ever willing to help us work through any of these. He has the solution to every problem — having overcome all things — and will help us overcome if we will come unto Him.

In addition to the scriptures, the following books are outstanding in helping to overcoming addictions:

- *He did Deliver Me from Bondage* by Colleen C. Harrison. It is available from Deseret Book.[63] It is a work book based on Book of Mormon Scriptures that when applied will bring a person to Christ, who will then deliver them from bondage. There are "Heart t' Heart" meetings being held weekly at various places across the country that are based on this book. There are also 'on-line' meetings.[64]

---

61) Gal. 4:19
62) Mosiah 3:19
63) The source may be found at Colleen's web site: *http://www.windhavenpublishing.com/*
64) *http://www. heart-t- heart.org/meetings.html*

- *Healing Hearts and Mending Minds* by Mark B. Kastleman has been referenced before in this article. It also is outstanding, and is written by a person who has overcome a serious pornography addiction that threatened to destroy his family. He is now free of this enormous burden of guilt. His book has also helped several others. Unfortunately, it is out of print and hard to find even used copies. In the Chapter 8, I outline the process he suggests for overcoming pornography addiction.

- *Clean Hands, Pure Heart: Overcoming Addiction to Pornography through the Redeeming Power of Jesus Christ*, by Philip A. Harrison, who is Colleen's husband and a cured addict from pornography.[65] It comes highly recommended. Colleen told me that church leaders are buying this book by the case. It is a Christ-centered book, helping to open the door of mercy to those with addictions, giving them hope. It is also designed for those who are working with those with addictions or who have family members with addictions.

Conclusions

Love is 'premièrement.' No opportunity or commandment is more important than to love. We see that as we come to understand and internalize true or pure love, harmony, joy, healing, and a oneness in our relationships both with God and with each other are the delightful fruits — especially with those who are near and dear. The message of gladness that the gospel brings into our lives is fundamentally a message of love. Learning to love is learning to live the gospel. As we live it, we receive and share the Lord's love, which fills our hearts, permeates our lives, and radiates to all within our influence.[66]

Faith in the Lord Jesus Christ and repentance are our path to receive this pure love, which entails coming to Christ so that He can change us and prepare us to receive His fullness of love, of light, and of truth. Doing so allows us to overcome all things because He overcame all things and has the solution to every problem and addiction. Coming unto Him brings the perfect peace from the Spirit that enlightens our lives and which is so critically needed in this world of wickedness.

---

65) Also available from *http://www.windhavenpublishing.com/*
66) Moroni 7:47-48

Chapter 4

Harmony and order exist in the ideal society filled with God's love. The confusion, addictions, hate, envy, immorality, etc. that we see in the world come from the dark side. Harmony of body, mind, and spirit come when we are one with the Lord. In contrast to the world, He helps us overcome the world. In contrast to the world, He helps us overcome the world in following three steps: 1) In our bodies, we are buried with Him by baptism — taking His name upon us, which baptism is symbolic of His death and glorious resurrection. 2) In our minds, we can choose to give Him the only thing that is really ours to give — our agency (we surrender our wills). When we truly do that, then, ironically, He can lead us to perfect and ultimate freedom. As He chose to totally submit to the Father's will, as our perfect exemplar, He was able to subdue all enemies under His feet and open the door to Eternal Life. 3) In our spirits we can — through full repentance and faith on the Lord Jesus Christ — receive the baptism of fire, which communicates to us the glorious truth that we are totally forgiven of our sins. Without this cleansing, we cannot enjoy that oneness and harmony with God that is essential to the greatest happiness here and in the world to come.[67]

In the arithmetic of the Lord, seven is basic. It takes seven thousand years to celestialize the Earth. There are seven days in the week to bring us to the Holy Sabbath, where we can better come to know the Lord and renew our covenants with Him.[68] There are seven archangels the Lord has appointed to head the seven main dispensations of the gospel to bring the inhabitants of the Earth to a restoration of all things promised by the Apostle Peter.[69] I have shown that there are seven dimensions to the Lord's atonement to bring us to perfection as explained in the link below.[70] There are seven learning centers in the brain, in which when fully utilized, we optimize our ability to love, to serve, and to best use our talents for personal growth.[71]

As we dig deeper, we find there are seven electron shells surrounding the nucleus of all atoms — making up all the elements in the universe. In Chapter 21 we share a new Unified Field Theory (UFT) being researched by the author in which there are seven spectral regions that the Lord uses to bring about the incredible harmony that we see throughout the universe. This

---

67) *http://www.allanstime.com/Spiritual/ born_again_body_mind_spirit.htm*
68) Ezek. 20:20
69) Acts 3:18-20
70) *http://ItsAboutTimeBook.com/spiritual-science-seven-steps-in-atonement/*
71) *http://www.allanstime.com/Spiritual/BookReports/magic_trees.htm*

universe interconnection is done via diallel-field-lines that connect everything to everything. We have been able to prove the existence of these diallel-field lines and that they have quantum states. Everyone has a direct diallel-field-line connection to the Father, the Son, and the Holy Spirit. Prayer uses these lines and is instantaneous, and is our greatest opportunity to ever be in touch with those who love us most. Time is a mortal limitation as we move along our eternal journey. We are limited by the four dimensions of space and time; the Lord is not, and we need not be. Love allows us to sense and move into the Eternity-Domain — the fifth dimension. Love is the power by which Father governs the universe and which motivates His great plan of happiness. Love is the power by which we may return to the Father and fully partake of a fullness of joy with Him, with the Lord, and with our loved ones. May we learn to love as Jesus and the Father love is my prayer. In so doing, we come to know them, which is eternal life — the greatest of all the gifts of God. One of my main purposes in writing this book is to help us all obtain that gift. This is my hope and prayer.[72]

4.2 Is Learning to Love Our Main Purpose in Life?

Love transcends the grave, and God's love is the primary motivation for all that He does. He so loved the world that He gave His Only Begotten Son to bring perfection to His plan of happiness.[73] He asks the same of us in the first two great commandments, and on these two hang all the law and the prophets.[74] In the commandment, then, to be perfect like our Heavenly Father and Jesus are perfect is our most important purpose — to learn to love like they love. Love is not only God's primary motivation for all that He does, but it is His primary characteristic; hence the scripture, "God is love."[75]

Every soul is precious to our infinitely loving Heavenly Father, and He wants us all to come back to Him and to be perfect like Him.[76] This brings Him the greatest honor and glory.[77] As we so choose, He promises us a fullness of joy and to be ONE with Him and His Beloved Son.[78] That path to perfection

---

72) *http://www.allanstime.com/Spiritual/In_Touch_with_Eternity.htm*
73) John 3:16
74) Matt. 22:36-40
75) 1 John 4:8, 16
76) Moses 1:39
77) John 15:8-9
78) John 17:20-23

Chapter 4

is through True Love, which perfection, ironically, we can never achieve by ourselves. But if we choose to follow Christ and inculcate His teachings in our lives, He promises to perfect us, and His promises are sure — leading us to the greatest happiness here and a fullness of joy hereafter.[79] His plan is the perfect plan of happiness and joy. There is no better plan. Why shouldn't each of us want to share it and follow it with all of our heart, mind, might, and soul?

As people observing all the hate, pain, and suffering in the world as part of His creation, many wonder, "Is what we observe a manifestation of God's love? How can this be, given all that we see?" All the problems in the world seem inconsistent with a " perfect plan." How can God, who is perfectly loving, allow or " create evil," as Isaiah says,[80] and all the badness we see. We find that in His infinite love " free choice — agency" is at the heart of the plan. Robert Burns so aptly said, "Man's inhumanity to man makes countless thousands mourn!"[81] Because of the importance of these question, I did a lot of research and wrote an article, which is on our web site entitled "Why does God, who is good, create (allow) evil?" We have already discussed this in some detail in Chapter 2.[82]

In Job's agony, he asked, "What is man?"[83] In contrast, the Psalmist, David, in his marveling at God's handiworks asks the same question.[84] We have both ends of the spectrum in our mortal journey: pain and suffering, but also joy. God created man — knowing of the great pain, heart ache, and suffering he would endure, and in contrast the ecstasy he could enjoy. Why would He do this? Is there a grand design behind it all? And surely there must be because God is God, who is Love and perfectly so. In His infinite capacity to love His children, would He as a loving parent knowingly subject his children to such as we see in the world if life's challenges were not necessary? As C. S. Lewis so profoundly stated "If they are unnecessary, then there is no God or a bad one. If there is a good God, then these. . .are necessary."[85] Many are led to atheism because they cannot believe that a "good" God would allow such as we see. They do not understand nor do they believe in a pre- mortal existence or an afterlife —leading in large measure to atheism, to immorality in the world.

---

79) Moroni 10:32-33
80) Isa. 45:7
81) Robert Burns *Man was made to mourn: A Dirge*, 1784.
82) (See http://www.allanstime.com/Spiritual/Why_Does_God,_Who_is_Good,_Create_Evil.htm
83) Job 7:17
84) Ps. 8:3-4
85) C.S. Lewis, A Grief Observed, p 43

The pre- mortal perspective seems quite different from what we perceive here on Earth. As I have shared before, we read from Job that we "shouted for joy" as we anticipated our opportunity to come to Earth.[86] Looking further into our premortal existence, we know from scripture that we were born as the literal spirit offspring of Heavenly parents.[87] Father explained His perfect plan of happiness, and we rejoiced in our anticipated opportunity to come to Earth. As this infinitely loving opportunity was extended to us to make this mortal journey in our path to godhood — to perfection and learning to love like our Heavenly parents, we looked at them, and we knew that any price was worth paying to be like them. We saw that any trial or challenge would be worth it; in fact, we saw that life's challenges were actually opportunities for growth in our lives and were necessary because of free choice. Further, we saw that because of free choice we would sin. The need for a Savior was clearly manifest, and Father's firstborn[88] offered to be our Redeemer — foreordained[89] to bring perfection to Father's plan because of His infinite love for us, as well. There we knew, and we further saw that here on Earth we would need to live by faith in the midst of life's challenges and that faith is to "hope for things, which are not seen, which are true"[90] — the most important being faith in Christ.

What other most important truths are there for us to come to learn and to know here below, as we exercise our faith? They are to come to know that God exists, that we are His children, that He loves us with an infinite love, that His plan provides a path based on free choice to come back to Him, to learn to love like He loves, which will bring us a fullness of joy, that He so loved the world that He gave His Beloved Son to bring perfection to the plan through the infinite atonement, and that He gives to every mortal child the Spirit of His Beloved Son[91] to know good from evil as a compass to find our way by faith — so that we may know in our hearts that these vital truths are true. This is why there are brain cells in the heart, and these cells are the most important part of our brains.[92,93] That is why we should listen to our hearts. We can build on these fundamental truths, which open the door to all truth.[94] This path of

---

86) Job 38:7
87) Acts 17:28-29, Heb. 12:9; D&C 76:24
88) Romans 8:28-29; Colossians 1:14-15
89) 1 Peter 1:19-20
90) Alma 32:21
91) John 1:9
92) *http://projectavalon.net/forum/showthread.php?t=1623*
93) *http://www.therealessentials.com/followyourheart.html*
94) D&C 93:24-28

righteousness leads to the greatest fulfillment, love, joy, peace and happiness in this life and a fullness of joy in the life to come, as we receive the greatest of all the gifts of God, Eternal Life.[95] A gift is not a gift unless it is appreciated.[96] Could eternal life, life with God, life like God, be appreciated if the recipient did nothing or there were no contrasts to perceive the enormity of the gift? Thus, opposites and opposition are a necessary part of the plan. Satan was purposely cast down to the Earth[97] in opposition to the Spirit of Christ. Otherwise, we could not enjoy and appreciate happiness here and a fullness of joy in the life to come.

This opposition creates some grand ironies in Father's perfect plan of happiness. Consider some examples: Those who suffer the most are rewarded the most, if they endure them well.[98] Of ourselves we are nothing,[99] but through the atonement we are of infinite worth;[100] each soul is precious to our Heavenly Parents, for we are their precious offspring. As another example, we gain total freedom by giving up the only thing that is truly ours to give — our free will. When we subject our will to the will of the Father, as Christ perfectly did, then as the Savior said, "If the Son therefore shall make you free, ye shall be free indeed"[101] — individually and as a society. And the one that seems the most ironic of all is that the Lord allows evil to bring about the greatest good; He is able to turn every evil into good. The capstone irony is that He allows hate and fear for us to enjoy the beautiful fruits of love. We know that " perfect love casteth out all fear."[102]

To many, life seems to be a great big test — to see how we will endure all of these trials and if we will be obedient to God's commandments in the midst of it all. It is taught that the first law of heaven is obedience. Rather than life being about love, they say it is about duty. Such a view seems so heavy. In addition, the commandments are often viewed as obligations, which also seems heavy. Is this what the scriptures teach?

The typical answer the Mormon missionaries give for our purpose in life

---

95) D&C 14:7; 59:23
96) D&C 88:33
97) Rev. 12:8-9
98) D&C 121:7-8; 122:5-9
99) Gal. 6:3; Moses 1:10
100) D&C 18:10-16
101) John 8:36
102) 1 John 4:18; Moroni 8:16

is to gain a body, to have Earthly experience, and to prove ourselves, which answer comes from Abraham 3:22-28. If we go to verse 25 of that scripture, it reads, "And we will prove them herewith, to see if they will do all things whatsoever the Lord their God shall command them;" And we know the greatest commandments are to Love.

In Doctrine and Covenants 59:1-5 we learn the beautiful truth that the commandments are opportunities for exaltation; a loving Father gives them as guidelines for our ultimate happiness. And as mentioned above, the greatest of these commandments is to love: to Love the Lord our God with all of our heart, might, mind, and strength, and the second is like unto it, to love our neighbor as ourselves (which includes our enemies), and on these two commandments hang all the law and the prophets.[103] John quotes the Savior saying, "If ye love me, keep my commandments." Later, in the same chapter, the Lord says, "He that hath my commandments and keepeth them, he it is that loveth me."[104] From logic and set theory we have the following profound truth: from John 14:15 we see that love implies obedience;

$$L \supset O$$

then in verse 21 we see that obedience implies love;

$$O \supset L$$

therefore, from set theory and logic being obedient is equivalent to the Lord's commandment to love:

$$O \equiv L$$

If we say, "He or she is an obedient person," it has a very different feel to us than if we say, "He or she is a loving person." That is because our telestial definition of obedience is not the same as Heavenly Father's. In other words, celestial obedience is bathed in love and celestial love is bathed in obedience. Noah Webster's definition comes very close[105]: "Voluntary [LOVING] obedience alone can be acceptable to God." Hence, the reason for Christ's answer to the lawyer as to what are the first two greatest commandments in the law and the Bible gives us the clarion message that everything in the Law and the Prophets hangs on these two Commandments.

---

103) Matt. 22:37-40
104) John 14:15, 21
105) 1828 Dictionary

Chapter 4

So, the most important thing we can learn to do is to learn to love like God loves. This we will naturally do as we come to know Him, and knowing Him, who is love,[106] is eternal life. Jesus said that eternal life is to know the Father and the Son and they are Love.[107] If we truly love, keeping the commandments is automatic because we want to please them and to bring honor and glory to their Holy Names.

How do we come to know them? Jesus further said, "If any man will do his will, he shall know of the doctrine, whether it be of God, or whether I speak of myself." In other words, if we live the Law of Love, we will find it to be the first law of heaven (bathed in obedience), because we want to do nothing but that which pleases our ever loving Heavenly Father and His Beloved Son.[108]

In the sermon on the mount we read,

> Not everyone that saith unto me, Lord, Lord, shall enter into the kingdom of heaven; but he that doeth the will of my Father which is in heaven. Many will say to me in that day, Lord, Lord, have we not prophesied in thy name? And in thy name have cast out devils? And in thy name done many wonderful works? And then will I profess unto them, I never knew you: depart from me, ye that work iniquity. Therefore whosoever heareth these sayings of mine, and doeth them, I will liken him unto a wise man, which built his house upon a rock: And the rain descended, and the floods came, and the winds blew, and beat upon that house; and it fell not: for it was founded upon a rock. And every one that heareth these sayings of mine, and doeth them not, shall be likened unto a foolish man, which built his house upon the sand: And the rain descended, and the floods came, and the winds blew, and beat upon that house; and it fell: and great was the fall of it.[109]

I love the French phrase in this scripture: 'les met en pratique' (put these teachings into practice). In other words, those who hear and do, have the

---

106) 1 John 4:8, 16
107) John 17:3
108) John 7:17; 14:21-24
109) Matt. 7:21-27

sure promise that they will "enter into the kingdom of heaven." We learn love by practicing love (the pure love of Christ). Those who truly love are on the path leading to the greatest of joy in this life and eternal life in the world to come — there to receive a fullness of joy and "all that the Father hath."[110]

By following the Spirit, listening to our hearts, and with our faith in Christ, we become Christ-centered; otherwise we are self-centered and are easily entangled in the sins of the world. The Lord has given the recipe of how we can always have His Spirit:

> Let thy bowels also be full of charity towards all men, and to the household of faith, and let virtue garnish thy thoughts unceasingly; then shall thy confidence wax strong in the presence of God;... The Holy Ghost shall be thy constant companion...[111]

Having the Spirit as our "constant companion" and treasuring up His word[112] gives us also the great guarantee that we will not be deceived by Satan and his minions. The Nephites understood this and prayed for it as should we — knowing the Spirit would lead them back to the Father through the Son.[113]

The prophet Mormon, who lived here in America during the fourth century, shares the following profound truths:[114]

> For no man can be saved, according to the words of Christ, save they shall have faith in his name;...How is it that ye can attain unto faith, save ye shall have hope? And what is it that ye shall hope for? Behold I say unto you that ye shall have hope through the atonement of Christ and the power of his resurrection, to be raised unto life eternal, and this because of your faith in him according to the promise. Wherefore, if a man have faith he must needs have hope... I say unto you that he cannot have faith and hope, save he shall be meek, and lowly of heart... and... he must needs have charity; for if he have not charity he is nothing; wherefore he must needs have charity. But charity is the pure love of Christ, and it endureth forever; and whoso is found possessed

---

110) D&C 84:38
111) Doctrine and Covenants 121:45-46
112) JS-M 37
113) 3 Nephi 19:9
114) Moroni 7:38-48; see also D&C 10:5

of it at the last day, it shall be well with him. Wherefore, my beloved brethren, pray unto the Father with all the energy of heart, that ye may be filled with this love, which he hath bestowed upon all who are true followers of his Son, Jesus Christ; that ye may become the sons of God; that when he shall appear we shall be like him, for we shall see him as he is; that we may have this hope; that we may be purified even as he is pure. Amen.

Following Mormon's counsel would solve all the world's problems, learning to love like they love us.

Because Father loves us, He has created this Earth for us as an essential part His perfect plan of happiness.[115] One of the grand ironies of God is that He can and will bring us to perfection out of all our imperfections through the infinite atonement of Jesus Christ if we will but come unto Him. That is why it needs to be infinite. As the Apostle Paul so poignantly said, we are His offspring and where we are born and when we are born are determined by a loving Heavenly Father to give us the best chance of coming back to Him.[116] The Fall of Adam was part of the grand design: " Adam fell that men might be; and men are, that they might have joy."[117] The atonement perfectly overcame the effects of the Fall. As each of us fully comes to Christ — internalizing His love in our lives — with all of our heart, might, mind, and strength, He can and will perfect us.[118] This is the Father's work and glory, "to bring to pass the immortality and eternal life of man,"[119] and He enjoys His work; in it He has a fullness of joy as may we.

The Savior gives us an additional enormous promise of "another Comforter" as we increase our love and keep His commandments:[120]

> If ye love me, keep my commandments. And I will pray the Father, and he shall give you another Comforter, that he may abide with you forever; Even the Spirit of truth; whom the world cannot receive, because it seeth him not, neither knoweth him: but ye know him; for he dwelleth with you, and shall be in you.

---

115) Abr. 3:24-25
116) Acts 17:26-29
117) 2 Nephi 2:25
118) John 17:23; Moroni 10:32-33
119) Moses 1:39; John 15:7-8
120) John 14:15-23

> I will not leave you comfortless: I will come to you. Yet a little while, and the world seeth me no more; but ye see me: because I live, ye shall live also. At that day ye shall know that I am in my Father, and ye in me, and I in you. He that hath my commandments, and keepeth them, he it is that loveth me: and he that loveth me shall be loved of my Father, and I will love him, and will manifest myself to him. . . If a man love me, he will keep my words: and my Father will love him, and we will come unto him, and make our abode with him.

Where do we observe the greatest joys in this life? Are they not in family, in sharing His eternal truths with His precious sons and daughters, in using our time and talents — in harmony and unity with others — to bless the lives of Father's children, and in coming to know Him through the Savior's atonement — learning of the mysteries of godliness. We have fun and pleasure in life's activities, but they typically don't bring lasting joy. Sharing and living the gospel brings deep and lasting joy — both individually and as a society.[121] Knowing that we are clean and forgiven of all our sins through what the Savior has done for us and knowing that we are promised a fullness of joy as glorious resurrected beings fashioned after His glorious body[122] indeed fills the soul with the greatest of hope and joy.

There can be no doubt these feelings of love and joy that we have in this life transcend the grave. A Loving God would not only allow this love to continue beyond death but would amplify it perhaps a 100-fold[123] in His Eternal realms and mansions on high to bring the fullness of joy He promises to the faithful.

Certainly, true love in marriage transcends the grave, and ultimately has the potential to bring about the greatest happiness in time and in eternity, as we are together as family with our most loving Eternal Parents and with our Beloved Savior. Well did Paul also say, ". . . neither is the man without the woman, neither the woman without the man, in the Lord."[124]

---

121) Luke 2:10; 1 Nephi 13:37; 19:11; Mosiah 3:3-4; Alma 22:15; Alma 38:2; 3 Nephi 12:10-12; D&C 18:15-16; 31:3; 128:19
122) Philippians 3:21
123) Matt. 19:29
124) 1 Cor. 11:11

Chapter 4

The Lord revealed:

> For strait is the gate, and narrow the way that leadeth unto the exaltation and continuation of the lives, and few there be that find it, because ye receive me not in the world neither do ye know me. But if ye receive me in the world, then shall ye know me, and shall receive your exaltation; that where I am ye shall be also. This is eternal lives—to know the only wise and true God, and Jesus Christ, whom he hath sent. I am he. . .[125]

We are able to partake of Their Great Plan of Happiness as we come to know them. This path leads to have eternal family relationships and a fullness of joy. The sealing ceremony in the temples of The Church of Jesus Christ of Latter-Day Saints is the crowning ordinance of the gospel. Conditioned on the faithfulness of the recipients, this ordinance seals together husband and wife and is for time (mortality) and for all eternity. As we live up to the covenants we make in His holy temples and learn to love as they love, we become heirs of all that they have. Can one imagine a more glorious doctrine?[126]

Understanding the importance of this glorious doctrine, one sees why Satan has so vehemently attacked the family. The celestial order is family, and he knows that. If he can destroy a family, he knows that this is the most significant thing he can do to work against God. We see Satan coming against the family in every direction — trying to destroy both the happiness that can be therein as well as its being the fundamental unit of a healthy society and the ultimate unit in heaven.

The home is the best place to learn and practice Christlike love as we serve one another in our homes and those around us. The Christ-centered home is the happiest home, for their love abounds in selfless deeds to one another. Satan does his best to introduce selfishness, but loving parents, who emulate the life of the Savior, can push him out the door. Children raised in such a home come to know that they can totally trust in the Lord and that practicing His love (unselfish heart-felt service) is the best-tasting fruit of the tree of life. Our home in Heaven is filled with this love. Let us pray with all the intensity of our hearts to be filled with this (His) love. It is the solution to every problem and the path to the greatest happiness here and hereafter.

---
125) D&C 132:22-24
126) Isa. 56:3-8

4.3 How Can We Live the Law of Love and Receive Its Enormous Blessings?

Knowing how to love like God loves is coming to know God, which is Eternal Life. (John 17:3) Can there be anything more important? Everything God does is motivated by love, and His capacity to love is infinite — beyond our capability during our mortal journey. So, how do we get there? From the scriptures and from the Lord I have learned what I call the great identity. The verb "is" is the identity indicator in English, and we learn that GOD IS LOVE IS LIGHT IS TRUTH IS LIFE IS SPIRIT IS "THE WAY" TO ETERNAL LIFE, the greatest of all the gifts of God.[127] One of the beautiful clarifications of the restoration is that we have Heavenly parents, a fact which has enormous ramifications and inherent within us is the potential to become like Them as Their offspring. The all-encompassing and infinite atonement makes that possible.

Doctrine and Covenants section 88 teaches us He is in everything; specifically verse 7: Which truth shineth. This is the light of Christ. As also he is in the sun, and the light of the sun, and the power thereof by which it was made.) As a specific example, I find it fascinating that the light of the sun triggers the remarkable photosynthesis process, which creates chlorophyll, making foods green and beautiful. The chlorophyll molecule, complex in structure, is identical to the hemoglobin molecule, except at the center of each is a unique atom (magnesium for chlorophyll and iron for hemoglobin), and these beautiful chlorophyll-laden foods optimize our health as a direct manifestation of His love. There is a great book in this regard, *The Green Pharmacy; New Discoveries in Herbal Remedies for Common Diseases and Conditions from the World's Foremost Authority on Healing Herbs*, by Dr. James A. Duke.

King Benjamin shares the profound perspective: "...that if ye should serve him who has created you from the beginning, and is preserving you from day to day, by lending you breath, that ye may live and move and do according to your own will, and even supporting you from one moment to another—I say, if ye should serve him with all your whole souls yet ye would be unprofitable servants."[128] It is interesting that the Church in the meridian of time called the

---

127) 1 John 4:8-16; D&C 84:45; John 14:6; D&C 14:7
128) Mosiah 2:21

church *The Way*.[129] He is the way, the truth, and the life...[130] in all aspects of our beings (emotions, body, mind, and spirit).

So as we come unto Him, who is the way, we learn that He is the way to learn to love, to be filled with light, to be taught truth, to receive the abundant life, and to enjoy the constant companionship of the Spirit. Christ did not receive a fullness at first but grew grace for grace, and He set the perfect pattern before us.[131] Following Christ's example, we grow line-upon-line, precept-upon-precept and are filled with eternal joy in the process as we come unto Him, and He moves us on the path to eternal life — life with God, life like God, where we learn to love like Him with an infinite love. Analogously, we can never perfect ourselves, but He can and will perfect us as we come unto Him.[132] As Immaculée Ilibagiza shares in her most inspiring book *Left to Tell*, "God will teach us how to love this intently, if we ask Him."[133] I will share more of her amazing story later.

The more we are motivated by pure love, the more we are like Him. Another way of looking at the identity aspect of it is that if we receive His light, we receive His love; if we receive His TRUTHs, we receive His love; if we receive His Spirit, we receive His love; and as He shows us "The Way," to live our lives as we ask, we are receiving of His love, and amazingly He is in the details. We see His love manifest everywhere and most poignantly in creation, the fall, and the Atonement. We are surrounded — for those who have eyes to see — by His love. If we observe through His loving eyes, we may see how He turns every evil into good in due process while not violating agency.

Even though loving God and our neighbor are the greatest commandments, the most oft'-repeated commandment is to pray; this action is our access to come to know Him, who is love and to learn how to love. Mormon's great counsel fits perfectly here:

> Wherefore, my beloved brethren, if ye have not charity, ye are nothing, for charity never faileth. Wherefore, cleave unto charity, which is the greatest of all, for all things must fail— But

---

129) Bercot *Will the Real Heretics Please Stand UP?*
130) John 14:6
131) John 1:16; D&C 93:12
132) Moroni 10:31-32
133) See http://www.allanstime.com/Spiritual/BookReports/Left_to_Tell.htm and http://ItsAboutTimeBook.com/left-to-tell-immaculee-ilibagiza/

charity is the pure love of Christ, and it endureth forever; and whoso is found possessed of it at the last day, it shall be well with him. Wherefore, my beloved brethren, pray unto the Father with all the energy of heart, that ye may be filled with this love, which he hath bestowed upon all who are true followers of his Son, Jesus Christ; that ye may become the sons of God; that when he shall appear we shall be like him, for we shall see him as he is; that we may have this hope; that we may be purified even as he is pure. Amen.[134]

Prayer is not only the path to find God's love, it is the path to conquer Satan: "Pray always, that you may come off conqueror; yea, that you may conquer Satan, and that you may escape the hands of the servants of Satan that do uphold his work."[135]

So we see that in our mortal journey, building loving relationships is our greatest opportunity in time and eternity since love is the first law of heaven. This we can best do by invoking constant, sincere, heart-felt prayer, and this is the path to come to know the Lord, to bring the greatest of joy in this life and a fullness of joy in the life to come. For if we love God and His children, we will keep His commandments and live the fullness of the gospel, which are designed to bring us into His presence — even in this life. As we do that, He can and will and wants to bless us to the fullest here and hereafter, as we make and keep covenants of faith with Him. We learn as we grow in the knowledge of God to become trustworthy as He is totally trustworthy; then He can share with us the mysteries of godliness and he will fill us with His love as we become ONE with Him.

How do we best build such loving relationships with God and His children? The Gentile physician Luke quotes the Savior saying: "Thou shalt love the Lord thy God with all thy heart, and with all thy soul, and with all thy strength, and with all thy mind; and thy neighbour as thyself."[136] The quote in Mark, which is the first gospel written, Chapter 12:30-31 is similar — giving us two witnesses of its importance. In other words, the Lord is asking us to love God in all four dimensions of our beings: heart (emotions), soul (spirit), strength (body), and mind (intelligence).

---

134) Moroni 7:48
135) D&C 10:5
136) Luke 10:27

Chapter 4

What are the blessings when we live this first and most important law? Are these blessings and promises tied to the mysteries of godliness? The following scripture is fundamental to our day:

> Listen to the voice of Jesus Christ, your Redeemer, the Great I AM, whose arm of mercy hath atoned for your sins;
>
> Who will gather his people even as a hen gathereth her chickens under her wings, even as many as will hearken to my voice and humble themselves before me, and call upon me in mighty prayer.[137]

The footnote below references all the scriptures regarding the "hen" gathering or desiring to gather "her chickens under her wings."[138] As these are studied, one sees a very interesting trend. In the meridian of time, Christ bemoaned, ". . . how often would I have gathered thy children together, as a hen her brood under her wings, and ye would not."[139] Because they killed the prophets and rejected the Savior, their "house is left unto you desolate." And so Israel was scattered because they would not accept Christ's (Jehovah's) invitation. Then "when the Gentiles. . . reject the fulness of my gospel," in our day[140] their "house is left unto [them] desolate." Then, which is now, He will "bring" the gospel back to Israel as explained in 3 Nephi Chapter 16. And His promise for our day is that He "will gather his people even as a hen gathereth her chickens under her wings, even as many as will hearken to my voice and humble themselves before me, and call upon me in mighty prayer" as cited above. In other words, the parable of the hen gathering her chicks was there for ancient Israel, and they rejected it. It is there for the Gentiles, who are now rejecting it. And so we have the incredibly important promise that as the Lord gathers the "pure in heart" in these last days, He "will" be successful in this gathering. Like the parable of the wheat and the tares, He will pull out the wheat from the midst of the tares before the tares are burned with fire, and the gathered will be as chicks under His wing — protected by loving arms and by Divine Providence. What an incredible promise to look forward to.

As He gathers his chicks under His wings in our day, we will receive of His promises:

---

137) D&C 29:1, 2
138) *http://ItsAboutTimeBook.com/spiritual-health-as-a-hen-gathers-chickens/*
139) Lk. 13:34-35
140) 3 Nephi 16:10

- Therefore the redeemed of the LORD shall return, and come with singing unto Zion; and everlasting joy shall be upon their head: they shall obtain gladness and joy; and sorrow and mourning shall flee away.[141]

- Thy watchmen shall lift up the voice; with the voice together shall they sing: for they shall see eye to eye, when the LORD shall bring again Zion.[142]

- Hear the word of the LORD, O ye nations, and declare it in the isles afar off, and say, He that scattered Israel will gather him, and keep him, as a shepherd doth his flock. . . Therefore they shall come and sing in the height of Zion, and shall flow together to the goodness of the LORD, for wheat, and for wine, and for oil, and for the young of the flock and of the herd: and their soul shall be as a watered garden; and they shall not sorrow any more at all. Then shall the virgin rejoice in the dance, both young men and old together: for I will turn their mourning into joy, and will comfort them, and make them rejoice from their sorrow.[143]

- Sing, O daughter of Zion; shout, O Israel; be glad and rejoice with all the heart, O daughter of Jerusalem. The LORD hath taken away thy judgments, he hath cast out thine enemy: the king of Israel, even the LORD, is in the midst of thee: thou shalt not see evil any more. In that day it shall be said to Jerusalem, Fear thou not: and to Zion, Let not thine hands be slack. The LORD thy God in the midst of thee is mighty; he will save, he will rejoice over thee with joy; he will rest in his love, he will joy over thee with singing.[144]

- Sing and rejoice, O daughter of Zion: for, lo, I come, and I will dwell in the midst of thee, saith the LORD. And many nations shall be joined to the LORD in that day, and shall be my people: and I will dwell in the midst of thee, and thou shalt know that the LORD of hosts hath sent me unto thee. And the LORD shall inherit Judah his portion in the holy land, and shall choose Jerusalem again.[145]

- [Jacob speaks of "things. . . to come," and quotes from Isaiah in doing

---

141) Isaiah 51:11
142) Isaiah 52:8
143) Jeremiah 31:10-13
144) Zephaniah 3:14-17
145) Zechariah 2:10-12

Chapter 4

so.] Therefore, the redeemed of the Lord shall return, and come with singing unto Zion; and everlasting joy and holiness shall be upon their heads; and they shall obtain gladness and joy; sorrow and mourning shall flee away.[146]

- [Alma, one of King Noah's priests, asks:] And it came to pass that one of them [Alma] said unto him [Abinadi]: What meaneth the words which are written, and which have been taught by our fathers, saying: How beautiful upon the mountains are the feet of him that bringeth good tidings; that publisheth peace; that bringeth good tidings of good; that publisheth salvation; that saith unto Zion, Thy God reigneth; Thy watchmen shall lift up the voice; with the voice together shall they sing; for they shall see eye to eye when the Lord shall bring again Zion; Break forth into joy; sing together ye waste places of Jerusalem; for the Lord hath comforted his people, he hath redeemed Jerusalem; The Lord hath made bare his holy arm in the eyes of all the nations, and all the ends of the Earth shall see the salvation of our God?[147] [Abinadi explains] And now I say unto you that the time shall come that the salvation of the Lord shall be declared to every nation, kindred, tongue, and people. Yea, Lord, thy watchmen shall lift up their voice; with the voice together shall they sing; for they shall see eye to eye, when the Lord shall bring again Zion. Break forth into joy, sing together, ye waste places of Jerusalem; for the Lord hath comforted his people, he hath redeemed Jerusalem. The Lord hath made bare his holy arm in the eyes of all the nations; and all the ends of the Earth shall see the salvation of our God.[148]

- Verily, verily, I say unto you, thus hath the Father commanded me—that I should give unto this people this land for their inheritance. And then the words of the prophet Isaiah shall be fulfilled, which say: Thy watchmen shall lift up the voice; with the voice together shall they sing, for they shall see eye to eye when the Lord shall bring again Zion. Break forth into joy, sing together, ye waste places of Jerusalem; for the Lord hath comforted his people, he hath redeemed Jerusalem. The Lord hath made bare his holy arm in the eyes of all the nations; and all the ends of the Earth shall see the salvation of God.[149] And

---

146) 2 Nephi 8:11
147) Mosiah 12:20-24
148) Mosiah 15:28-31
149) 3 Nephi 16:16-20

there shall be gathered unto it out of every nation under heaven; and it shall be the only people that shall not be at war one with another. And it shall be said among the wicked: Let us not go up to battle against Zion, for the inhabitants of Zion are terrible; wherefore we cannot stand. And it shall come to pass that the righteous shall be gathered out from among all nations, and shall come to Zion, singing with songs of everlasting joy.[150]

- And the blood of that great and abominable church, which is the whore of all the Earth, shall turn upon their own heads; for they shall war among themselves, and the sword of their own hands shall fall upon their own heads, and they shall be drunken with their own blood. And every nation which shall war against thee, O house of Israel, shall be turned one against another, and they shall fall into the pit which they digged to ensnare the people of the Lord. And all that fight against Zion shall be destroyed, and that great whore, who hath perverted the right ways of the Lord, yea, that great and abominable church, shall tumble to the dust and great shall be the fall of it. For behold, saith the prophet, the time cometh speedily that Satan shall have no more power over the hearts of the children of men; for the day soon cometh that all the proud and they who do wickedly shall be as stubble; and the day cometh that they must be burned. For the time soon cometh that the fulness of the wrath of God shall be poured out upon all the children of men; for he will not suffer that the wicked shall destroy the righteous. Wherefore, he will preserve the righteous by his power, even if it so be that the fulness of his wrath must come, and the righteous be preserved, even unto the destruction of their enemies by fire. Wherefore, the righteous need not fear; for thus saith the prophet, they shall be saved, even if it so be as by fire. Behold, my brethren, I say unto you, that these things must shortly come; yea, even blood, and fire, and vapor of smoke must come; and it must needs be upon the face of this Earth; and it cometh unto men according to the flesh if it so be that they will harden their hearts against the Holy One of Israel. For behold, the righteous shall not perish; for the time surely must come that all they who fight against Zion shall be cut off. And the Lord will surely prepare a way for his people, unto the fulfilling of the words of Moses, which he spake, saying: A prophet shall the Lord your God raise up unto you, like unto me; him shall ye hear in all

---

150) D&C 45:69-71

Chapter 4

things whatsoever he shall say unto you. And it shall come to pass that all those who will not hear that prophet shall be cut off from among the people. And now I, Nephi, declare unto you, that this prophet of whom Moses spake was the Holy One of Israel; wherefore, he shall execute judgment in righteousness. And the righteous need not fear, for they are those who shall not be confounded.[151]

A Zion society is motivated by the pure love of Christ. What incredible blessings await those who seek to establish Zion and are filled with this love! These are they who will "hear [and hearken to] His voice," " humble" themselves, "calling upon [God] in mighty prayer," and be gathered as a hen gathereth her chicks under her wings — protecting us from the evils of this wicked world. Indeed, we will rejoice and sing the songs of Zion. Let us, therefore, learn how to love the Lord with all of our heart, soul, body, and mind, and our neighbor as ourselves that we may be the recipient of these joyful and incomprehensible blessings. How do we do that in the detail of the four dimensions of our beings?

Let us think back to our pre- mortal existence where we were intelligences.[152] There our intelligence was clothed with a spirit body as we were born spirit offspring of our loving Heavenly parents;[153] hence, we are literally the offspring of God and have the potential to become like them. We are gods in embryo. When we were born into this life, our intelligence was downloaded to our brain cells, which exist in our physical mind, in our heart, and in our bowels.[154]

It is interesting that only recently have scientists discovered brain cells in the heart and the gut, but it has always been in the scriptures as the Lord references on many occasions about the "thoughts of the heart," and His "bowels" are filled with "mercy" that He may "succor his people." This remarkable connection between the brain-cells in the mind, the heart, and the gut gives us our full range of emotions needed during our mortal journey and for us to have joy in our path to godhood. When we are motivated by the pure

---

151) 1 Nephi 22:13-22
152) D&C 93:29
153) Acts 17:26-29; D&C 76:24
154) *Brain Cells Located in the Heart,* from an interview with Joseph Chilton Pearce 9/12/2008 by Chris Mercogliano and Kim Debus from *Journal Of Family Life Magazine*, Vol. 5 #1 1999.
Also: *Not Only Does Our Gut Have Brain Cells It Can Also Grow New Ones, Study* http://www.medicalnewstoday.com/articles/159914.php

love of Christ, then we choose with our brain to give our hearts to God, and to let our "bowels" be full of charity toward all men. We let our "hearts be full, drawn out in prayer unto him continually for [our] welfare, and also for the welfare of those who are around [us]."[155] We let "virtue garnish our thoughts unceasingly," and our bowels are full of charity toward all men and to the household of faith, then the Holy Ghost will be our constant companion.[156] In so doing, we move in the direction of the heavenly society called Zion.[157] Then we can "quench all the fiery darts of the wicked."[158]

When Alma the younger was undergoing his great change, as he turned to God with all of his heart, might, mind and soul, he said:

> Now, as my mind caught hold upon this thought, I cried within my heart: O Jesus, thou Son of God, have mercy on me, who am in the gall of bitterness, and am encircled about by the everlasting chains of death. And now, behold, when I thought this, I could remember my pains no more; yea, I was harrowed up by the memory of my sins no more. And oh, what joy, and what marvelous light I did behold; yea, my soul was filled with joy as exceeding as was my pain![159]

The Lord tells us that he wants our hearts and a "willing mind."[160] "And the Lord called his people ZION, because they were of one heart and one mind, and dwelt in righteousness; and there were no poor among them."[161]

Because our first relationship is with God and our second with His children and because the first law of heaven is love, it makes total sense why the first and second greatest commandments are to love the Lord thy God with all thy heart, soul, strength, and mind and then to love our neighbors as ourselves. He wants to make of us gods and goddesses — filled with His love and having fullness of joy. Learning to live the Law of Love gets us there.

---

155) Alma 34:27
156) D&C 121:45
157) JST Psalm 14:7
158) D&C 27:15-18
159) Alma 36:18-20
160) 1 Chronicles 28:9 and D&C 64:34
161) Moses 7:18

Chapter 4

How To Love God With All Thy Heart, And With All Thy Soul, And With All Thy Strength, And With All Thy Mind

Love God with all of our Heart

"Let the affections of thy heart be placed on the Lord forever."[162] "And thou shalt love the LORD thy God with all thine heart, and with all thy soul, and with all thy might. And these words, which I command thee this day, shall be in thine heart: And thou shalt teach them diligently unto thy children, and shalt talk of them when thou sittest in thine house, and when thou walkest by the way, and when thou liest down, and when thou risest up. And thou shalt bind them for a sign upon thine hand, and they shall be as frontlets between thine eyes. And thou shalt write them upon the posts of thy house, and on thy gates."[163] Hence, we see why the Jews put the Mezuzah with the above words in it on their door post.

" And it shall come to pass, if ye shall hearken diligently unto my commandments which I command you this day, to love the LORD your God, and to serve him with all your heart and with all your soul, That I will give you the rain of your land in his due season, the first rain and the latter rain, that thou mayest gather in thy corn, and thy wine, and thine oil. And I will send grass in thy fields for thy cattle, that thou mayest eat and be full. Take heed to yourselves, that your heart be not deceived, and ye turn aside, and serve other gods, and worship them;"[164] This deceit is a major problem of our modern world — worshiping the works of men's hands. "For they have strayed from mine ordinances, and have broken mine everlasting covenant; They seek not the Lord to establish his righteousness, but every man walketh in his own way, and after the image of his own god, whose image is in the likeness of the world, and whose substance is that of an idol, which waxeth old and shall perish in Babylon, even Babylon the great, which shall fall."[165] "Wherefore, my beloved brethren, pray unto the Father with all the energy of heart, that ye may be filled with this love, which he hath bestowed upon all who are true

---

162) Alma 37:36
163) Deut. 6:5-9
164) Deut. 11:13-16
165) D&C 1:15-16

followers of his Son, Jesus Christ; that ye may become the sons of God; that when he shall appear we shall be like him, for we shall see him as he is; that we may have this hope; that we may be purified even as he is pure. Amen."[166] "And again, verily I say unto you, blessed is my servant Hyrum Smith; for I, the Lord, love him because of the integrity of his heart, and because he loveth that which is right before me, saith the Lord."[167, 168] In the last days, the Lord will write His law [of love] in our hearts. He will be our God and we will be His people.[169] This will be Zion motivated by our pure love of God in all four dimensions of our beings and our neighbor as ourselves. For on these first two commandments "hang all the law and the prophets," and we will fully understand what this means. Then we will be of one heart and one mind with no poor amongst us and with the Lord in our midst, the glorious day of Zion.

Love God with all of our Soul (Spirit)

Having our spirit coupled with the Holy Spirit is our greatest opportunity in our mortal journey; the Nephites recognized this concept as the Savior visited and taught them the truths of eternal life.[170] The Lord gives us the recipe for how we can have the Holy Ghost as our constant companion,[171] which says, "Let thy bowels. . . be full of charity towards all men, and to the household of faith, and let virtue garnish thy thoughts unceasingly;. . . The Holy Ghost shall be thy constant companion. . ." This charity is the "pure love of Christ."[172] Our Heavenly Father and His Beloved Son are most pleased, when we are filled with this love; we are motivated to share the truths of eternal life with our neighbors as guided by the Spirit. They rejoice, the angels rejoice, and then they and we are "edified and rejoice together."[173] This is the Lord's work, "to bring to pass the immortality and eternal life of man,"[174] because He loves us.

---

166) Moroni 7:48
167) D&C 124:15
168) See also Moroni 8:25-26; D&C 4: and 12: for additional attributes, emotions, and characteristics of a heart that is right before the Lord.
169) Jer. 31:31
170) 3 Ne. 19:9
171) D&C 121:45, 46
172) Moroni 7:47
173) D&C 50:13-25
174) Moses 1:39

Chapter 4

Love God with all of our Body

"Know ye not that ye are the temple of God, and that the Spirit of God dwelleth in you? If any man defile the temple of God, him shall God destroy: for the temple of God is holy, which temple ye are."[175] "What? Know ye not that your body is the temple of the Holy Ghost which is in you, which ye have of God, and ye are not your own?"[176] "And all saints who remember to keep and do these sayings, walking in obedience to the commandments, shall receive health in their navel and marrow in their bones; And shall find wisdom and great treasures of knowledge, even hidden treasures; And shall run and not be weary, and shall walk and not faint. And I, the Lord, give unto them a promise, that the destroying angel shall pass by them, as the children of Israel, and not slay them, Amen."[177] "For whoso is faithful unto the obtaining these two priesthoods of which I have spoken, and the magnifying their calling, are sanctified by the Spirit unto the renewing of their bodies."[178] We love God as we use His priesthood to bless the lives of His children and serve them with love — helping them to come to know Him and of His love. This is magnifying, and the Lord promises a renewing of our bodies as we do so. What a great promise!

Love God with all of our Mind

"Look unto me in every thought; doubt not, fear not."[179] With our mind we choose to follow God or not — to do His will, as did the Savior perfectly. The Spirit of Christ is given to every soul at birth to know good from evil. If we choose to follow that Spirit, He will lead us back to the Father. We love Him with our mind as we so choose to give our will to Him, the only thing we really have to give. The Earth is the Lord's and the fullness thereof. "Now we have received, not the spirit of the world, but the spirit which is of God; that we might know the things that are freely given to us of God. Which things also we speak, not in the words which man's wisdom teacheth, but which the Holy Ghost teacheth;. . .the things. . . of God. . . are spiritually discerned. [Which allows us to]. . .have the mind of Christ."[180] A Zion people are of one heart and one mind with no poor among them; their main motivation is love.[181]

---

175) 1 Cor. 3:16
176) 1 Cor. 6:19
177) D&C 89:18-21
178) D&C 84:33
179) D&C 6:36; Alma 37:36
180) 1 Corinthians 2:12-16
181) Moses 7:18

Love Our Neighbor as Ourselves

The difference between the first and second commandments is "all." God wants us to love Him with "all" of our heart, soul, body, and mind, for He knows that as we do that, we are on the path to a fullness of joy. He further knows that we cannot enjoy celestial spheres unless we also learn to love ourselves and our neighbor in all four dimensions of our beings.

In the grand irony of Father's perfect plan of happiness we find the realization of our "nothingness" compared to God and our infinite worth as His child. The way to be truly humble is to compare ourselves to God. We are His precious offspring, and the one fundamental truth I know is that God loves us. When we know and feel of His love for us, we will love ourselves like unto His infinite love. This process will bring faith, hope and charity into our bosoms and fill us with joy. We will want to share those feelings of love and joy. How then can we best learn to love ourselves and our neighbor with our heart, soul, body, and mind?

How to Love Our Neighbors as Ourselves
Loving our neighbor and ourselves — with our heart

We love our neighbor with our hearts when we listen to them with our hearts — with a desire to understand their desires, needs, and goals in life — their emotions, their feelings. Passing no judgment as we listen with love, we help them to feel our love. We help them understand the heart of God and His Divine characteristics, so that by emulating these Divine characteristics they and we will come closer to God and have the peace, love and joy in our hearts and their hearts, which He wants for all of His children. Their joy is His joy.

The art of listening with love is almost non-existent in our busy, noisy, and entertainment-oriented society. The average attention span of adults is about 20 seconds. Hence, that is how long advertisements last. Because of its importance and because it is pretty much a lost art, I have written an article available from the link below to this extremely important topic.[182]

The Apostle Peter succinctly proclaims: "Seeing ye have purified your souls in obeying the truth through the Spirit unto unfeigned love of the brethren, see

---

182) *http://ItsAboutTimeBook.com/six-magic-skills-in-the-art-of-listening/*

that ye love one another with a pure heart fervently: Being born again, not of corruptible seed."[183]

The most important love in human relationships is husband-wife love. The Lord says: "Thou shalt love thy wife with all thy heart, and shalt cleave unto her and none else."[184] The love and oneness in the husband-wife relationship are among the richest of blessings the Lord offers during our mortal journey, and He has the greatest of blessings to give a righteous couple in the life to come, for He says: "and they twain shall be one flesh, and all this that the Earth might answer the end of its creation; And that it might be filled with the measure of man, according to his creation before the world was made."[185] As designed and planned in premortal, our greatest desire and ultimate goal was to be like our Heavenly Parents. This is the primary reason the Earth was made. Hence, developing celestial-loving relationships between husband and wife are the most important of all human relationships — with all of our hearts.

The heart-felt love of a mother for her children approaches the Divine. It is pure and without bound. The heart-felt love of a father provides protection and security for his children. It approaches that of our perfectly-loving Heavenly Father. The father's pure love for their mother and hers for him sets a perfect example for the children when they get married. That love along with helping them to "understand" and internalize the beautiful truths of the gospel of Jesus Christ will give them the best foundation possible in this life.[186]

In such a loving environment, children naturally develop a deep and abiding love for their parents. They naturally want to honor them consistent with the commandment of the Lord to do so throughout their lives. In Father's perfect plan of happiness, such a love extends and is greatly amplified into eternity in Father's eternal realms of joy. Again, to this end was the Earth created, for family is the order of heaven.

---

183) 1 Pet. 1:22
184) D&C 42:22
185) D&C 49:16-17
186) D&C 68:25

With our Soul (Spirit)

We love our neighbor with our soul when we share with them by the Spirit the truths of eternal life — the fullness of the gospel — that we and they may rejoice and be edified together in partaking of those truths.[187]

"And how great is his joy in the soul that repenteth! Wherefore, you are called to cry repentance unto this people. And if it so be that you should labor all your days in crying repentance unto this people, and bring, save it be one soul unto me, how great shall be your joy with him in the kingdom of my Father! And now, if your joy will be great with one soul that you have brought unto me into the kingdom of my Father, how great will be your joy if you should bring many souls unto me!"[188]

The worth of a soul is worth more than a lifetime of our labor. Let us reach out in love and share these glorious truths of eternal life.

With our Body

We love our neighbor with our body when we delight to serve him or her in a way to best help them in their lives to come to know the Father and the Son. There are all kinds of loving acts of service we can do for our neighbor.

Using body language is a major part of communication, and letting it be loving is a very effective way to share our love with our neighbor. Healthy tactile interaction is critical in human development as well in showing God-like love. Mariana Caplan succinctly says, "Touch, when done with heart, is always healing — period."[189] Every language has its nuances and lacks in pure communication — except for the Adamic language. Since communication is critical in human relationships, we see the importance in truly loving our neighbor of communicating with love as the Savior has asked with our hearts, our souls, our strength, and our minds. Let our language, especially of our bodies, be the language of love.

---

187) D&C 50:13-24
188) D&C 18:13-16
189) Mariana Caplan, *Untouched: The Need for Genuine Affection*

Chapter 4

With our Mind

We love our neighbors with our minds when we think about them as the Father does as His precious sons and daughters of infinite worth. The Earth was created by Him to the intent that we may have the best opportunity to return to Him — there to receive a fullness of joy and to be like Him as His offspring. We keep in our minds that it "is [His] work and glory to bring to pass the immortality and eternal life of [family]."[190] We desire that His work becomes our work. This we do as we love our neighbors with our minds as we think of ways that we can share with them truth that their minds may be enlightened by the same. In that process both the giver and the receiver are edified and rejoice together.[191] This Divine process brings about eternal friendships that run to the deepest levels and bring great joy and peace to the souls in this sharing. The Spirit bears record of the truths of God, and this two-way communication becomes divinely three-way.

---

190) Moses 1:39
191) Doctrine & Covenants 55:22

Chapter 5

Mary's Miracle; God is Closer Than You Think; Documented Miracles Showing Evidence of God

Major Miracle on the Mountain: Our grand-daughter, Mary Owen, was found on Mt. Hood after falling and injuring herself and spending six nights on the mountain. We have in her experience a most inspiring and documentable combination of miracles, messages, and direct evidence of the Hand of the Lord.

When my wife and I got home late Thursday night (28 March 2013), we had received the following e-mail from our oldest daughter, Shelli Owen:

Please pray with us that we will be able to find where Mary is soon, or that she would contact us or her roommates… regarding her whereabouts… soon. We have officially filed a missing person's report with the police, because all the people who should know where she is, don't know where she is, and haven't heard from her since Sunday. She told a friend [that she] was planning on being back in Newberg Sunday evening or Monday morning and she was supposed to be in contact with her roommate and others by now. We are not sure that she didn't try to hike Mt. Hood. If you know where she might be or have information that might help us to locate her, please contact us right away. Thank you!

Bruce & Shelli

Friday, the next day, they found a truck at Timberline Lodge, which she had borrowed. We learned later that she had some friends, who had been planning to summit Mt. Hood with her, the previous Saturday, but the climb was called off by the leader of the group, who is an experienced guide and climber, because the snow and weather conditions were expected to be bad that day. However, we learned later that this cancellation was viewed by Mary as another of several 'failed' attempts to find people to climb with her, and so when these plans fell through, she decided to climb Mt. Hood alone the next day (Sunday 24 March 2013).

After talking with Search and Rescue (SAR) on the phone on Friday (29 March), her father, Bruce, posted the following at 10:33 p.m. on his Facebook wall:

Here is the latest information I have from SAR. They have had two crews out since this afternoon and a plane. One crew checking the Pacific Crest Trail [Mary had hiked the PCT three years before] where it runs by Mt Hood... Newberg police [have] sent a computer crimes specialist out to hack Mary's laptop which they found in the truck. They are hoping it might have information on it to help narrow the search. They have traced her debit card activity to several places and reviewed video in each which showed her to be alone at the time she used her card. They have identified what she bought (shoes and another ax) and are figuring that stuff in when they look at what was left in the truck. So far, we still do not have any info on the people she was planning on climbing the mountain with, so we are still trying to find out who they are.

SAR says they will have upwards of 50 trained people out there beginning at 4 am tomorrow. They plan to have one crew brought to the summit to do a top down search of the mountain...

Please keep praying and spreading the word. Please share this post to your wall so more people will see it. We are continuing in prayer and faith. We hope to find Mary soon.

Chapter 5

The next day, Saturday morning, the day before Easter, I sent out the following e-mail to family and many friends:

Our grand-daughter, Mary Owen, has been missing since last weekend on Mt. Hood. Please focus your prayers with us. I believe she will be found today alive. Mary is one of the most competent mountaineers that I know. She did the Pacific Crest Trail — a 2,700 mile, five-month hike in rugged country from Mexico to Canada, and she did extremely well. She is also one of the most spiritual and loving people you will ever meet.

Shelli, called me at about 11:50 a.m. that Saturday morning saying they had found Mary. A National Guard helicopter in the shortest search they had ever done (nine minutes) had found her, and she was alive. They were on the way to the Portland hospital. I could not contain my tears of gratitude and joy; I was sobbing. I fell to my knees in gratitude and prayer. Shelli asked me to let everyone know, so I did that through every channel I had.

At 2:35 p.m. that day I talked to Ruth (Mary's younger sister) and got an update and sent out the following:

Mary is doing okay. Ruth said she was using her ice axe and slipped and fell — gashing her leg and spraining her ankle — apparently making her immobile. Ruth said that the helicopter followed some tracks from the top of the mountain that led right to her, and that is how they found her, but they were not hers. We gratefully acknowledge Divine intervention and an answer to massive prayers in her behalf. It is interesting that she was found today, as many of us felt she would be, the day before Easter…

Our hearts are full of gratitude to the Lord for His tender Mercies and for hearing our prayers. Thank you for your faith and prayers; they obviously helped big time.

Saturday, 3:23, I just talked to Bruce. He said for prayer purposes she is in the burn unit at Legacy Immanuel Hospital in Portland, OR, for dealing with the frost bite. Bruce thought she would recover from that, but please keep your prayers focused in that direction. Bruce said that if you call the hospital they will not acknowledge that she is there because of all the publicity. She was able to light a fire one night, and she tried to set a tree on fire so people would know where she was, but didn't succeed in that.

If you were to look at her black toes and then to see them a few days later, you would say, "It is another miracle!" Our hearts are so full of gratitude for answering our prayers in her behalf. [1]

This is what she wrote a week later of the experience:

Mountain Time-out
By Mary Owen (Notes) on Sunday, April 7, 2013 at 4:36am

It's been over a week since they pulled me off the mountain, and I know I haven't written much as far as updates on here [Facebook]. Part of it has been that there is so much to write. Part of it has been trying to figure out how to say what needs to be said. I've stayed away from a lot of the negative coverage on my own story. I've never believed in living a life in regret. But one of the most cutting comments has been the simple question from a much loved friend, "Well, did you learn anything?"

And my answer is, 'Yes. Yes I have.'

This experience for me has been first and foremost a very stern though gracious rebuke from God. I feel as though I was put in timeout for four days. The other day I was reading Isaiah 30 and it is basically the story of [my] "accident." It begins, "Ah, stubborn children," declares the Lord, "who carry out a plan, but not mine. . ."

Looking back at my own behavior on Sunday, preparing to go up the mountain, I made one bad decision after the next. I feel as though God purposefully removed His Spirit from my reason. Without Him I am just a dumb blond who stubbornly walks up a mountain after several warnings not to (two from experienced mountaineers); carries a GPS, but doesn't use it; has the opportunity to turn back, but doesn't take it. God never removed His hand of protection from me. Believe me, during the whiteout I was literally walking through a graveyard - places where others more experienced than I have died on that mountain. But I do believe He allowed me to end up where I did on Monday morning, curled up in the snow with one leg torn open and the other foot sprained, unable to just get myself out of the situation. "therefore I will call her 'Rahab who sits still.'"[2]

---

1) You may see some pictures at this link: *https://www.google.com/search?q=Mary+Owen+Mt.+Hood*
2) Isaiah 30:7

Chapter 5

On Monday I noticed that God wasn't talking to me and it scared me a lot. I don't believe in fear. I wasn't afraid of death, or the elements, or whatever I would experience out there, but I have learned now that there is one fear that I must believe in - you will fear God.

Sitting out there with nothing else to do, I began to realize how very carelessly I have carried my own life, treating death as though it was just between myself and God. The immenseness of this foolishness hit me on Monday. You may think that your life doesn't count, that it's not a big deal if you disappear - or like myself - that you know where you're going when you die - life security - Jesus and I are buddies. You are wrong. I was wrong. Your life is inextricably tied to the lives of millions of other people. There is no one-person who can be removed from the tapestry without leaving a gaping hole behind; and such a momentous decision - to remove a life - does and must belong solely to God. I realized while I was out there that the way I carried my own life, with such carelessness and so little regard for those who loved me, was equivalent to committing suicide.

So to all of my dear friends and family, I am truly sorry.

God cut me off from His own voice and from the intercessory prayers of others including my own mother. He didn't talk to me again until good Friday morning, when I woke with a verse in my mind: "Do not be anxious, saying 'What shall we eat?' or 'What shall we drink?' or 'What shall we wear?' . . . your heavenly Father knows that you need them all. But seek first the kingdom of God and His righteousness, and all these things will be added to you."[3]

I also woke with the overwhelming knowledge that thousands and thousands and thousands of people were praying for me.

"And though the Lord give you the bread of adversity and the water of affliction, yet your Teacher will not hide himself anymore. But your eyes shall see your Teacher. And your ears shall hear a word behind you saying, 'This is the way, walk in it.'"[4]

This is the way, walk in it: If you call yourself a Christian, it is more

---
3) Matthew 6:31-33
4) Isaiah 30:20-21

than life security; it is more than a nice relationship that you can depend on to make you feel better when you're feeling discouraged or threatened; if you call yourself a Christian, your identity is now in Christ. You have been crucified, buried, and resurrected with the Living Savior and your life no longer belongs to you! You and I no longer have the freedom to walk in our own will and according to our own desires and understanding. This is the God of the Universe who has purchased your life with His own Blood! It is not a thing to be taken lightly or frivolously. You are now a part of the Kingdom Mission - Reconciling the World to the God that Loves it.

It is hard to believe, looking back, that I almost gave up everything that God has planned for my life and all of the precious relationships that He has given me for the foolish pride of standing on a mountain peak and saying I had summited.

God is good. So good. That is my conclusion to every message, every conversation. He graciously protected and provided for me while I was out on that mountain. He sternly corrected my foolish and careless attitude towards the Life that He has given to me. He has given me another chance to be a part of His kingdom work; and I will take it with joy, a new humility, and a greater fear of the Lord.

To all of you who prayed and hoped and waited, thank you!

And to the search and rescue teams and black hawk crew that risked their own lives to find me, thank you!

And to all who are praying for my recovery, thank you!

I am stuck in the hospital for a while longer - the doctors are being very careful with the wound in my left leg - it is humbling to be here, but I'm not complaining. I know that God truly does work all things together for the good of those He loves and for the good work of the Kingdom. If you do not yet know that God loves you, believe me, HE DOES! And He will shake the world and move mountains, to bring you to Himself - to bring you to wholeness and freedom.

If you claim the title " Christian" but live as though your life belonged to you, STOP. You have no idea what you are messing with. This is the Holy

One, the God of the Universe, and if you give your Life to Him, you had better be prepared for Him to take it - all of it.

I am blessed that I still get to be a part of Kingdom work. I am blessed that I still get to live in relationship with each of you who are reading this post. I am blessed to be alive and to be a servant of the King. And if my life does not show it, please, please remind me of what He has brought me through, saved me from, and purchased me for.

"The grace of the Lord Jesus Christ be with your spirit."

My God is good. So good.

I love you all.

Mary Owen

I wrote the following on 1 April 2013

The Power and Purpose of Prayer

There is a fascinating and very inspiring story behind the story of our grand-daughter, Mary Owen's surviving six nights on Mt. Hood. It made international news as she was rescued last Saturday morning. Most would not have survived with the inadequate provisions she had — having planned to be on the mountain only one day. Descending the mountain in a white-out, she got disoriented and took a bad fall when her ice axe broke loose at about 4:00 am Monday morning. As she fell some 40 feet into a ravine a tree limb gouged her leg causing an eight by one inch wound with splinters and bark inside, which were later removed by a series of operations. One of the splinters barely missed the femoral artery — another miracle.

Mary is a brilliant lady, very spiritual, and very competent in the mountains... She is also a free spirit and it is not uncommon for her to go on one or two day hikes alone. Her room-mates knowing this didn't let Bruce and Shelli know she was missing until Thursday. Bruce immediately called the police. Shelli couldn't get hold of us, as we were in Orem, so she sent me an e-mail. When I read it upon arriving home, I had this heavy feeling that Mary was in trouble and that she had had an accident. At the same time Shelli was sending out prayer requests, I was doing the same to our family and friends, and the response was fantastic.

Mary had filed a hiking plan and expected that they would start looking for her Monday, since she indicated on her plan that she would be back by then. She had this empty feeling that no one was looking for her or praying for her all day Monday, Tuesday, Wednesday, and Thursday, and that was really hard on her — given that she is a very spiritually sensitive person. Friday morning, she woke up and felt peace and that now thousands were praying for her. She saw search planes but could not catch their attention, but she then knew she would be found, that God had a work for her to do yet on this Earth. Her feelings corresponded exactly with the prayer requests that went out. We are all interconnected!

One may ask, "Couldn't the Lord have answered her prayers earlier in the week without the rest of us praying?" Certainly He could. She is full of faith. But praying for one another is one of the main ways we reach out in love to one another, and this was a great opportunity for the Lord to teach this principle to a large number of people. I am sure that Mary having to suffer those six days and nights alone without help is a price she does not regret paying seeing so many people touched for good through her ordeal. The faith that was developed, the love that was felt, and the heart-felt prayers offered in our family were deeply touching to my soul. I can remember exactly where I was when Shelli called me with that great news that she had been found alive and my profound prayer experience as I poured out my gratitude through sobbing tears of joy to my loving Heavenly Father. So prayer is our great opportunity to grow in love and to be filled with His love. We have a great witness in her ordeal of the Divine purpose and power in prayer. It is our shield against the adversary, our way to show our love to the Lord and for His children, and our way to know His will for each of us in our personal lives; each of us has a different mission given to us by our loving Heavenly Father in our pre- mortal sphere. Fulfilling that mission brings the greatest of love, happiness, joy and peace in time and eternity. Ultimately, this is the ideal society, where we love the Lord with all of our heart, soul, strength, and mind and our neighbor as ourselves.[5]

---

5) Luke 10:27

Chapter 5

An interesting aside, we introduced Mary to chia seed for her PCT hike. We were so glad she had some with her on this hike. Not only is its nutrient value excellent, but it staves off hunger pains and has excellent hydration retention. It is the richest plant based food in omega-3 fatty acids. You may see its benefits at the following link.[6] She couldn't have had a better supplemental food for this ordeal — another miracle.

Saturday (13 April 2013), just got the following e-mail from our daughter, Shelli. It has been two weeks today since they found Mary on Mt. Hood.

From: Bruce & Shelli Owen

Sent: Saturday, April 13, 2013 11:13 AM

Mary is home! They put her under yesterday (Friday, Apr. 12th) and stapled her leg wound closed; she has to leave it untouched under the bandage until the 15th, and then have the stitches removed after the 19th if all goes well and it remains uninfected, etc. . . . They let her go around lunch time. She was able to walk - though slowly - out to our car in the hospital parking garage (no wheelchair for the distance). She has to wear an ankle brace and shoes that protect her toes from getting damaged while she still is getting feeling back in them. The toe guards look like a cow nose.. so 'moo-shoes'. They asked her about pain. She says 'no pain'.. They say "miraculously no pain?". She says, yes, no pain, and the reason being that so many people are praying for her!

Below is what Bruce shared on his Facebook page that evening, It sums things up well:

So tonight, after a week lost on Mount Hood, and two weeks in the hospital after her rescue, Mary is home. She is just sitting on the couch talking with us and emailing friends. In a few minutes we will gather at the table for dinner and give thanks. So much to be thankful for, for all of you who prayed that Mary would be found alive, for a God who hears and answers prayers and is so good. We will also be thankful for Search and Rescue workers who answered the call, and brought Mary home to us, and local police who worked so hard to find out she was on the mountain in the first place. This could get long . . . Thanks again!

---

6) *http://ItsAboutTimeBook.com/chia-seed-a-super-food/* and *http://www.allanstime.com/Health/Chia_seed.htm*

With grateful hearts for your prayers, love and encouragement,

And prayers that God richly bless you for your goodness to us, and just because He is good and loves to bless those who pursue Him,

Bruce & Shelli

We ask for your continued prayers; thank you all and thank you Lord.

The Rest of the Story

Mary's experience has deeply touched many lives — seeing the hand of the Lord and His matchless miracles. On Wednesday (1 May 2013) our daughter Celeste was up from Cortez to attend Women's Conference at BYU. We had a chance to have dinner with her and were discussing the angelic tracks that led down off the mountain directly to Mary. I learned from Celeste, "The rest of the story:" I document it as follows. On March 30th at 7:31 a.m. — the day Mary was found — I received a text from Celeste, "Any news?"

I responded at 7:32 a.m., "Yes. I am preparing an email.

At 7:36 a.m. Celeste responded back, "I had a big feeling of something this a.m. at 6. Then at 7 my hands involuntarily clasped together and went to my chest."

At 7:53 a.m. I responded, "Thx for sharing. Is significant. Love u."

I then sent out the e-mail mentioned above that I believed she would be found today alive. I also learned that Shelli and Bruce had been praying similarly — for angelic intervention in finding Mary. I thank the Lord for the power of prayer. Celeste is very close to Mary, and told me Wednesday that she had been praying specifically for angelic assistance in finding Mary. I believe we saw manifest in those tracks that the National Guard helicopter followed directly to Mary a direct answer to the focused faith of the thousands who were praying for her coupled with the specificity of Celeste and Shelli and Bruce's prayers. The tracks could not have been Mary's; she never summited!

When Shelli called me later that morning and told me that just as they had arrived at the parking lot of Timberline Lodge, they were informed that a search and rescue helicopter was hovering over someone who fit Mary's description. Shortly afterwards they were informed that it was in fact Mary, and that she was alive, though somewhat injured and was being flown to the hospital in Portland, the feeling that swept over me of Divine intervention was so profound, and as I said before, I was dissolved to tears of gratitude and fell to my knees sobbing and thanking the Lord for His tender mercies and for answering our prayers in such a miraculous way. The tracks from the top of Mt. Hood going directly to Mary were a documented miracle.

For those of you who don't know the details since, Mary continues to have a remarkable recovery — given the extent of her injuries. THANKS AGAIN for all of your prayers. How we need each other in this incredible mortal journey and most especially the Lord.

We had a family reunion at Lake Tahoe (22-29 June 2013) and we got to be with Mary there. When I saw her, we gave each other a big hug. We have always been close, and it was so wonderful to see her and with no apparent side-effects from her traumatic and life-threatening accident. She was barefoot, as is her usual walking style. They call her "Barefoot Mary" on campus. I asked her if she could lift her foot, which she had not been able to fully do because of some nerve damage incurred during her fall and she lifted her big toe for the first time right there in front of us. We rejoiced together. During the reunion she shared a lot more details of the experience on the mountain:

She was close to summiting — a goal she had set for herself when she walked by it three years before on the PCT. She estimates that she was within about 1,000 to 1,500 feet of the summit before the white-out got too intense, she knew she had to abort the attempt for the safety of her own life. She also knew there was a technical part left ahead of her.

One fascinating detail she shared with us was that Saturday night — just before her hike — she asked the Lord to wake her up when it was time. He woke her at 2:30 a.m. She was tired and rolled over and went back to sleep. Had she gotten up, I and she are convinced that she could have summited and been off the mountain before the white-out rolled in. The Lord honors our free agency. Being a perfect God, He honored her request, but also knew she would do what she did providing an opportunity for Him to teach her and all of us who will pay attention an extremely important lesson so poignantly stated in her inspired writing of "Mountain Time-Out." I thought to myself, "How many times have I ignored the ' still small voice,' and done my own thing?"

From the web one can learn the following interesting details about hiking Mt. Hood:

Mount Hood (11,239 ft.), Oregon's highest peak, forms a prominent backdrop for Portland and most of Northern Oregon and Southern Washington. Located 50 miles east of Portland, Mt. Hood rises nearly seven thousand feet higher than any other peak within 70 miles. Due to its prominence, proximity to an urban center and the well maintained road that runs high onto its south side, Mt. Hood is a popular destination for climbers.

Mt. Hood is considered a dormant volcano by most climbers, though as anyone who has climbed the south side will tell you it is not dead. Active fumaroles continue to spew sulphur gas into the air,…

Mt. Hood is currently one of the most climbed glaciated peaks in the world, some even claim it is second in the world only to Mt. Fuji in Japan, though this claim is mostly speculative because climbing registration is optional and self-regulated. Current estimates put the number of climbers who attempt to summit Hood at 15,000 to 20,000 annually. It is common to see a nearly solid line of headlamps stretching from Crater Rock to the summit in the early hours of a clear spring morning.

The South Side Route, which begins at Timberline Lodge parking lot (5,924 ft.), is the shortest and by far the most popular route to the summit. Mt. Hood's popularity and dubious distinction as an "easy non-technical hike" or as a "walk up" is a misconception that tends to increase the number of deaths on the mountain. It is a technical climb requiring at least an ice axe. These inexperienced climbers in conjunction with severe weather (which can move in quickly) or the ever present danger of avalanches, rock and ice fall account for most accidents.

Mary went up this most popular route on the south side and encountered these sulphur-emitting fumaroles. Because they emit heat, they are tempting

Chapter 5

to hikers to get close to get warm. Mary was similarly tempted but ignored it knowing the consequences. More than one hiker has been succumbed to the oxygen-depleted sulphur gas fumes. More than 130 people have died on that mountain.

When she encountered the intense white-out she knew she had to get off the mountain. Upon her descent she felt impressed to stop at one point and turn on her head lamp. As she did so, she was on the edge of a crevasse. She had a near full moon to navigate by. She had not replaced her batteries in her headlamp; being a college student working her own way through and trying to stay out of debt, she had bought only what she felt were the critical essentials, and so had been conserving battery life.

During her descent and fearing some of the bad places she had heard about to her left, she gravitated too much to the right. When she finally saw the lights of Portland, she knew where she was and that she was on the wrong side of the mountain. But the snow was so deep, she could only effectively descend. She said that she was almost swimming in the snow. She saw some lights on the west side of the mountain and designed to go to them. When she got down to timberline, she was in a canyon and felt to climb up out of it. She built a fire to warm up and to dry out before trying to climb up a canyon wall on the south side of the canyon — going toward the lights she had seen. It was during this climb that she fell as she was scrambling using tree roots and her ice axe to make the ascent. After falling, she was on the snow, out of reach of any fuel for Monday, Tuesday, and Wednesday. It wasn't until Thursday, because of falling snow and rocks threatening avalanche, that she, with great difficulty, moved herself over to a stretch of ridge with exposed Earth and some trees and brush. Then she was able to get fuel to build a fire again, warm her hands, and dry herself out. We have another miracle here. When she had built her first fire on Monday, she had used all of her matches and all the fuel in the cigarette lighter she had along for that purpose. She felt to try the lighter again, and it had fuel in it to start her fire. Thanks again, Lord, for your continual watch care.

She was wearing her poncho she had on the PCT trail, the whole time she was out there, except after her fall when she built herself a snow shelf (not a snow cave as some have said). She was then inside the poncho, sitting on a secure shelf of snow with enough of a wall to create a wind shield but not much more. She was still exposed to rain and snow and sun Monday through Thursday. She had some food and some chia seed that kept her going, but since she had been planning to be out only the day, her food and clothing were way short of her needs with this accident.

As I mentioned before, she fell at about 4:00 a.m. Monday morning, as she was climbing this canyon wall. When her ice axe broke loose, she let go of the axe to reduce chances of injury from it and positioned herself to stop her fall as soon as possible. After falling about 40 feet, she stopped herself just before going over a ledge into the bottom of the canyon with a running stream under the snow. Had she gone over that ledge that could have been fatal — falling through the snow and into the stream. At our family reunion, she described her feelings as she fell. My heart wrenched as she shared this horrific experience.

At that point, realizing that she could no longer hike with her badly sprained right ankle and the eight-inch gouge in her left thigh — making that leg immobile as well — she built her snow shelf and snow windbreak using her good foot and her elbows. Her hands were very wet and cold. She had four layers on her torso and two on her legs. She looked at her wound and it was not bleeding, so she felt to deal with it later. She could not afford good gloves, and the ones she had were not adequate for the challenge facing her. She sat on her climbing gear — helping her to keep up off the snow to preserve body heat. She curled up in her windbreaker jacket and her poncho in a fetal position. She was able to cover everything except her feet. Her design goal was to keep as warm as possible, and she spent the rest of the night that way.

It was really hard on her for those four days (Monday through Thursday) knowing that the Lord was not talking to her and no one was praying for her. My heart wrenched for her as she shared the loneliness she felt, but I found this particularly fascinating because of the Unified Field Theory work we had done over a decade ago showing our interconnectedness to each other and to God through diallel-field lines.[7] What she felt was exactly the case; no one knew, and no one was praying or looking for her that whole time because we didn't know and the Lord had His loving purposes for

---

7) For details of this research and the experiments substantiating the theory see: *www.allanstime.com/UFT_private* and the following article that I wrote for the Wisdom Society in San Diego: *http://www.allanstime.com/Spiritual/In_Touch_with_Eternity.htm*

being silent. Seeing no evidence of search efforts puzzled her because she had filed a hiking plan and thought that when she did not return at the time indicated on it they would start a search. We learned later that hikers typically don't sign out when they come off the mountain and they had filed away her hiking plan after a few days.

She woke up Friday to a totally different feeling; she knew thousands were praying for her, which again was exactly what was happening at the time, and the Lord was talking to her. And she also knew that she would be found! This whole experience marks clearly the reality of a loving God in our lives and that He is in the details in His infinite ability to love His children. I believe the Lord wants us all to learn the very important lesson she learned on the mountain.

Another miracle, not noted by most, is the manifestation of the Lord's love by making that week unseasonably warm. Had she lost more body heat the frostbite damage might have been irreparably bad or she might very well have lost her life. As it was, the Lord preserved her life and has worked and is working remarkable healings of all her injuries. She still has some nerve damage, but that too is healing amazingly well.

On the 27th of June, 2013, she and I hiked Mt. Tallac overlooking Lake Tahoe during our family reunion — a hike we were both looking forward to doing together. This was her first hike after her fall at the base of Mt. Hood. Mt. Tallac is labeled a difficult climb and is 9 miles round-trip with 3,500 feet vertical. It is known as the jewel in the crown of mountains which ring Lake Tahoe, Mt. Tallac promises unparalleled views from its summit 9,735ft above sea level. There were some rock scramblings in a couple of places, but not nearly as hard as Mt. Timpanogos or Mt. Nebo, which I have climbed in Utah. We went up in four hours, had a spectacular view during our lunch hour on top, and came back down in three hours. I am a good hiker — even at age 76, being an avid mountain biker — and she let me set the pace. She kind of gently pushed me up the mountain.

We stopped and visited several times with fellow hikers, and I had to tell them Mary's miraculous story. She is a walking miracle given all she had gone through on Mt. Hood, her incredible rescue, and subsequent healings. The scenery and mountain flowers were gorgeous. From the top we had a spectacular panoramic view: to the west was the Pacific Crest Trail she had hiked three years before; to the north and north- east are breath taking views of Emerald Bay (reportedly the most photographed bay in the world) and Lake Tahoe. Mary and I are not only family, but also kindred spirits and had nearly non-stop conversations up and down — a hike I will ever remember most fondly. Thank you, Mary, and THANK YOU, LORD, for teaching us all such valuable lessons through this beautiful daughter of Thine.

As a fun afternote: Mary graduated magna cum laud from George Fox University 3 May 2014 — a year after the accident — and marched across the platform "barefoot!" The next day, she married Benjamin Grimm on Mt. Tabor east of Portland, OR, and again she was barefoot! They are a great match. We are excited for them. Ben is fully aligned with her goal of translating the Bible for a culture that has no written language. Now Ben will be able to share with Mary this wonderful experience of giving a people God's word and a written language.

Chapter 6

The Theory of Evolution is Pervasive. Have We Been Deceived?

The theory of evolution, as generally held in the world, includes man as being evolved from lower forms of life — contrary to the Biblical account. We have a lot of data for evolution within a species, and we see this as God's design in helping any species adapt to its environment. Yet we have no reliable data for cross-species evolution. What is the truth? If the evolution of man is false, this has enormous implications. Most of the world believes it to be true. If false, this is a massive world-wide deceit.

Signature in the Cell by Stephen C. Meyer is a great example showing scientifically that behind DNA is Intelligent Design (ID). George Gilder, author of *Wealth and Poverty and Telecosm*, commenting on Dr. Meyer's book shares the following great insight:

> Meyer demolishes the materialist superstition at the core of evolutionary biology by exposing its Achilles' heel: its utter blindness to the origins of information. With the recognition that cells function as fast as supercomputers and as fruitfully as so many

factories, the case for mindless cosmos collapses. His refutation of Richard Dawkins will have all the dogs barking and the angels singing.

Having read the book, I fully concur with Gilder's comment.

Stephen Meyer says of Darwin's work:

> *On the Origin of Species* seized the attention of the scientific community like a thunderclap. Darwin's analogy to artificial selection was powerful, his proposed mechanism of natural selection and random variation easily grasped, and his skill in dispensing with potential objections unrivalled. Moreover, the explanatory scope of his argument for universal common descent constituted something of a tour de force. By the close of the Origin it seemed to many that Darwin had dispensed with every conceivable objection to his theory but one.[1]

As Meyer further states, " Darwin was puzzled by a pattern in the fossil record that seemed to document the geologically sudden appearance of animal life in a remote period of geologic history..." Darwin gave a copy of the Origin to Louis Agassiz — considered to be the greatest natural scientist of the time — and asked him to read it with an open mind. As Meyer reports, Agassiz "concluded that the fossil record, particularly the record of the explosion of Cambrian animal life, posed an insuperable difficulty for Darwin's theory." In other words, the data contradict the theory. Yet, the masses adopted the theory. In 1909, Cambridge University had an enormous 100 year anniversary celebrating Darwin's birth. But now time has proven the erroneous nature of his theory and substantiated Agassiz's conclusion.

In 1980, Harvard paleontologist Stephen Jay Gould stated, "that neo-Darwinism 'is effectively dead, despite its persistence as textbook orthodoxy,' the weight of critical opinion in biology has grown steadily with each passing year."[2] In 2009, Dr. Stephen C. Meyer was asked to testify before the Texas State Board of Education, who was considering a provision "that would encourage teachers to inform students of both the strengths and weaknesses of scientific theories." He had a binder with one hundred peer-reviewed

---

1) Steven Meyer, *Darwin's Doubt* p. 6
2) Gould, *Is a New and General Theory of Evolution Emerging?* p. 120

scientific articles in which biologists described significant problems with the traditional theory of evolution.³

The world ignores the above information as Dr. Meyer points out in his 2013 book, *Darwin's Doubt, The Explosive origin of Animal Life and the Case for Intelligent Design*. He goes on to share the alarming fact that several "official scientific organizations. . . routinely assure the public that the contemporary version of Darwinian theory enjoys unequivocal support among qualified scientists and that the evidence of biology overwhelmingly supports the theory." And "The media dutifully echo these pronouncements." ⁴ This they do in spite of the data to the contrary.

Meyer shares the alarming information that atheism coming out of Darwinian evolution is "a materialistic worldview in which entities such as God, free will, mind, soul, and purpose [play] no role."⁵ Whereas, the biological and paleontological data when properly understood "brings science and faith into real harmony. . . The evidence of purposeful design behind life. . . offers wholeness, and hope."⁶ Reflective of the last century's deceitful trend, Richard Dawkins wrote his bestselling book, *The God Delusion* (over 2 million copies sold). As a counter-attack to Dawkins' book and several others like it, David Berlinski wrote, *The Devil's Delusion: Atheism and its Scientific Pretensions*. Ben Stein in his excellent movie and documentary, *Expelled; No Intelligence Allowed*, interviews both Dawkins and Berlinski and several others on both sides of the issue: Darwinian evolution versus intelligent design (ID). ID has implicitly in its fabric that God is the author of creation. Now the large amount of ID data flies in the faces of the atheists. As a result, many of the atheists resort to "name-calling" of the ID folks. They have in their narrow-minded paradigm that good science excludes anyone who supports creationism as well as any religious beliefs. Richard Dawkins's book is a direct and frontal attack on traditional religions. Another implicit assumption in this battle of words on the part of the evolutionist's camp is that those who believe in religion are deleterious to a healthy society and religions are counterproductive to science. In this book we show just the opposite — using data and empirical results. We show that true science and true religion perfectly harmonize. In other words, using an improved and enlarged scientific method we demonstrate this harmony.

---

3) Steven Meyer, *Darwin's Doubt; The Explosive Origin of Animal Life and the Case for Intelligent Design*, p. xi
4) Steven Meyer, *Darwin's Doubt; The Explosive Origin of Animal Life and the Case for Intelligent Design*, p x, xi
5) Steven Meyer, *Darwin's Doubt; The Explosive Origin of Animal Life and the Case for Intelligent Design*, p 409
6) Steven Meyer, *Darwin's Doubt; The Explosive Origin of Animal Life and the Case for Intelligent Design*, p 412

It's About Time

In the interview of Richard Dawkins by Ben Stein, Dawkins said that anyone who doesn't believe in evolution is either stupid or ignorant. He said that evolution is one of the surest facts in science — totally ignoring the "fact" that Darwinian evolution is a theory and has no solid data to support cross-species evolution.

I have found it interesting that those in the evolutionary camp seem to feel that by labeling and putting down the ID people, they will win the day. Lying is an acceptable strategy, since they believe morals come of religious dogmas. Dawkins argues in his book that believing in evolution doesn't make a person immoral, that altruistic genes may prevail leading a person to do good. Historical data show the opposite! They admit that believing in evolution leads to atheism, and we have seen over the last century the immorality coming out of that influence. As I see it, an advantage that atheists have — if you want to call it that and I don't — is that there are no morals to live by; it is survival of the fittest mentality. Hence, we see this lie within Dawkins's book.

In his documentary, *Expelled: No Intelligence Allowed*, Ben Stein does a good job of sharing how atheists lie about how they mistreat those who are "expelled" for advocating and/or teaching creationism. I find it a common practice among them to name call, belittle and tell falsehoods about the intelligent-design people also. It is like the person who feels that if he yells louder than his opponent, he will win the argument; whether he is right or wrong is immaterial.

Taking Dawkins' statement that those who don't believe in evolution are "stupid and ignorant," this list would include some of the most brilliant minds I know: Isaac Newton, George Washington, Thomas Jefferson, C.S. Lewis, and Stephen C. Meyer, and the list could be made very long of those who are devout followers of Christ and are brilliant. The list of those believing in intelligent design is growing rapidly because of the data and the goodness one feels in seeing this evidence come forth.

David Berlinski, as a secular Jew, is an interesting example of a brilliant and non-religious person whose intellectual integrity has convinced him of the validity of intelligent design. In his book, *The Devil's Delusion*, he shows Dawkins and his camp have no basis for fighting against religions and against intelligent design. Berlinski's subtitle to his book is apropos: *"Atheism and Its Scientific Pretensions."* In this regard, Ben Stein's interviews in his documentary, *Expelled: No Intelligence Allowed*, with Berlinski and Dawkins are most revealing and show the emptiness and flagrancy of atheism. Meyer's books brilliantly document the non-validity of Darwinian evolution, i.e. that life came about by chance.

For me, David Berlinski's credentials are pretty amazing: He received his Ph.D. in philosophy from Princeton University and was later a postdoctoral fellow in mathematics and molecular biology at Columbia University. He is currently a Senior Fellow at Discovery Institute's Center for Science and Culture. Dr. Berlinski has authored works on systems analysis, differential topology, theoretical biology, analytic philosophy, and the philosophy of mathematics, as well as three novels. He has also taught philosophy, mathematics and English at such universities as Stanford, Rutgers, the City University of New York, and the Universite de Paris. In addition, he has held research fellowships at the International Institute for Applied Systems Analysis (IIASA) in Austria and the Institut des Hautes Etudes Scientifiques (IHES) in France.[7, 8]

In his book, *The Devil's Delusion: Atheism and its Scientific Pretensions*, he asks some interestingly articulated and poignant questions:

- Has anyone provided a proof of God's inexistence? Not even close.

- Has quantum cosmology explained the emergence of the universe or why it is here? Not even close.

- Have the sciences explained why our universe seems to be fine-tuned to allow for the existence of life? Not even close.

- Are physicists and biologists willing to believe in anything so long as it is not religious thought? Close enough.

---

7) Recent articles by Dr. Berlinski have been featured in *Commentary*, *Forbes ASAP*, and the *Boston Review*. Two of his articles, *On the Origins of the Mind* (November 2004) and *What Brings a World into Being* (March 2001) have been anthologized in *The Best American Science Writing 2005*, edited by Alan Lightman (Harper Perennial), and *The Best American Science Writing 2002*, edited by Jesse Cohen, respectively

8) Dr. Berlinski is author of numerous books, including *A Tour of the Calculus* (Pantheon 1996), *The Advent of the Algorithm* (2000, Harcourt Brace), *Newton's Gift* (The Free Press 2000), *The Secrets of the Vaulted Sky* (Harcourt, October 2003), *A Short History of Mathematics for the Modern Library* series at Random House (2004), and *The Devil's Delusion: Atheism and Its Scientific Pretensions* (Crown Forum, 2008).

- Has rationalism in moral thought provided us with an understanding of what is good, what is right, and what is moral? Not close enough.

- Has secularism in the terrible twentieth century been a force for good? Not even close to being close.

- Is there a narrow and oppressive orthodoxy of thought and opinion within the sciences? Close enough.

- Does anything in the sciences or in their philosophy justify the claim that religious belief is irrational? Not even ballpark.

- Is scientific atheism a frivolous exercise in intellectual contempt? Dead on.

Berlinski does not dismiss the achievements of western science. The great physical theories, he observes, are among the treasures of the human race. But they do nothing to answer the questions that religion asks, and they fail to offer a coherent description of the cosmos or the methods by which it might be investigated. William F. Buckley Jr. said of his book, " Berlinski's book is everything desirable: it is idiomatic, profound, brilliantly polemical, amusing, and of course vastly learned. I congratulate him."

Militant atheism is on the rise. However, Berlinski and many more like him, who are increasing in number, are countering this rise. Even though he is a secular Jew, Berlinski delivers a biting defense of religious thought. As an acclaimed author who has spent his career writing about mathematics and the sciences, he turns the scientific community's cherished skepticism back on itself, daring to ask the above rather embarrassing questions. Some of the above difficult questions Berlinski poses we answer in this book — not because I claim brilliance, but because I have sought out sources of Light and Truth revealing answers to these soul-searching questions.

Toward the end of the movie, *Expelled; No Intelligence Allowed*, Ben Stein goes directly to Richard Dawkins and interviews him. It is a fascinating dialogue. As Ben concludes the interview, he asks Dawkins:

What if after you died you ran into God and He said, "What have you been doing, Richard? I've been trying to be nice to you. I gave you a multimillion-dollar paycheck over and over again with your book, and look at what you did?"

Chapter 6

Richard: Bertrand Russell had that point put to him, and he said something like, "Sir, why did you take such pains to hide yourself?"

Ben: But if the intelligent-design people are right, God isn't hidden. We may even be able to encounter God through science, if we have the freedom to go there. What could be more intriguing than that?

Howard Storm was an atheist, and then had a near-death experience. I suggest you read his book, *My Descent Into Death and the Message of Love that Brought Me Back*. Scientific integrity leads us to learn a great deal from the some 13 million documented NDEs — many of them with irrefutable evidence as to their authenticity. Storm's near-death experience, as an atheist, was such an one. This experience totally changed his life.

He died from complications when his duodenum perforated — opening a hole allowing the hydrochloric acid in his stomach to leak into his abdominal cavity resulting in ever increasing and searing pain until he passed through the veil of death. There his spirit was in indescribable darkness with other spirits tormenting him. He says, "There was a profound sense of timelessness… A terrible sense of dread was growing within me. This experience was too real… it was Hell… and it was more horrible than anything I could possibly have imagined." Before he had thought that life was every man for himself — survival of the fittest — and he was about building his ego. He thought that when you died that was it — nothing, and he found something very different.

He heard a voice that sounded like his voice saying, "Pray to God." He remembered thinking, "Why? What a stupid idea… I don't believe in God." He heard this three times. From his childhood Sunday school experiences he tried reciting the 23rd Psalms, the Star Spangled Banner, the Lord's Prayer, the Pledge of Allegiance, and God Bless America and whatever phrases came to mind. The tormenting increased as the evil spirits tormenting him hated to see him trying to pray. Then, again, from his childhood he heard his voice singing full of innocence, trust, and hope, "Jesus loves me…" He said, "I desperately needed someone to love me, someone to know I was alive. A ray of hope began to dawn in me, a belief that there really was something greater out there. For the first time in my adult life I wanted it to be true that Jesus loved me. I didn't know how to express what I wanted and needed, but with every bit of my last ounce of strength, I yelled out into the darkness, 'Jesus, save me.' I yelled that from the core of my being with all the energy I had left. I have never meant anything more strongly in my life."

And then, like the thousands if not millions of others who have had NDEs from all parts of the world, he met the Savior of mankind and felt of His infinite and profound love.

This loving, luminous being who embraced me, knew me intimately. He knew me better than I knew myself. He was knowledge and wisdom. I knew that He knew everything about me. I was unconditionally loved and accepted. He was King of Kings, Lord of lords, Christ Jesus the Savior. I experienced love in such intensity that nothing I had ever known before was comparable. His love was greater than all human love put together… He was indescribably wonderful goodness, power, knowledge, and love.

This experience — being as life changing as it was — caused Howard to change his career from being a university art instructor and painter to helping people come to know God's love. He said, "Creating art had been the driving passion in my life," and that egocentrically. His new life had a new focus, "Personal relationships have become my artistic expression," and to teach them what he knew about God's love and the importance of loving one another. He saw his old world centered in "egocentric pride." He had learned that if we want to know God, "we have to surrender our individual and collective pride/ego if we are ever to know God's love." He had learned that "God loves everyone beyond anything we can imagine. God loves atheists, agnostics, murderers, prostitutes, thieves, drunks, drug addicts, homeless people, and liars." He abhors much of our behavior, but He always loves us because of His infinite capacity to do so.

He became an avid reader of the Bible and found the scriptures to harmonize beautifully with his NDE experience. The evolutionary theory has now led many people to not believe the Bible story of creation and of the flood and that these are "fairy-tells" — as Dawkins would say — and are out of harmony with science. For the Christian community, the implications of this disharmony are enormous. If there were no Adam and Eve and the fall of mankind as described in the Bible, then there is no need for an atonement — a Savior to overcome the effects of the fall and the sins of the human family. In other words, organic evolution, as believed by most scientists, implicitly teaches that there is no need for a Christ, and Biblical morality is out of vogue. In other words, organic evolution is Satan's subtle anti-Christ with enormously far-reaching effects as it has infiltrated most of the societies of the world.

Since Darwin published his book on the theory of evolution in 1859, *On the Origin of Species by Means of Natural Selection, or the Preservation of Favoured Races in the Struggle for Life*, atheism has increased significantly. By the 1930s to the 1950s — triggered in large measure by the Scopes Monkey Trial in 1925 — a broad consensus developed in which natural selection was believed to be the basic mechanism of evolution and the origin of life. Darwin's work became the accepted theory of the life sciences, and his theory became the explanation for the diversity of life. In contrast, all the great prophets throughout the ages have preached the creation, the fall, and the atonement (three pillars of eternity) as being fundamental to Heavenly Father's plan of happiness and eternal life.

The people of the world seem to have lost track that the greatest scientist we have ever known, Isaac Newton, along with many other early scientists were devout creationists. Most do not know that Newton spent more time in the Bible than doing science as is documented in his personal writings. Louis Agassiz (1807-1873) — considered to be the "greatest natural scientist of his day" said, "It is the job of prophets and scientists alike to proclaim the glories of God." Agassiz was greatly respected by Charles Darwin. In fact, Darwin said of him, "What set of men you have in Cambridge! Why, there is Agassiz — he counts for three." At that time, Agassiz showed that Darwin had errors in his extrapolations of how life came to be. As we examine the data with integrity we are led to the conclusion that "natural selection" and "survival of the fittest" coming out of Darwin's theory of evolution are not only wrong, but are devastating to a healthy societal structure.

A good scientist will not proclaim a theory to be true unless there is data to support it. There is no data to support the atheist's view that there is no God, yet the effects of organic evolution are evident in that a recent survey shows that 93% of the leading scientists in America are atheists or non-believers. Furthermore, and this will surprise most, there is no data to support organic evolution across species; it is an unproven theory, and obviously it is widely accepted, but not true. Evolution within a species is common knowledge and illustrates again divine design as any species can adapt to its environment. As Agassiz told Darwin at the time, he was extrapolating too far in believing that evolution could go across species and from a chance beginning.

As a devout Christian and as a scientist, I find that my thoughts align with Newton and Agassiz, and I see evidence of God's love and of the beauty and harmony of His creations in every direction I turn. The past half century has seen the fruits of science without God, as society has spiraled down in moral decay. Fortunately, this century is delightfully seeing numerous examples of God being brought back into science. Stephen C. Meyer in his 2009 book *Signature in the Cell: DNA and the Evidence for Intelligent Design*, gives irrefutable probabilities for intelligent design in all DNA. The book *Thousands. . .Not Billions*, by Dr. Don DeYoung (copyright 2005) documents the work of a team of experts showing that radiometric dating yields thousands of years not billions of years for the age of the Earth. Two separate experiments using DNA to estimate the age of Mitochondrial Eve (the first mother of all men) give her origin date to be about 6,000 years ago. You will see in the following footnote the link to good scientific data, which outlines the work described in Dr. DeYoung's book how integrating God into science brings harmony to science.[9]

A dear friend and colleague introduced me to the spear-head point called the Clovis point and the extremely important information around it. He has a whole chapter in the book he is writing on this subject (68 pages; here, I am writing a paragraph) and his work will go a long way in bringing science and religion into harmony. He wants his name and his work to be kept anonymous until his book is completed. Respecting his request, I have used just information publicly available.

You can find out on the internet that the Clovis point is a real stumbling block to archeologists and to other sciences as well. The first Clovis point was discovered in Clovis, New Mexico in 1929, hence, the name.[10] Subsequently, tens of thousands have been found with the maximum density in North America but extending as far south as Venezuela and into SW Europe. We will see significant implications of these findings.

---

9) *http://www.youtube.com/watch?v=AMy2IUeXJRI*
10) One can find some useful information at the following link: *http://en.wikipedia.org/wiki/Clovis_point*.

Chapter 6

The Clovis point is the most sophisticated and cleverly made point ever found. We are challenged to replicate it today. Many experts now believe that the Clovis people are regarded as the first human inhabitants of the New World — yet with the most sophisticated point technology. We see this is exactly the opposite of evolutionary theory. Furthermore, archeological data indicate that the Clovis-point culture ended cataclysmically as with Noah's flood. This understanding aligns profoundly with the scriptures and supports the fact that these spear-head points were made by the families descending from Adam and Eve.[11]

It is also fascinating that Clovis points have been found in southern France as well. That fact makes perfect sense because the family of Adam lived before the Earth was divided. A study of the Earth's plate tectonics shows that both France and Venezuela were attached to North America before the Earth was divided — again aligning beautifully with the Bible.

It is also interesting to note that they have found no Clovis-points in eastern Asia. Since the Clovis points have been dated pre- flood and are the oldest points, the dearth of Clovis points in East Asia directly contradicts the first man coming up out of Africa and up across the Bering Strait as is promoted by evolutionists. What has been found fully harmonizes with the scriptures, which is fundamental to my basis of truth.

To my knowledge, my good friend and colleague is the first one to uncover the true origin of the Clovis culture people as the family of Adam and Eve over their 1,600-year history until they were destroyed in the deluge. He, like Newton and Agassiz, believes in the Bible and the words of the prophets, and using those as his basis of truth sorted out this very fundamental information. In contrast, since the first Clovis point was discovered in 1929, archeologists, as well as other scientists, have not been able to make sense of the Clovis culture because of their belief in evolution. I use this knowledge as a beautiful example of how when true science and true religion come together, then the TRUTH is revealed and its harmony resonates with our souls. For additional information on the Clovis Culture follow the link below.[12]

My good friend has also done some fascinating experiments simulating the conditions of the flood using an autoclave. He has demonstrated that

---

[11] See Doctrine and Covenants 29:41 and 116:1
[12] *http://ItsAboutTimeBook.com/as-old-as-adam-science-and-religion/*

fossils, rocks, and crystals — in the conditions of the flood — can be grown in months and don't need millions of years to form. He took us to the biggest rock show in the world held in Tucson, Arizona during February of 2013 and shared with us how many of these were formed in the flood. I purchased an enhydro quartz crystal, which means it has one or more bubbles of water in it, that was formed during the flood. Geologists have a hard time explaining how these are formed, but my friend's model fits very nicely the scriptural account and aligns with his experimentation. Enhydro quartz crystals can be found across the globe — an enigma to geologists.

My friend also demonstrates that dinosaurs died during the flood. I have a nice picture of a bone as a sample from eight humans buried in the same stratum with dinosaurs found in Montrose, Colorado. All of this will be documented and explained in detail when my friend's book is published — we hope soon.

As mentioned, these data directly contradict the first man coming up out of Africa and up across the Bering Strait as is promoted by mainstream evolutionary science. What does the evolutionary camp do with these data? Ignore the data and stick with the theory in order to justify their actions. You will see in the following footnote the link from the work sponsored by the Creation Research Society in Seattle, Washington — following good scientific procedure — which outlines the work described in Dr. DeYoung's book, how integrating God into science also brings harmony to science:[13]

It is also interesting that in contrast to the last century where science and religion were at odds, several ancient civilizations had science and religion fully integrated and had technologies and construction concepts that baffle our best experts today. The example I shared before is a good one.[14] And watch the DVD, *Lost Civilizations of North America*. In this DVD you will learn of ancient cities that existed in North America that were larger than London or Rome and with advanced technologies.

It is an exciting time as we see true science and the scriptural account coming into harmony — witnessing to all who have hearts to feel, eyes to see, and ears to hear the beautiful truths of God's love for each of us. He so loved the world that He gave His Only Begotten Son that we may

---

13) *http://www.youtube.com/watch?v=AMy2IUeXJRI*
14) *http://www.allanstime.com/Spiritual/ Hopewell_Civilization_Great_Octagon/index.html*

be forgiven of our sins as we believe in Christ and His great gospel plan, which offers love, joy, peace and happiness in time (in the midst of a world awry) and in eternity a fullness of joy with God and our loved ones. What a glorious message.

Professor Chauncey Riddle, who wrote the Foreword for the book, was kind enough to share his perspective on the conflict resolution between science and religion, so called. I felt it to be of extreme value and include it here:

Professor Riddle's notes on the conflict between science and religion:

Science is the art of creating descriptions which correctly characterize the natural world in which we live. It has five levels, four of which are assertions. To describe these levels we will use an example prominent in this book, that of organic evolution.

The first level is phenomena, the sensations we humans have as we relate to the world around us. Example: We see lots of plants and animals in the world. (No assertion necessary here, but all things which are true science must begin with phenomena.) The second level is facts, assertions we invent to interpret the phenomena we perceive. Example: We see a fossil and call it a "trilobite," an ancient marine arthropod. This fact is a combination of what we see as a phenomenon and our general understanding of the world and what that world contains. The third level is laws, assertions we invent which are reliable generalizations about the facts we invent and which are based upon masses of facts. Example: Trilobites evolved slowly from the Early Cambrian era to the Permian era when they became extinct. This law is well grounded in observation and no one challenges it. The fourth level is theories which we invent to explain the laws we invent. Theories are our imagination at work, proposing the existence of things which are unobservable. Example: The theory of organic evolution: Trilobites evolved from more simple forms of life, as did everything else that presently lives as a plant or animal, and life originally came about spontaneously. It is important to realize that there is absolutely no physical evidence for this theoretical explanation of how trilobites came to be. It is only a hope, wishful thinking on the part of some scientists. The fifth level is postulates we invent to regulate our theory construction. One such postulate is the idea of causation: every event must have a cause sufficient to the result. Another postulate is that of least action: The simplest explanation that suffices for sufficient cause is preferred. Another possible postulate is that of naturalism: All that exists is the matter that physics describes, and there is no spiritual realm nor God nor spirit in man, and every human ceases to exist at mortal death. Perhaps you can see that it is in the postulates that govern scientific theory construction that the fundamentals of any scientific outlook are established.

Organic evolution is an unproved, unprovable theory based on the postulate of naturalism. Its only real support is the opinion of those scientists who do not want a God or spiritual existence as part of their thinking. The unfortunate part is that the many solid physical evidences for the law of evolution are dishonestly used by some to attempt to prove the theory that all life descended from primitive forms created spontaneously by natural forces. The law in no way proves that theory. But some want to believe that theory anyway, so they misappropriate the evidence for the law and apply it to the theory, and most people are unaware of that dishonesty.

And there are some people of good will who are good Christians who do support the theory of organic evolution. Apparently they are able to do so because they do not have in their minds a clear distinction between law and theory in science and particularly between the law of evolution and the theory of evolution. If they could be brought to see better the structure of science, especially the importance of the usually unstated original postulates behind the theory of organic evolution, they would be wiser about the matter. Unfortunately, the intellectual structure of science is seldom taught or mentioned. Law and theory are usually run together. And thus many are confused without knowing it. That confusion is carefully conserved by some who do understand the difference in order to win converts to the cause of the theory of organic evolution.

The bottom line is that the conflicts between science and religion are not necessary but are invoked by those persons who do not want to have a God or immortality. The conflicts arise from the unproved and unprovable postulates which individual scientists adopt to create their theories. Then they pretend that science demonstrates that there is no God nor spirit, nor immortality, nor any judgment of human conduct by a higher being. Therefore, in their thinking, there are no moral standards and anything is permissible for "enlightened" persons (persons who share the premise that there is no God and no right and wrong). Thus the same people who believe in the theory of organic evolution also tend to believe in abortion, sexual promiscuity, euthanasia, that might makes right, that lying is justified, and that ' religion' is a mark of a weak mind.

It so happens that actually every normal human being has a " religion." Each person's religion has three parts: First, beliefs about the universe and who and what is in it (variously called "theology", or "metaphysics", or "reality," depending on with whom you are speaking). Second, standards of conduct, what should and should not be done (variously called "ethics", or "morals", or " wisdom.") Third, rituals: things people do to reinforce or amend their religious beliefs and practices. Some rituals are prayer, conferring with others to compare notes, and meditation.

The theory of organic evolution is a personally chosen theology, the metaphysics for some persons, part of their personal religion. Those same persons tend to believe there is no right or wrong and no such thing as sin. One of their rituals is to get together with persons who think as they do and proclaim how enlightened they are and how unenlightened are the people who believe in traditional religions such as Christianity.

But not every writer sees a conflict between science and religion. They do not see one because all such conflicts are man-made, quite unnecessary to good science and good religion. If one avoids the common approach in our modern times to blindly worship science and current scientific hypotheses as "the truth" and to denigrate religion as being associated only with backward, non-intelligent teachings of some church, then there need be no conflict.

It's About Time

Chapter 7

Food, Toxins, and Nourishment to Body, Mind & Spirit

7.1 How to be Healthy, How to be Sick — the American Way, or What's on My Plate? Does it Affect My Body, Mind and Spirit?

Truth is Light

We love and honor our parents. However, just because a tradition came to them and on to us does not make it right. Satan, in his great chicanery, has infiltrated our society with traditions which destroy. We need to examine every tradition in the light of truth. The following are documented in the medical/health literature and/or in the writings of the prophets.

Did you know that:

- The healthiest people that have ever lived or that live now on this planet follow(ed) a life style with little or no meat and/or dairy in their diet?

- American meat consumption has more than doubled over this century and most of the major killing diseases have followed this trend?

- Most of the transmittable diseases are carryable by meat and dairy products?

- A high-protein diet (meat and dairy centered) causes calcium leaching from the bones, regardless of how much calcium is ingested?

- Osteoporosis is epidemic in American culture with 200% consumption over protein need?

- The grain and soybeans fed to U.S. livestock would feed more than 6 times the population of the U.S.?

- The high cancer rate correlates with a meat and dairy centered diet causing the body to be too acidic?

- It is almost impossible to suffer from protein deficiency on a wide variety, plant-based diet?

- Most of the world's people are lactose intolerant.

- Most Americans have bought into believing the BIG LIE that we have to have meat and dairy to fulfill our protein and calcium needs?

These and many other startling distortions have lead the masses down the disease laden path – also leading to memory loss, deteriorated immunity systems as well as a desensitization of our spirits.

The American diet on average is a disaster. The average American diet and life style are the driving reasons for the epidemics we now have of obesity, diabetes, heart attacks, cancer, dementia, eyesight problems, and other health challenges, as you will learn through Section II of this book.

Let us start off with a positive approach. I call it the ideal meal. Consider the environment of the ideal meal. It is one of the beautiful pleasures of life when we share it with precious family and friends – not rushed, but greatly enjoyed.

Over the centuries and even millennia, the "breaking of bread" together has been a tradition of great importance in almost every culture. With the proper food, properly and lovingly prepared, the ideal meal provides nutrition to the body, the mind, and the spirit. Not only does it provide nutriment to all three, but it can provide healing for all three as well.

Chapter 7

The joke that Americans "inhale" their food because of their busy-ness is not funny. The French say that Americans eat to live, and that they live to eat. I have eaten with them on many occasions, and they may take two hours or more to eat their evening meal. The French cuisine is outstanding, and they mostly don't get fat. They take a bite; savor the flavor and chew it well; then they visit.

I am so incredibly blessed to have a Sweetheart, who is also my wife, who prepares meals this way. We have learned some very valuable lessons to bring us to this point. As a result, I have enjoyed personally the reversal of several different "Western diseases" over the last two decades or so. I write this in celebration of our fifty-five years together, and to let the world know of some of the very important lessons we have learned from experience and from experts.

The Prophet Joseph Smith received a health code revelation in 1833 known as the Word of Wisdom.[1] It agrees, in a profound way, with scientific data gathered since regarding health. Gaining a better understanding of that Section has helped us a lot. The Lord promised both physical and spiritual blessings to those who would follow it.

New York Times #1 best seller author, Michael Pollan, has written several, very informative books toward knowing what is the ideal meal.[2] We have three of them: *Omnivore's Dilemma*, *In Defense of Food*, and *Cooked*.

Pollan documents the disaster our industrialized food system has created – being almost the total cause of all the epidemic diseases that afflict our western society. He shows that we get over half of our calories from tax-subsidized corn: corn-fed beef, chickens, turkeys, pigs, and processed corn-based foods, etc. It is mostly about money with empty words promising nutrition. He shows that corn-feed is not the best-feed, and the whole system is supported by fossil fuels in contrast to renewable, healthful, solar based food chains that can potentially come from our own gardens.

Pollan further shows that high fructose corn syrup (HFCS) is pervasive in the American diet, and it is in almost all soft drinks. It has become the preferred sweetener in a large and increasing percentage of processed foods. It is the cheapest sweetener and prolongs the life of that to which it is added.

---

1) D&C 89
2) *http://michaelpollan.com/books/*

From pure economics it was a major breakthrough when it was developed in the 1960s. Its use started in the 1970s, and now it is a challenge to buy food without it. But if you care about your health, you better look at labels and avoid it like a plague.

From a health standpoint, it is subtle disaster. HFCS is linked as one of the main causes of the obesity epidemic as well as to the diabetes epidemic. It tends to suppress the satiation sense, so that when we have eaten enough, we don't know it and overeat. It also tends to be addictive. Its abundance in the American diet is often causing liver problems. As a synthesized sugar, the body doesn't know how to digest it and turns it into fat.[3]

When I learned from Pollan that HFCS "is the most valuable food product refined from corn, accounting for 530 million bushels" [in 2006], I was aghast. This is a major conspiratorial activity where the rich get richer at the expense of your health, and you are paying for it through your taxes and because it is a fossil-fuel-based industry, as Michael proves so profoundly. The middle-man folks doing the processing, who make most of the profits from this cheapest and most pervasive of all sugars in America, would not let Michael review their processing procedures. It was discovered in 2009 that some HFCS had mercury in it at harmful levels. It seems the manufacturers have responded to this.[4] They, however, don't bother to tell you that it is still a poison to the body.

The Lord knew this was coming and warned: "In consequence of evils and designs which do and will exist in the hearts of conspiring men in the last days, I have warned you, and forewarn you, by giving unto you this word of wisdom by revelation" for "the temporal salvation of all saints in the last days."[5]

Pollan's book, *Omnivore's Dilemma*, tracks how we get our food in America, and it is a real eye opener. Interestingly, Cargill and ADM have a significant influence on the FDA, which has approved HFCS – regardless of HFCS being harmful.

Michael Pollan says, "Eat food [by this he means "whole fresh foods rather than processed food products]. Not too much. Mostly plants." This statement reminds me of Thomas Jefferson, who was basically vegetarian using meat

---

3) http://articles.mercola.com/sites/articles/archive/2008/11/06/corn-syrup-s-new-disguise.aspx
4) http://en.wikipedia.org/wiki/High_fructose_corn_syrup
5) D&C 89:4, 3

more as a condiment. He believed one should always rise from the table just a little bit hungry. This belief is in total contrast to the typical American diet, which is meat and dairy-centered – resulting in the large variety of "Western diseases" that are the major killers of Americans.

The FDA has also approved MSG – a flavor enhancer often used in restaurants – which is also harmful to the body and has been shown to do damage to the brain.

If you take the French attitude of relax with family and friends and enjoy, add the whole plant-based foods, prepared with love, eaten with gratitude and with conversation that is uplifting – providing spiritual nourishment to the soul, then you have the IDEAL MEAL. You will be properly and enjoyably nourishing body, mind, and spirit. Couple this with keeping the commandments of the Lord – loving Him and your neighbor – then you have His sure promise that you "shall receive health in [the] navel and marrow to [your] bones; And shall find wisdom and great treasures of knowledge, even hidden treasures; And shall run and not be weary, and shall walk and not faint. And I, the Lord, give unto [you] a promise, that the destroying angel shall pass [you by], as the children of Israel, and not slay [you]. Amen."[6] This I know and am most grateful to the Lord for the blessings of health to body, mind, and spirit in my life.

We recommend that you eat from your own garden and get away from the chemicals and poisons that are all too common in commercial, highly processed foods. We have had good luck with green houses for growing food year around. (We bought ours from www.growingspaces.com and you can see our solar home and the geodesic dome we bought online.[7]

The *Back to Eden* film was released August 2011.[8] It has been viewed over two million times in 211 countries. It shares the inspiring story of Paul Gautschi, as he asks God how He did His garden in creating the Earth. Paul shares some remarkable skills learned and has developed an organic gardening technique which has been sustaining his family – and lots of others – for about 40 years.

---

6) D&C 89:18-21
7) http://www.allanstime.com/SolarHome/index.html
8) It is available on line http://www.backtoedenfilm.com/ and as a DVD

My wife and I have been using Paul's techniques since we learned about them two years ago. It is a lot of work to get started, but the fruits of our efforts are paying off. Growing your own food has so many benefits for a healthy body, mind, and spirit. We highly recommend you watch Back to Eden. The DVD has, in addition to the film, extra tips in dealing with weeds and insects and in how to prune trees.

Buying organic foods helps. Experts also recommend, "Daily supplements of vitamin B12, and vitamin D for people who spend most of their time indoors and/or live in [high latitudes – like the USA] are encouraged."

The Lord makes it very clear that by using conspiring men, Satan has successfully inflicted our society with disease, early death and misery because of the false traditional diets which have found their way onto our plates.

The Lord has also said, "There is enough and to spare."[9] Given the current trends on the earth, most people perceive that there is not enough and to spare. Is the Lord wrong, or is the fault at our own feet? A meat and dairy centered diet is about ten times more taxing of the Earth's resources than is a plant-centered diet.

Perspectives from the Prophets

The following quotes from the scriptures and statements of the prophets provide food for the soul.

> "The earth mourneth and fadeth away, the world languisheth and fadeth away, the haughty people of the earth do languish. The earth also is defiled under the inhabitants thereof; because they have transgressed the laws, changed the ordinance, broken the everlasting covenant. Therefore hath the curse devoured the earth, and they that dwell therein are desolate: therefore the inhabitants of the earth are burned, and few men left." (Isa. 24:4-6) "Shake thyself from the dust; arise, and sit down, O Jerusalem: loose thyself from the bands of thy neck, O captive daughter of Zion." (Isa. 52:5; 2 Ne. 8:25; 3 Ne. 20:37) Indeed, it is time for us to awake and arise; take off the shackles of our minds, our

---

9) D&C 104:17

Chapter 7

hearts and our souls, and remove the false traditions which bind us down. "And it came to pass that Enoch looked upon the earth; and he heard a voice from the bowels thereof, saying: Wo, wo is me, the mother of men; I am pained, I am weary, because of the wickedness of my children. When shall I rest, and be cleansed from the filthiness which is gone forth out of me? When will my Creator sanctify me, that I may rest, and righteousness for a season abide upon my face?" (Moses 7:48)

As we look toward a Zion (pure in heart) society and being a millennial people, we can have great hope. We have His promise that the knowledge of the Lord will cover the earth as the waters cover the seas, and He will be in our midst. We will be His people and He will be our God, and surely this will be the happiest people who have ever dwelt on the face of the earth. "The wolf and the lamb shall feed together, and the lion shall eat straw like the bullock: and dust shall be the serpent's meat. They shall not hurt nor destroy in all my holy mountain, saith the LORD."[10]

Here are some statements regarding the millennial covenant:

> "And in that day will I make a covenant for them with the beasts of the field, and with the fowls of heaven, and with the creeping things of the ground; and I will break the bow and the sword and the battle out of the earth, and will make them to lie down safely." (Hosea 2:18) "And I, God, said unto man: Behold, I have given you every herb bearing seed,. . .and every tree in the which shall be the fruit of a tree yielding seed; to you it shall be for meat. And to every beast of the earth and to every fowl of the air, and to everything that creepeth upon the earth, wherein I grant life, there shall be given every clean herb for meat" (Moses 2:29-30).

Josephus tells us regarding Daniel's superior diet:

> "Now Daniel and his kinsman had resolved to use a severe diet, and to abstain from those kinds of food which came from the king's table, and entirely to forebear to eat of all living creatures; . . .but gave them pulse and dates for their food and anything else

---

10) Jeremiah 31:31-34; Isa. 11:9; 65:25; 2 Ne. 21:9; 30:15

beside the flesh of living creatures that he pleased for that their inclinations were to that sort of food, and that they despised the other. . . ..those that were with Daniel looked as if they had lived in plenty, and in all sorts of luxury,. . .they had their souls in some measure more pure, and less burdened, and so fitter for learning, and had their bodies in better tune for hard labour;" "As for these four children, God gave them knowledge and skill in all learning and wisdom: and Daniel had understanding in all visions and dreams. Nebuchadnezzar . . .found none like (them). . . in all matters of wisdom and understanding" (Dan.1:17-20).

The Prophet Joseph Smith, during Zion's Camp, prevented the brethren from killing three rattlesnakes, saying, "Leave them alone. Don't hurt them! How will the serpent ever lose its venom, while the servants of God possess the same dispositions, and continue to make war upon it? Men must become harmless before the brute creation and when men lose their vicious dispositions and cease to destroy the animal race, the lion and the lamb can dwell together, and the suckling child can play with the serpent in safety."[11]

Brigham Young taught, "You mothers and daughters in Israel . . . You may think that these things are not of much importance . . .but let the people observe . . . and lay the foundation for longevity . . . Do you think they will stuff themselves then with . . . beef, pork . . . No; you will find they will live as our first parents did, on fruits and on a little simple food, and they will never overload the stomach . . . The strength, power, beauty, and glory that once adorned the form and constitution of man have vanished away before the blighting influences of inordinate appetite and love of this world. Then let us not trifle with our mission, by indulging in the use of injurious substances. These lay the foundation of disease and death in the systems of men, and the same are committed to their children, and another generation of feeble human beings is introduced into the world . . ."[12]

Joseph Fielding Smith added, "This is my answer to you in relation to Brigham Young's statement that mothers should not feed their small children meat. Yes! Small children do not need the flesh of animals . . . .when the Millennium reaches us, we will live above the need of killing dumb, innocent animals and eating them."[13]

---

11) *History* V.2, p.71 & TPJS p.71
12) JD V.12 pp 37,118,119
13) Letter, 30 Dec.1966

Chapter 7

Joseph F. Smith shares an interesting perspective, "The reason undoubtedly why the Word of Wisdom was given — as not by 'commandment or restraint' was that at that time, at least, If it had been given as a commandment it would have brought nearly every man, addicted to. . . these noxious things, under condemnation; so the Lord was merciful and gave them a chance to overcome, before He brought them under the law."[14] If the Lord gave the millennial commandment now, most Americans would be under condemnation.

Summary

The Lord asks us to "serve him with all your heart, might, mind and strength . . . with an eye single to His glory"[15] He will not force; it is our choice. The diet we choose to feed body, mind and spirit determines, in significant measure, what we are. Three times a day we sit at the altar of our table. Who do we worship? "Our whole social order could self-destruct over the obsession with freedom disconnected from responsibility, where choice is imagined to be somehow independent of consequences."[16]

The Lord says, "If ye are not one ye are not mine."[17] The spirit of oneness includes the spirit of love and harmony with all the Lord's creations; it is the spirit of Zion; "And the Lord called his people Zion, because they were of one heart and one mind, and dwelt in righteousness; and there was no poor among them."[18] The scriptures are clear that there will be a Zion society "when men should keep all my commandments;" then can the City of Enoch come down to greet them.[19] "The earth will tremble with joy" when man realizes there is "enough and to spare" as we respect, enjoy and thank the Lord for all that comes forth from the earth "for food and for raiment, for taste and for smell, to strengthen the body and to enliven the soul."[20] He is the vine; we are the branches. As we abide in Him, we can bring forth much fruit, for of ourselves we are nothing.[21] When we are obedient to His voice, "then shall thy light break forth as the morning, and thine health shall spring forth speedily: and thy righteousness shall go before thee; the glory of the Lord shall be thy reward"[22]

---

14) JD 12:118
15) D&C 4:2, 5
16) President Boyd K. Packer, General Conference, April 1996
17) D&C 38:27
18) Moses 7:18
19) JST Gen. 9:21-22
20) D&C 59:19
21) John 15:5
22) Isa.58:8

"... your body is the temple of the Holy Ghost... therefore glorify God in your body, and in your spirit,"[23] "Come unto Christ, and be perfected in him, and deny yourselves of all ungodliness;...and love God with all your might, mind and strength, then is his grace sufficient for you,...that ye become holy, without spot."[24]

7.2 One of the Major Causes of Hormone Imbalance

Dr. David Williams writes the monthly health newsletter *Alternatives, For the Health Conscious Individual*. His May 2014 issue was titled *Achieving Hormone Balance*. My wife and I have found his information to be among the best in the world on the subject of optimum health. Several years ago we learned from him the importance of chia seed in our diets as the richest plant on the planet in omega-3, and most are deficient in this essential fatty acid.[25] Generally, this May issue information is not known and is a profound contribution on the part of Dr. Williams. Since it is very important to our health, It's About Time for me to share a summary of this information for you. Because of the importance of the professional details and insights shared, I recommend that you order this issue. The information below is only a summary.[26] Synoptically, Dr. Williams documents that over the last 50 years there has developed a large imbalance of the estrogen/ progesterone ratio. With the large increase in this ratio from our exposure to estrogen-like compounds, he reports, "We now live in an estrogenic, feminizing fish bowl. These compounds, called xenoestrogens, are sometimes referred to as 'gender benders' and can be 10 to 100 times more powerful than natural estrogen." This "gender bender" problem correlates with the decrease in the age of puberty for an average of 14.2 years for girls in the 1900s to now girls as young as 7 enter puberty. For males, we see more and more female characteristics. Serious cancer problems arise from this imbalance as well. Fortunately, he gives some combative measures to counter this problem.

The problem "has become endemic" because common "xenoestrogens include BPA, PCBs, phthalates, flame retardants, pesticides, herbicide, and DDT residue, all of which have been found in our food and water supplies."

---

23) 1 Cor. 6:19-20
24) Moroni 10:32-33
25) *http://ItsAboutTimeBook.com/chia-seed-a-super-food/*
26) This issue can be ordered from custsvc@drdavidwilliams.com or call 1-800-527-3044, and ask for Volume 17, No 5, May 2014 of Alternatives.

Chapter 7

Many plastic products contain BPA. "Phthalates are used in PVC products to make them softer and more flexible. They show up in everything from toys to food packages, flooring, and shower curtains, as well as nail polish, hair spray, and shampoo." BPA shows up in the lining of many cans. "Campbell's uses BPA free cans."

Dr. Williams points out that it would be impossible to list all the xenoestrogen compounds that are legally added to our foods, cleaners, and personal care products. However, he suggests some very important steps to limit the effects of this endemic hormone ratio problem and the exposure to excess estrogen and estrogen like compounds and lack of sufficient progesterone:

- Go organic and have a good multivitamin supplement;
- Avoid pesticides and herbicides;
- Use natural home cleaning products, e.g. vinegar and baking soda;
- Opt for chemical-free personal care products;
- Don't touch credit card and cash register receipts from thermal printers; if you need the receipt for record keeping, have them place it in your shopping bag; then handle it with tongs or tweezers until you have recorded the information; then discard it;
- Don't use plastic containers that contain BPA;
- Studies have shown the following are useful: turmeric, cilantro, and spirulina;
- An herbal combination called Myomin "can reduce excess estrogen levels:
- Avoid using the following: processed soy products, sunflower, safflower, cottonseed, and canola oils, licorice and red-clover; olive and avocado oils are good to use and the occasional use of fermented soy is good.
- Fat cells produce estrogen, so our obesity epidemic is a large contributor;
- It has been shown that a vegan diet will move a person away from obesity, and Dr. Williams sites some specific beneficial vegetables:

broccoli, cabbage, bok choy, Brussels sprouts, cauliflower, cress, kale, radishes, watercress, arugula, mustard greens, horseradish, turnips, rutabaga, and kohlrabi;

- Healthy liver function to get rid of toxins and excess estrogen is essential; alcoholic men have impaired liver function; Tylenol is toxic to the liver;

- Bowel transit time is important; constipation causes the toxins to not be eliminated in a timely fashion and the " estrogen is reabsorbed and placed back into circulation." Dr. Williams likes "fresh ground flax, hemp, or chia seeds in [his] morning protein shake;" and eat more fermented foods.

- "A quality probiotic on a daily basis is… one of the best disease-preventing and longevity-promoting supplements ever;

- Supplementing with natural progesterone in a crème or orally is very helpful and he gives some suggested sources.

Dr. Williams points out the disturbing situation that there are a lot of misinformation both with doctors and the public regarding estrogen and progesterone, because of all the synthesized products produced by the pharmaceutical industry. "Most doctors will be quick to say that oral contraceptives contain both estrogen and progesterone. More specifically, they contain synthetic forms of these hormones." This has provided enormous profits to be made from, for example, the estrogen-based Pill, which causes a fertilized egg to be aborted, and the increasing estrogen-like to progesterone ratio imbalance is causing ever increasing serious disease issues.

Defining conception as when the sperm enters the egg, estrogen-based contraceptives are actually abortive. Progesterone is a natural "true" contraceptive, but since it occurs naturally, it cannot be patented and marketed, so the pharmaceutical industry has no motivation to develop this natural contraceptive. No big money to be made here.

Further, Dr. Williams shares the " brain is very responsive to progesterone. Its concentration in the brain is around 20 times higher than in the blood. Studies have shown that it not only protects brain, but it also improves memory, relieves anxiety…" and more. "Progesterone has a calming effect on the brain as opposed to estrogen's excitatory influence." In addition, " progesterone stimulates new bone formation, which is needed to actually reverse osteoporosis."

I am deeply grateful for Dr. David Williams continued help for us toward optimum health. Our health — and several others of our family and friends — has significantly benefited since we learned from him of the importance of chia seed in our diet on a regular basis. As you can see from the article,[27] it helps with both the cardiovascular and digestive systems along with helping to lose weight. So it also addresses some of the other issues caused by the harmful increase in the estrogen/ progesterone ratio imbalance.

7.3 Harmony in Music Toward Good Health to Body, Mind, and Spirit

It's about time to share my feelings here. There may be some who wonder what music has to do with It's About Time. Our division in Boulder, Colorado, was called the "Time and Frequency Division." Man's time is impossible to generate without a frequency standard. Frequency is music, but much richer, bringing together a harmony of many frequencies in some most pleasing ways. Resonances in the Universe are part of the workings of God and His angels. The angels singing at the birth of the Savior was a climax of harmonious frequencies blended together to praise and honor Him who would make perfect the Plan of the Father. We were probably all there blending our voices with that heavenly choir and rejoicing in anticipation of the glorious opportunities His infinite atonement would open for us.

Over the course of my 79 years on this planet, I have witnessed an enormous overall deterioration in the music. As the Christian nation of America came into being, music was an important part of people's lives. They could sing the lyrics to hymns to different tunes having the same meter, and they could do this in their homes as well as in their worship services. Almost no one can do that now. Now singing in the home is almost unheard of unless it is rap or rock music. The American culture has lost that sacred song ability. Harmony and musical pitch are foreign to most of our children today. Typically, the music in our homes comes from the media. I'm not sure a lot of it could be considered music!

---

27) *http://ItsAboutTimeBook.com/chia-seed-a-super-food/*

I know several people who have had near-death experiences and have heard heavenly choirs sing. It seems that with the moral decay of society we are forgetting the harmony of heavenly music that has come through inspired and great composers. Handel's *Messiah* is the most performed piece of music ever and for very good reason. The Mozart effect is well known — how as you listen to Mozart's music while you are learning something, you retain it better. There are seven learning centers in the brain and music is one of them. They complement each other and when exercised properly they help to grow neurons in the brain at any age, a fact which was proven by Dr. Marion Diamond in the 1960s.[28]

In contrast, experiments have shown that hard rock and the like destroys. It has been demonstrated that such stimulates cancerous growths in the brain stems of mice. Over the course of Darwin's life he lost interest in classical music and beautiful art and a dark feeling came over him. He said, "A loss of these tastes is a loss of happiness, and may possibly be injurious to the intellect and more probably to the moral character, by enfeebling the emotional part of our nature."[29] Toward the end of his life he sadly shared, "Often a cold shudder has run through me, and I have asked myself whether I may not have devoted my life to a phantasy."[30] Following Darwin's introduction of the theory of evolution (24 November 1859) atheism became more and more common, and is becoming more common today—especially among scientists.

Beautiful harmonious music is associated with light and truth. I have personally witnessed angelic assistance in inspiring music performances. It uplifts the soul, and brings wonderful feelings of peace, love, and joy. Music has always been a very important part of our home environment.

I share another important dimension of beautiful music in Chapter 13 entitled *Growing the Brain and Fullness of Joy*. But here I would like to share two profound personal experiences.

The Washington D. C. " Mormon" temple is an amazing architectural edifice.[31] Its inspiring architectural design can be seen as one drives west at night on the north end of the Washington D. C. beltway, which heads right for it. The temple is on the top of a wooded hill — looking as if it is suspended

---

28) Read her book *Magic Trees of the Mind*, and see the link: *http://www.allanstime.com/Spiritual/BookReports/magic_trees.htm*
29) Charles Darwin, *The Autobiography of Charles Darwin*, p. 139
30) Darwin, Francis ed., *Charles Darwin: His Life Told in an Autobiographical* Chapter, and in *a Selected Series of His Published Letters*, p. 213
31) *http://www.ldschurchtemples.com/washington/*

Chapter 7

from the heavens as the shafts of light go up into the sky.

The Church knew the temple would be seen by millions, as people come from all over the world to Washington D. C. They wanted the design to illustrate our faith. Elder Keith W. Wilcox — one of the architects for the temple — visited us in Boulder in conjunction with a Stake Conference we were having. I was then serving as Stake President and he shared the following inspiring story with my family as we visited before the conference.

President Hugh B. Brown was coordinating the architectural design at the time. He asked Keith and the other architects to share their designs on a particular date. The others shared their elegant designs. President Brown then said, "Keith, where is yours?" Keith had been working on another Church project that had occupied all his time, and told President Brown. This was a Thursday and President Brown responded, "I would like to meet again next Tuesday, and Keith you can share yours." Keith called his wife and said, "Turn off the phone! Cancel all my appointments!" And he began an intense fasting and prayer time — seeking divine help.

About 2:00 the next morning, his mind went back to when he had done his Master's Thesis at the University of Oregon. His thesis professor had asked Keith to come up with "One word or phrase to describe the spirit of [your] church." Keith repeated the request in surprise.

"Yes," the professor replied, "and then include in your thesis a design of a church building that would demonstrate that word or phrase."

The professor's statement was completely unexpected. This thesis was his last hurdle before graduation. He asked several church leaders for help and arrived at nothing satisfactory. He finally decided to take it directly to the Lord. He had done all in his power to find an answer but had not been able to find an adequate "word or phrase" that he felt good about. He truly needed direct help from the Lord.

He reported the following experience,

> I found a quiet, private place to pray, and there I knelt and poured out my heart to my Heavenly Father. As I concluded my prayer, a word flashed into my mind: enlightenment. Then

the phrase light and enlightenment followed. Joy swept through me. My prayer had been answered. I thought of how light and truth have been restored in our day through the Prophet Joseph Smith. As prophets, seers, and revelators, our Church leaders continue to offer light and truth to all who will listen. Our missionary efforts truly bring enlightenment to the world. Our temples glow with spiritual light. Eternal truths are taught and enlighten all who enter therein. Suddenly it was easy to envision a meaningful architectural design for one of our Church buildings. I decided to design a building that would allow light to penetrate from the heavens all day long and that would radiate light heavenward each evening.

After his mind had been turned back to this experience, a vision opened to him of what the Washington D. C. temple should look like. What you see coming from the east as you approach the D. C. temple, is exactly what he saw. The vertical lines remain unbroken indicating the connection between Earth and heaven. Where windows were needed, the white translucent marble they located in Missouri was shaved to 5/8 of an inch retaining the beauty of unbroken marble on the exterior and emitting a warmly glowing light inside.

I asked Elder Wilcox to share that story with our members at stake conference, which he did, as we were getting ready to have a temple in Denver. Many were touched with his experience, and I told him afterward that we have no hymn in our book that describes this unusual transcendent event.

My wife and I had saved our money so that we could serve a mission in 1997, which we did for a year and a half in West Africa (Cote d'Ivoire). Try learning French at age 61; it was a stretch. Our oldest son, Sterling, is an excellent pianist and can compose music as well, and as they held a church service farewell for us, I asked Sterling if he would be willing to prepare some music that would fit this visionary event that Elder Wilcox shared with us. He agreed to do so. I told him I would prepare the lyrics from the scriptures. It came together in a miraculous way — reminding me of Frederick Handel's miracle in bringing forth the *Messiah*. We had people from four states helping us perform it as part of our farewell send-off. We recorded it, and we listened

Chapter 7

to it "n" times — often with tears of joy in our eyes — while we were in Africa, where n is a large number. We had a great experience there.

Here are the lyrics, and my wife gave it the title. You can listen to the composition on Sterling's web site.[32] Some of our grandchildren sang the Isaiah part.

A Message Of Gladness

Righteousness will I send down from heaven;
Truth will I send forth out of the Earth;
To bear testimony of mine Only Begotten;
To gather mine elect unto the Holy City.
Looking forth for the time of my coming.[33]

(Go to Chorus)

In my name shall they do many wonderful works;
In my name they shall cast out devils;
In my name shall they heal the sick;
Open the eyes of the blind; Unstop the ears of the deaf;
And the tongue of the dumb shall speak, and sing Hallelujah![34]

My law I will write in their hearts;
I'll be their God, and they my people.
They shall all know me—
From the least of them, unto the greatest of them.
I will forgive their iniquity, and remember their sin no more.[35]

Jehovah is my Strength and my Song; He is become my salvation;
Praise the Lord, All-mighty! Declare His doings among the people.
Call upon His name; make mention his name is exalted.
Sing unto the Lord; for He has done excellent things.
Cry and shout thou inhabitants of Zion;

---

32) *www.greaterthings.com*
33) Moses 6:72
34) D&C 84:66-70
35) Jer. 31:33-34

For great is the Holy One of Israel, in the midst of thee, Alelujah![36]

The Lord hath brought Zion down from above;
The Lord hath brought Zion up from beneath;
A Message of gladness; rejoice all ye Earth!
I will trust and not be afraid;
I will trust in the Lord![37]
Chorus:

Endowed from on high; Truth from the Earth;
Hope in our hearts; Faith in the Father;
Redeemed by His Son.
Endowed from on high; Truth from the Earth;
Hope in our hearts; Light of our lives;
Perfected in Him; Perfected in Jesus Christ, the Lord, Amen.[38]

Here we mean by Zion the pure in heart and a people of one heart and one mind centered in Christ as prophesied by Isaiah and the Apostle John. Such a society will be prepared as the Bride to meet the Bridegroom at His coming. So, the above song is really a millennial song as well.

A second musical experience that has brought about a lot of good in many people's lives occurred also while I was serving as Stake President for the Boulder Colorado Stake. I had been asking the Lord how we could do a better job of befriending our neighbors in the Boulder area. At that time I attended a performance of Handel's *Messiah* in St. John's Episcopal Church in downtown Boulder. They have a lovely pipe-organ, which brought a real richness to the performance. The Spirit whispered, "Put a pipe-organ in the new stake center you are building in Louisville and perform Handel's *Messiah*." " And it came to pass" we did as the Spirit directed.

To do this, we needed an extra $65,000 to acquire the desired organ. We asked 30 of our families living in Boulder County if they would help us raise the needed monies. They did, and the first major performance in the stake center was Handel's *Messiah*. We put an invitation out to the community to attend the performance and about half of the nearly one thousand people who

---

36) Isaiah 12:2-6
37) D&C 84:99-100; 128:19; Ps 85:11
38) Moroni 10:32-33

came to the performance were our friends and neighbors. Miracles occurred while we pulled together this challenging and inspiring piece of music and the Spirit warmed many hearts — especially mine as it all came to pass.

We became known as the "Musical Stake" in the Denver area. Our Stake Music Director, Delta Bement, was asked to coordinate the music when the Denver Temple was dedicated. The rippling effects have been wonderful over the years as many souls have been touched in Boulder County and in the Denver area with beautiful-inspiring music. Again, I have witnessed many miracles in conjunction with these rippling effects, and I thank the Lord for the inspiration given in blessing the lives of His children. His tender mercies are ever present.

7.4 Water, Proper Hydration — A Panacea and Healing Back Pain

Dr. Fereydoon Batmanghelidj, Dr. B as he is known by many of his associates, has had a remarkable career path. Born in Tehran, Iran, in 1931, he was schooled in Edinburgh, Scotland, and at St. Mary's Hospital Medical School of London University. Immediately before the revolution in 1979, he was engaged in the completion of a family charity medical center, the largest medical complex in Iran. The revolutionary government put him in prison. It was there, in prison, that he began his research on the benefits of proper hydration in the human body. He had an unusual clientele with the high stress level of the prisoners — many of them suffering from abdominal pains, peptic ulcers, and numerous other problems. Because of his excellent success in treating his fellow prisoners, his execution was delayed. When his trial came, he presented the judge with an article on water treatment of peptic ulcers. This saved his life so that he could continue his research, and he was released from prison. The article was sent to a professor of gastroenterology at Yale University, and later published in the *Journal of Clinical Gastroenterology*, and then in the science section of the *New York Times*.

In 1982, he was released from prison, and later escaped from Iran and went to the United States to further his medical research. He has written several important books on the importance of body hydration. I was so impressed with his book, *The Body's Many Cries for Water*, that I wrote a book report for our web site.[39] He called me one day — having read my book report — and thanked me for it. We became e-mail friends.

I was doubly grateful for his work because as I read *The Bodies Many Cries of Water* I learned of his book, *How to Deal with Back Pain & Rheumatoid Joint Pain*. Having a significant back problem, I anxiously read this book. It is the only book I know of that teaches how to rebuild the back. Since there is no blood circulation in the spine, it is nourished by osmotic pumping. He details the physiology of the back and how through proper exercises, diet, and hydration, the back will heal. I shared with him my personal experience of how this second book helped me. He asked me to write him a letter telling of my experience.

I have shared Dr. B's information and counsel with many other people, and we have seen some of them benefit as well. Because of the importance of this message, I share this letter here:

Healing My Back Pain

26 February 2003

Dear Dr. Batmanghelidj,

I would like to thank you and share with you the significant progress I have enjoyed in the healing of my back. While reading your book, *Your Body's Many Cries for Water*, I learned of your other book, *How to Deal With Back Pain and Rheumatoid Joint Pain*. Following your suggested exercises - plus a few of my own - and drinking the amount of water you suggest as well, I am pleased to report that I have enjoyed much more strength and a lot less pain in my back.

I had a back injury well over twenty years ago, while helping to lift a piano for a friend. It seems that one of the disks in my lower back was crushed or badly damaged. It had gotten progressively worse over the years to a point a couple of years ago when it would go into spasms - rendering me unable to walk. The pain was excruciating. This would usually be triggered by a slight amount of lifting - especially front on. If I bent over and lifted something much over 10 pounds, I would be in trouble for some days.

Upon starting your suggested back exercise program about a year and a half

---

[39] *http://www.allanstime.com/Health/ water.htm*

Chapter 7

ago, I saw slow but consistent progress. Last summer was the acid test. Ten-and-a-half years ago, I retired as an atomic clock researcher in Boulder, Colorado, and bought a farm in central Utah - near my home town and family. My wife and I have pretty much had others run the farm until this last summer, but circumstances arose that we needed to run it. This challenge included moving 40-feet-long irrigation pipes, baling and hauling three crops of hay. The farm is not big enough to be modernized, so my sons and I with some hired help were hand-lifting 90-pound bales of hay up on a truck and hay wagon - with hay leaves down my sweating neck and back. We had to also lift them into the hay barn — harvesting well over a thousand bales. Having been raised on a farm, I have never minded hard work and this was hard.

There is no way that I could have done that the year before, and it actually felt good to be able to do this hard work. I have always tried to keep myself in good physical condition, but some years ago I had to even stop jogging because it hurt my back too much. Now I can also run again without serious pain. Considering, also that I am 66 years old, I feel that the whole thing is pretty remarkable, and I thank you for the tremendous assist in my recovery. I am far from all the way better, but the trend is right and my back is getting stronger all the time. This morning and yesterday I shoveled about a foot of snow off our walk with no significant back pain. I hadn't been able to do that for years.

You suggested extra weights on the feet as one does the back exercises. I just wear shoes and have a three-pound weight in each hand to strengthen my upper body while I am at it. The extra exercises that I do in addition to yours are all toward strengthening the back muscles and for general cardiovascular and lymph system circulation improvement. These include: 1) lying on each side and doing scissors with my free arm and leg in circular motion as well as up and down; 2) sit-ups and leg lifts while lying on my back; 3) push-ups from the knees and the toes with particular emphasis on flexing and using the lower back muscles - activating the osmotic pumping of water action as you suggest for the lower back; 3) jumping jacks, touching the floor with the weights in my hands, and cross- touching the right hand to the left foot while straddled, and vice versa; 4) doing a variety of jumping exercises on a re-bounder, which is excellent for the lymph system circulation; and 5) then a variety of stretching exercises to keep the whole body limber. I add to these walking with my wife, cross-country skiing, mountain bike riding every day when the weather permits. We can usually do one of these most days of the year.

It feels so good to feel good. Life has been really good to me, and good health allows me to better serve and do those things that I feel are most important. I love to express my gratitude by serving, and you can't do that very well lying on your back in pain. Thanks again for helping so significantly to take that pain away. I was so impressed with your first book, that I did a book report and have it posted on my web site.[40] Your second book is saving my back. My wife has found significant benefit from your water cure as well.

Gratefully yours,

David W. Allan, President
Allan's TIME, Inc.

The following are the most important points I gleaned from Dr. B's first book. It is taken in part from the book report on our web site.

Water, the Remarkable Panacea!
By David W. Allan September 10, 2001 and updated 21 May 2014

It seems too good to be true that such a simple procedure (drink eight eight-ounce glasses of water a day) could provide such astounding benefits. From a preventative as well as curative point of view, the list is most impressive — digestive system problems, pain from arthritis, lower back, neck, angina, and headaches; stress-related problems and depression; high blood pressure and hypertension; excessive cholesterol levels and many heart problems; excess body weight and over-eating problems; asthma and allergy problems; nerve disorders, including multiple sclerosis (MS); both insulin-dependent and insulin-independent diabetes and improper tryptophan balance; the latter may tie to the prevention and cure of AIDS; and lastly, but not the end of the list, is the Big C (cancer).

If this seems overly simplistic, I understand. I felt exactly the same way before reading Dr. F. Batmanghelidj's book, *Your Body's Many Cries for Water* with subtitles: You are not sick, you are thirsty! and don't treat thirst with medications. I would suggest that you read his book. Even though the book is written for a lay audience, it is quite technical. Quoting from one of a large number of inspiring testimonials in this book (several of whom are doctors),

---
40) *http://www.allanstime.com/Health/water.htm*

M.D. Dan C. Roehm says, "... it is very apparent that this work is revolutionary and sweeps nearly all diseases before it. As an Internist/Cardiologist I find this work incisive, trenchant and fundamental. This work is a God-send for all."

In his research, he has also uncovered the inappropriate activities of the pharmaceutical industries and the American Medical Association (AMA) to promote medicine at the expense of health. For example, the typical low salt diet often recommended is more a problem of lack of water to help the body to work well electrically.

Here, I will give only brief synoptic information from the above-cited book. One really needs to read this book to better appreciate his work. This synopsis is written with the intent to help promulgate health, happiness and a better understanding of the marvelous miracle the body is. We learn from Dr. B that we cannot have optimum health without proper hydration, and if we are sick or diseased, it is a lot harder to be happy. Billions of dollars are spent on health, yet in water and its proper utilization for our bodies we could avoid much of these expenditures by understanding how our bodies work and their critical need to be properly hydrated. He further points out that most drugs and medications treat only the symptoms of illnesses and do not provide a cure. This up-side-down treatment is often because the medical community does not recognize that the proper cure could come from proper hydration of the body in many instances.

He points out the doctors take an oath to serve mankind, but the "business of America is business — making money. They have no right, but some of them are doing it anyway, to obstruct the simple message of 'you are not sick, you are thirsty,' from reaching a wider cross-section of the public."

> This marvelous mechanism we call the body has a priority sequence in how it uses the water we give it. This [divinely designed system] is so that it can function as well as possible in the state of water shortage, which exists for most people — not being aware of their bodies' need for more water nor of being aware of their many "cries" indicators of their being in a state of dehydration. The first priority is the brain. Though the brain is only 2% of our body weight, it gets about 20% of the blood flow and is the first to be hydrated so that it can properly sense and prioritize the

utilization of water for the rest of the body.

> ...there is a mechanism for establishment of priority for circulating blood to any given area.. The order is predetermined according to a scale of importance of function. The brain, lungs, liver, kidneys, and glands take priority over muscles, bones, and skin in blood distribution... When we do not drink enough water to serve all the needs of the body, some cells become dehydrated and lose some of their water to circulation... In water shortage and body drought, 66 % is taken from the water volume normally held inside the cells; 26 % is taken from... outside the cells; and 8 % is taken from blood volume.

The long-term consequences of this dehydration is various kinds of disease. These illnesses and the pains associated therewith are then really our body crying for water.

For complete health our bodies need proper hydration, proper diet, proper rest and environment, and proper exercise. He also points out how stress in conjunction with dehydration brings on a variety of diseases — many of them digestive-system related.

In regard to diet, he further points out the large misunderstanding we have is "[in] advanced societies, thinking that tea, coffee, alcohol, and manufactured beverages are desirable substitutes for the purely natural water needs of the daily "stressed" body is [a]... catastrophic mistake... [They]... contain dehydrating agents. They get rid of the water they are dissolved in plus some more water from the reserves of the body! One of the diseases of dehydration is to confuse the sensations of thirst and hunger. We think we are hungry and eat when we are actually thirsty. Aspartame and saccharin have similar effects —tricking the brain into thinking it has sugar when it doesn't. For example, the diet and/or caffeine drinks confuse the body into thinking it is hungry when it is actually simply asking for water. If when we go for a snack, we would get a drink of water instead, we would undoubtedly be a far healthier people — a simple but far-reaching change of behavior. This would help greatly with the obesity epidemic.

Also, "Caffeine has diuretic properties. It... acts on the kidneys and causes

Chapter 7

increased urine production. It is physiologically a dehydrating agent. This characteristic is the main reason a person is forced to drink so many cans of soda every day and never be satisfied. Thinking they have consumed enough ' "water" that is in the soda, they assume they are hungry and eat more than their bodies need. . . causing a gradual gain in weight. . ." Alcoholic beverages do basically the same thing — put the body in a state of dehydration and confusion. Have you noticed how many people carry around those big soft-drink mugs who are also fat! Of course, they drink diet Pepsi or diet Coke! Guess what the sweetener is?

In regard to exercise, he says, "Walking two times a day — every 12 hours —will maintain the activity of the hormone-sensitive fat-burning enzyme (. . . lipase) during day and night and help clear away the excess lipid deposits in the arteries." The walking can be replaced, if desired, with an equivalent aerobic exercise. He has several testimonials where people have accomplished significant weight loss (proper "fat-burning") and great health gains by simply cutting out all sodas, by drinking enough water, and by having proper exercise. Of course, a reasonable diet needs to be followed as well, but the body moves in the right direction when it is treated as it was designed to be.

In regard to diet, he treats the body chemistry in general terms as well as in specifics as dehydration affects the chemistry. This dehydration imbalance is the cause of many of our main diseases. In this context, the book becomes quite complex, but is interesting and inspiring as you see the wonderful harmony of how all the body functions work together when in a properly hydrated state — bringing about the prevention of most diseases.

As is well known, there are 20 amino acids (AA) needed by the body. Twelve are manufactured by the body — three in limited quantities. The other eight need to be imported from food intake. A critical one of these eight is tryptophan. "The brain tryptophan content, and its various by-product neurotransmitter systems, are responsible for maintenance of the "homeostatic balance of the body." Normal levels of tryptophan in the brain maintain a well-regulated balance in all functions of the body — what is meant by homeostasis. With a decrease in tryptophan supply to the brain, there is a proportionate decrease in the efficiency of all functions in the body.

Dr. B. makes the important point, "Depression and some mental disorders are the consequence of brain tryptophan imbalance. Prozac used in some mental disorders. . . is a drug that stops the enzymes that break down serotonin, a byproduct of tryptophan. When more serotonin is present, all nerves function normally. However, Prozac cannot replace the indispensable role of tryptophan itself. . . Hydration of the body, exercise and the intake of right foods help replenish brain tryptophan reserves. . . Chronic dehydration causes its loss from the pool of different amino acids held in the body." He also points out that stress can cause a dumping of the AAs and that "tryptophan seems to be one of the most important" ones that gets dumped.

As an important aside, Ann Blake Tracy in her book, *PROZAC, Panacea or Pandora?*, says that the side effects from using PROZAC can be devastating, listing: Agitation, Akathisia (need of movement, suicidal tendencies, and tendencies of violence), Atazia (defective muscular coordination), Asthenia (loss of strength), Depersonalization (the belief that one's own reality is temporarily lost), Hypesthesia (lessened sensitivity to touch), Hypoglycemia (deficiency of sugar in the blood), Hypomania (mild mania and excitement with moderate change in behavior), Hysteria, Mania, Neuralgia (severe sharp pain along the course of a nerve), Neuropathy (disease of the nerves), Paranoia, Psychosis, and Tremors. This is one of many drugs approved by the FDA, and which treat symptoms — and provide no cures — but which can have very serious side effects on one's own health. As you can see, it is so important that we take our own health into our own hands.

"One must consume the full range of AAs to build the "reserve pool" in due time. . . The best proteins are those stored in the germinating seeds of plants, such as lentils, grains, beans, etc." From my perspective, the consumption of raw foods from a plant-based food selection approaches the ideal diet for optimum health. He says that, "Carrots (for their beta-carotene content) are an essential dietary requirement. Beta-carotene is a precursor for vitamin A and absolutely essential for liver metabolism, apart from its need by the eyes." Potassium is also essential; but not too much. It can be obtained from orange juice, bananas, or chia seed. In addition, he discusses the importance of the omega-3 and omega-6 fatty acids. These are needed for the manufacture of cell membranes, hormones and nerve coverings in the body. Chia seed, flax seed, and flax seed oil are excellent sources for these — along with "wild caught" salmon, which we love to supplement with from time to time.

Chapter 7

In the book, the various chapters work through the body chemistry and the effects of dehydration and proper hydration as it relates to sickness and health. For example, there is a chapter on the digestive system; one on rheumatoid arthritis, lower-back pain, neck pain, angina pain, and headaches; one on stress and depression; one on high blood pressure and hypertension; one on cholesterol and why people tend to overeat; one on asthma and allergies; one on both insulin-independent diabetes and insulin-dependent diabetes — showing the very important relationship to the AA tryptophan in the body; one on AIDS and that HIV is a misnomer; and a general chapter on dealing with insomnia, fainting, heart attacks, the fallacy of a salt-free diet, our health-care systems and the enormous savings that could be appreciated if everyone understood and applied this new " water-cure" paradigm. In the beginning of the book, he makes the intriguing statement that, "Morning sickness of the mother is a thirst signal of both the fetus and the mother." Hence, expectant mothers that have been continually drinking caffeine, sugar-free, or alcoholic beverages, put themselves in a state of dehydration —thus increasing the probability of morning sickness and of exacerbating the problems associated therewith. Independently, we have learned that making a ginger drink is often helpful in dealing with morning sickness.

Interspersed throughout the book are other very important discussions around cancer, Alzheimer's, osteoporosis, puffiness of the eyes and swelling of the ankles, muscle cramps, the balance of salt and water for healthy cell regeneration, hydration, and sustenance, and providing the proper balance for all aspects of the reproductive process. He says that, "If bulimics begin to rehydrate their [bodies] well and drink water before their food, [their] problem will disappear."

He has some astounding bullet statements in the book:

- Excess cholesterol formation is the result of dehydration.

- The present way of treating hypertension is wrong to the point of scientific absurdity.

- Dehydration is the number one stressor of the human body —or any living matter. Alcohol will suppress the secretion of vasopressin from the pituitary gland. Lack of vasopressin in circulation will translate to general dehydration of the body —even the brain cells.

- It is. . . more prudent to attend to our daily water intake. . . than to what foods we eat.

He recommends eight to ten eight-ounce glasses of water per day — depending, of course, on body size. For each ten glasses of water, he recommends one-half teaspoon of salt. We recommend iodized sea salt. It is better to drink the water away from meals (½ hour before eating and about 2 hours after eating). It takes a lot of water to help properly digest food, but the body needs to be well hydrated before. Often, people drink their water with their meals — feeling thirsty mid-way through. Then, he says, "It is already too late, because the damage is registered by the cells lining the blood vessels." He says, "If we begin to appreciate that for the process of digestion of food, water is the most essential ingredient, most of the battle is won. If we give the necessary water to the body before we eat food, all the battle against cholesterol formation in the blood vessels will be won."

He has an extensive and fascinating discussion of the importance of water filtration into each of the cells in their myriad, miraculous, and diverse functions in a healthy, well-balanced, hydrated body. He says, ". . . water regulates the volume of a cell from inside. Salt regulates the amount of water that is held outside the cells — the ocean around the cell. There is a very delicate balance process in the design of the body in the way it maintains its composition of blood at the expense of fluctuating the water content in some cells of the body. When there is a shortage of water, some cells will go without a portion of their normal needs and some others will get a predetermined rationed amount to maintain function (. . . the mechanism involves water filtration through the cell membrane). However, blood will normally retain the consistency of its composition. It must do so to keep the normal composition of elements reaching the vital centers."

Current traditional medical thinking is that water acts as the conveyor of various essential minerals in solution to the sundry parts of the body. Dr. B. says, "This is where the 'solutes paradigm' is inadequate and goes wrong. It bases all assessments and evaluations of body functions on the solids content of blood. It does not recognize the comparative dehydration of some other parts of the body. All blood tests can appear normal and yet the small capillaries of the heart and the brain may be closed and cause some of the cells of these organs a gradual damage from increasing dehydration over a long period of time."

He goes on to say, "When we lose thirst sensation (or do not recognize the other signals of dehydration) and drink less water than the daily requirement, the shutting down of some vascular beds is the only natural alternative to keep the rest of the blood vessels full. The question is, how long can we go on like this? The answer is, long enough to ultimately become very ill and die. Unless we get wise to the paradigm shift and professionally and generally begin to recognize the problems associated with water metabolism disturbance in the human body and its variety of thirst signals, chronic dehydration will continue to take its toll on both our bodies and our society!"

7.5 Is Sugar a Poison?

There are many books written around this question. I have room to touch only on some highlights here. There are about 200 different kinds of sugars and only a few of them are really good for you. It's about time to learn that, unfortunately, table sugar, $C_{12}H_{22}O_{11}$, is a poison to the body and is contributing to a lot of societies health challenges. Read the books, *Sweet and Dangerous* by John Yudkin and *Sugar Blues* by William Dufty. Dr. Lorraine Day[41] has a CD *Eye See; Prevent and Reverse Eye Diseases* in which she documents the causes and the cures for many eye diseases. This is well worth listening to. She shares how sugar consumption is a principle contributor to cataract formation and other eye problems as well. I used to wear glasses; now I don't. My story can be found on our web site.[42]

Specifically, regarding eye-sight, I had to get glasses in the mid 1990s in order to read. Following Dr. Robert O. Young's program of moving the body to a more alkaline state to get rid of the Rouleau, which happened within two weeks, I noticed after a while that I could see better. Marie Dahlen, who is one of Dr. Young's microscopists (a person who can look at a drop of your live blood and tell you your health problems and the cures), was able to reverse macular degeneration for her mother using Dr. Young's program — including bio- light, a special herbal product that Innerlight sells. After I had my eye-sight reversal, I listened to Dr. Lorraine Day's excellent CD called *Eye See* that tells how to reverse many eye diseases and what causes them. Dr. Day's information augments Dr. Young's program nicely. There is a pioneering work of Dr. William H. Bates, *The Bates Method for Better Eyesight without Glasses*

---

41) http://drday.com/
42) http://www.allanstime.com/Health/Reversal_of_Health_Challenges.htm

— a tested method for improving vision by the doctor who originated and perfected it. The book was published by his wife nine years after Dr. Bate's passing in 1940. It is now available in paperback.

Typically, as with highly processed foods, so it is with highly processed sugars; they are not good for you. As nature provides them, they are usually very good for you. As examples, we like stevia, raw honey, pure maple syrup, raw cane sugar, dates and xylitol as our sweeteners. It is interesting that table sugar is historically a relatively new sweetener. It was used only by the nobility anciently, and has come into common usage over only the last two centuries. Is there a significant correlation with the diseases inflicting us over that same time period? Sugar is only one of many causes of society's diseases. We know from experience that the body can tolerate a little sugar. It is the excess use in our society that is the problem. For example, we know that excess sugar consumption weakens the body's immune system. Hence, we see a significant increase in sickness after Halloween and during the Christmas holidays.

7.6 How Serious Are the Vitamin-D and K2 Deficiency Problems?

When I came across the work of Dr. Kate Rheaume-Bleue, who is the world's expert on vitamin K2, I learned that the North American diet provides only 10% or less of the vitamin K2 needed for good health. I felt It's About Time to help get the word out, because this deficiency contributes to several health problems — including osteoporosis, heart-attacks, and cancer, strokes, and diabetes, and stunted growth, women's inability to conceive and to have children as healthy as they could be, and the list goes on. The proper amount of K2 works in harmony with vitamin-D causing the body to put calcium where it is needed and to remove it from where it is not needed.

Dr. Kate Rheaume-Bleue authored the book entitled Vitamin K2 and the Calcium Paradox. She explains why vitamin K2 is so important and why it is not well known. Dr. Mercola interviews her, and you may find it on his web site.[43]

Most people don't know; I didn't, that vitamin K1 and K2 act very differently in the body and come from very different food sources. K1 comes mainly from leafy greens — cooked kale, spinach and collards being three of

---

43) *http://articles.mercola.com/sites/articles/archive/2012/12/16/vitamin-k2.aspx*

the best sources. K2 comes from fermentation process and cooked Natto is an excellent source, which is common in Japan, but not in the USA. Grass-fed-animal meats (especially the livers — goose liver pâté being the best) and free-range eggs are high in K2. The fermentation process in cheeses provides an excellent source — brie and Gouda being some of the best. The following footnote contains a link that gives an excellent explanation of vitamin K and its different forms as well as good food sources for both,[44] but, unfortunately, the information in this link has the erroneous message that we don't have a deficiency in K2. This is because this information has only recently been discovered. I learned from the Wikipedia article, however, that the gut converts some of our consumed K1 into K2, but the current research indicates not nearly enough — especially for the typical North American diet.

Dr. Rheaume-Bleue says that you need about 200 micrograms of K2 per day along with at least 2,000 IU of vitamin-D-3. Without supplementation most people at higher latitudes (North American, for example) are vitamin-D deficient during the fall, winter, and spring.[45] Only one ounce of cooked natto is needed to get 200 micrograms of vitamin K2, but as you will see from the table in the Wikipedia article, it takes a lot of the other foods to get enough K2 — brie and Gouda cheeses being convenient North American sources. It takes about 10 ounces of those! That's a lot of cheese.

She makes the very important point that Vitamin K2 will guide calcium into the bones and teeth, where we want it to be. And it will get calcium out to prevent the deposition and even remove calcium from areas where we don't want it to be, like soft tissues and arteries, for example, where calcium can cause hardening of the arteries. K2 is really critical for keeping our bones strong and our arteries clear. Those are majorly important health concerns, and K2 is a very important nutrient for those health issues.

In 2005 I learned the importance of vitamin-D. Subsequently, I learned some additional important lessons from Dr. John J. Cannell, who heads the Vitamin-D Council, and from Dr. Joseph Mercola. In addition to the information available at the above link please note the following: Vitamin-D deficiency is typically a high-latitude problem. For example, there is no MS on the equator — a devastating autoimmune disease. The UVB band of sunlight

---

44)   http://en.wikipedia.org/wiki/Vitamin_K
45)   See the following link for the importance of vitamin-D and how to get adequate amounts: *http://www.allanstime.com/vitamin_D_and_sun.htm*

energy, which generates the vitamin-D-3 in your skin, bounces off the atmosphere, when the sun angle gets down to low angles. That occurs at high latitudes on the Earth and when you are two or three months or more away from the summer solstice — when the sun angle starts getting low in the sky. During that time, my wife and I supplement with 5,000 IU of vitamin-D-3 per day.

An interesting aside that Dr. Mercola points out is that the vitamin-D generated by the sun on our skin is actually an oil. So if one takes a shower after sunbathing with soap, it gets washed off. If you wait a day for it to be absorbed or take a shower without soap then you retain the benefits of the sunlight.

In the following footnote see the article.[46] I have an edited transcript of Dr. Joseph Mercola's (DM) interview with Dr. Kate Rheaume-Bleue (KB).

I love Ockham's Razor: simple is often the best. The story of Dr. Andrew Weil, who is often referred to as America's doctor, traveled around the world after he finished his medical training to gather the best of medical modalities. Coming home he found some of the best medical advice from a friend, who taught him the importance of deep breathing exercises for our overall health. It is simple and easy and doing it daily is a significant asset to being healthy.[47]

Another simple health rule I learned from Dr. Fred Bohman, who is also a dear friend. He says, "Drink your food, and chew your drink!" In other words, he says to chew each bite 29 times until the solids turn into liquid. This process intermixes the saliva properly in the food and signals the stomach's digestive system to be prepared for that food. This doesn't work very well for Americans who "inhale" their food at a fast- food restaurant. I am aghast at how fast most Americans eat. Then, when you drink something other than water, take a swallow and work it around in your mouth as if you were almost chewing it. Then again the saliva properly mixes and signals the stomach for optimum digestion of that drink. Simple!

I am inspired and amazed at the body the good Lord created for us to enjoy. We know proper exercise is key to good health. The simple rule here is to enjoy your exercise; if you do, you will persist. In 1982 Wayne Pickering warned, "If you don't find time for exercise now, you will have to find time for

---

46) http://ItsAboutTimeBook.com/benefits-of-vitamin-k2-heart-health/
47) http://www.drweil.com/drw/u/ART00521/three-breathing-exercises.html

illness later!"[48]

The next simple rule is to enjoy and feel good about what you eat. Let it be your medicine. The American diet reminds me of what Thomas Moffet profoundly stated back in the 1600s: "men dig their graves with their own teeth and die more by those fated instruments than by the weapons of their enemies." What's on your plate affects your body, mind and spirit.[49] Perhaps you have heard the quote: "We spend the first half of our lives wasting our health to gain wealth. And the second half of our lives spending our wealth to regain our health."

We are fortunate to live in a time when a lot of excellent health of body, mind, and spirit modalities are coming forth. My wife and I take no pharmaceutical drugs, but we are "big time" into natural ways to nourish body, mind, and spirit. In Chapters 10 and 11 we will share some of these modalities, where Dr. Lissa Rankin teaches us to "listen to your inner pilot light." It turns out that YOU are your best doctor!

Dr. Bruce Lipton shares a similar and far reaching message in his book, *Biology of Belief*, which ties the mind, body, and spirit together in a scientifically documented way and in contradiction to current teachings of biology accepted by the medical and pharmaceutical communities. In addition, Michael Brook has a fascinating new book, just ready to come off the press, *New Dimensions in Health; Simple Secrets to Creating Optimal Health*.

7.7 Is Yeast Good for You and What Is the Proper pH Balance?

When Marie Dahlen — a Nutritional Microscopist trained by Dr. Robert O. Young — found Rouleau in my live-blood analysis — as I mentioned earlier — my suspicions were correct that something was out of balance in my body. I was in reasonably good health but was about 15 pounds overweight and felt something amiss. Dr. Young introduced live-blood analysis using dark-cell microscopy technique, and it has blossomed into an extremely useful practice. Our second daughter, Karie, when she heard I was going to have my live- blood (she took only one drop from my finger!) analyzed, said she was

---

48) I have an extensive write up on our web site on the benefits of different kinds of exercise: *http://www.allanstime.com/Health/run_not_be_weary.htm*

49) *http://www.allanstime.com/Health/plate.htm*

fascinated by this technique and asked if she could come along.

Dr. Young can look at your live blood and tell the condition of your health and what to do to make your health better. Marie video-recorded my live-blood analysis showed me the symptoms of Rouleau — a clumping of the red- blood cells — which, of course, makes it more difficult for the blood to flow through the capillaries. She said I could get rid of it by going off yeast and sugar, by eating lots of green salads with a variety of vegetables and nuts (avoiding peanuts because of their tendency to have fungus). She also suggested I use Dr. Young's SuperGreens (a combination of 49 specially-prepared, herbal-blend of dried greens, leaves, vegetables, and sprouts — for long shelf life). The SuperGreens are sold by Innerlight. My daughter, Karie, was so impressed, she immediately signed up to be trained by Dr. Young as a Nutritional Microscopist.

I did as Marie suggested and within two weeks, I lost 15 pounds — felt great — and have discovered new flavors and tastes in salads that I never dreamed possible. It took me an hour to eat a salad, because the variety of flavors amazed me. After a couple of weeks, Karie, who was newly trained, had her microscope and equipment to re-examine my live- blood. It was totally clean, and she said my blood looked like a teenager's — not today's teenager's!

Marie told us that her mother, who had macular degeneration, had a reversal using Dr. Young's SuperGreens. A few months later, I noticed I could see better and read without my glasses. I took them off and haven't put them on since. That was over ten years ago.

I learned later from Dr. Matt McClean that the introduction of yeast into our diet is problematic. Charles Fleishchmann introduced yeast cakes at the world's fair in the US in 1876. Dr. McClean reports that, "The decreased leavening time, improved reproducibility and increased loaf volume catapulted his compressed yeast into modern fame." Dr. McClean further shares the very important dimension of natural leavening uses the natural bacteria to bring about the leavening process, which is much slower, but in the process the natural yeast digests the gluten. So for folks with gluten intolerance this can be extremely important. He further says,

> Current research is rediscovering why and how the ancient ways

of using natural leavening aid in human health and how natural leavening can prudently help many people to consume grains without deleterious effects. Natural leavening is not only beneficial with gluten containing grains such as wheat, barley, and rye, but it can also be used with all gluten-free grains such as rice, quinoa, teff, amaranth, oat, millet, sorghum, and buckwheat to improve nutrient availability and decrease inflammation.[50]

Natural leavening can be used to improve nutrition in any grain recipe — from batter breads to flatbread and loaf breads.

This yeast problem may be part of the cause for the significant increase in celiac disease. He has found that some persons with this disease can tolerate to some degree eating breads made with natural leavening.

Dr. Young has written and extremely informative book called *Sick? And Tired*. It documents the importance of pH balance in the body. This was a life-changing book for me. It was here I learned the importance of an alkaline diet. Putrefaction, decay, and the spoiling process are all acidic, and are all part of the natural process of returning organic-living material back to "dust." Acidity moves us toward death, while alkalinity moves us toward life — in simple terms. A meat-and-dairy centered diet tends to move our bodies more acidic, while a plant-based diet tends to move the body more alkaline. He recommends 95% foods that move the body alkaline and 5% that move the body acidic for optimum health. He has observed the reversal of almost all diseases in people's lives — including the big "C."

He like Dr. Batmanghelidj knows the importance of proper body hydration. He says,

> The importance of water is more than obvious since we are a gelatinous material in a body of water. As babies we were 90% water; as we age it decreases to 70%; and approaching complete cellular disorganization (death) we are 50% water. Water helps to maintain alkalinity in the blood, lymph, and intracellular and extracellular fluids by diluting excess acidity born out of cellular metabolism and acidic lifestyles, diets and thinking.

---

50) Dr. Matthew McClean, *Grain the Staff of Life? A Religious Perspective*.

So where is all the excess acid coming from? We get it from our inverted lifestyles, diets, drugs, pollution, emotional and physical stress, lack of exercise, lack of spirituality, immorality, "acid" music, "acid" books, our inverted thoughts, deeds and traumas, just to name a few.[51]

He recommends up to a gallon of water a day. If you need to move your body more alkaline, you can add 3 to 5 drops of chlorine-dioxide, which is highly oxidative according to Dr. Young.[52] Stanford University studies of shown significant health benefits by using ClO_2 with no side effects.

By having the body in a proper alkaline state then the immunity system is strong and disease goes to zero. Another amazing fact shared in Dr. Young's book is that the disease model of Louis Pasteur is flawed, which is the basis of the medical establishment and the pharmaceutical industry. Antoine Béchamp, the brilliant Professor of Medical Chemistry who had so many other achievements that it took eight pages of a scientific journal to list them when he died (1816-1908), knew Pasteur was wrong, but politics won the day.

Originally, the giant Antoine Béchamp (1816-1908) and later Professor Günther Enderlein (1872-1968) established that disease was a terrain-environment issue. In other words, it is not some disease bacteria that necessarily makes you sick, but more importantly what is going on in your body, mind and spirit from what we have learned since their original research. Louis Pasteur, who was six years younger than Béchamp, worked to piggyback on Antoine's great work by plagiarizing and politicizing and basically won the day. Dr. Robert O. Young, who picked up Béchamp's work in America in recent times points out that, "The concept of specific, unchanging types of bacteria causing specific diseases became officially accepted as the foundation of Western medicine and microbiology in the late 19th century."[53]

In her book, *Béchamp or Pasteur? A Lost Chapter in the History of Biology*, Ethel Douglas Hume cites an observation by Florence Nightingale as she observed people's symptoms, she noticed them changing spontaneously from one "disease" to another. As a result of her observations, she wrote: "The specific disease doctrine is the grand refuge of weak, uncultured, unstable minds, such

---

51) *Sick and Tired*, p.. 58
52) *Sick and Tired*, p. 101
53) Dr. Robert O. Young, *Sick? And Tired*, p23

as now rule in the medical profession. There are no specific diseases; there are specific disease-conditions."

Even though the Pasteur model "won the day" and has been the basic model of medicine and pharmaceuticals over the last century and to this day, in this century we are seeing the truth come out. Claim your terrain by having a healthy body, mind, and spirit where no disease can live!

Later, in Chapter 10, we will discuss the work of Dr. Lissa Rankin and her book *Mind over Medicine*, which harmonizes beautifully with what Florence Nightingale said over a century ago and which was contrary to the medical community then as it is now. In Chapter 11, we will discuss also the groundbreaking work of Dr. Bruce Lipton and brand new book of Michael Brook, which also harmonize with Dr. Rankin's unification of body, mind, and spirit for healing and optimum health.

7.8 Unusual Cause of Heart Attacks — Not Well Known

A very good friend of ours, Keith Chandler, had a heart attack due to exposure to phthalates, which are sometimes listed on ingredient's labels as "fragrance." A few years ago, my wife had a heart attack. When we learned the cause of Keith's heart attack, we realized that just before Edna had hers, she had extensive exposure to paint fumes while helping our daughter, Celeste, paint their home in Prescott Valley, Arizona. Those fumes had phthalates in them!

It's About Time

Chapter 8

Can All Addictions Be Overcome?

Christ can and will heal us of all addictions if we will truly come unto Him. We often think of the atonement as being given only to cover our sins as we repent, but it is so much more than that. The woman with the issue of blood, who in faith reached out and touched the hem of Jesus' garment, was healed. He sensed that virtue had gone out of Him and knew He had been touched and she healed, in which all rejoiced. Did He have any less virtue after? No, He has an infinite supply which we can access as well for all of our ills, as well as for addictions. We all have issues, and like her, we can reach out in faith, and guaranteed, He will be there for us, whether they be mental, emotional, spiritual, or physical. The challenge we often have is the healing will come in His timetable, not ours. In His infinite wisdom it will be done in the way that is best for us in the eternal perspective. This perspective is hard for most of us to see.

April 11, 1970, at 19:13:00 UTC Apollo 13 was launched from Cape Canaveral, Florida. The year before the world had been astounded by the first lunar landing. This was to be a follow-up mission. However, 55 hours into the flight a major accident happened with the oxygen supply system for the three astronauts. This mishap aborted their opportunity to land on the moon, and made it look as if they would be lost in space.

Perhaps you have seen the movie: Apollo 13. There is a main message that is missed in that movie; it is that the world united in prayer. The movie portrays the incredibly creative things that were done for them to survive — using very cleverly the limited resources they had available. I am confident, as were many at the time that those repairs and how they used manual navigation means augmented by astronomical measurements to re-enter the Earth's atmosphere that they were guided by the inspiration of heaven.

I still remember the emotional relief and thrill we all had when they safely splashed down just off the Cook Islands within sight of the recovery ship. They had endured 142 grueling hours. The world shed tears of joy. The joy the world experienced then will be nothing like the thrill we will have with the coming of our Lord and the tears of joy we will then shed. Like with the Apollo 13 crew, the united prayers of the righteous will be effectual, and nothing is too hard for the Lord.

The last days will include a purging time. Like the Apollo crew, we will be stretched to our limits and need to use every creative skill, but most importantly, we will learn to rely on the powers of Heaven. The key is to trust in Him. All of these experiences will be for our good, and will bring us ever closer to the Lord as we reach out to Him.

Rather than being on Apollo 13, visualize that you are on a space ship called Earth traveling through the heavens. The Earth has traveled from somewhere nigh unto Kolob in our Galaxy, and because of the Fall, it is here in its current orbit. Through the infinite atonement of Jesus Christ, it will be redeemed from the Fall and will return to the presence of the Father, celestialized. We know its destiny, and to be in harmony with our Creator is the greatest opportunity of our existence. We can overcome any problem encountered during our space travel if we will totally trust in him.

Now, Satan knows it is his last hour, and as the world has submitted to the arm of flesh and to Satan's seditious influences we see the world spiraling down in moral decay. As we reach out in faith and touch the hem of the Lord's

Chapter 8

garment, as it were, Satan has no power over us as we anchor our souls in Him and in His eternal truths. We can put on the breastplate of righteousness and the whole armor of God as is so eloquently outlined by the Apostle Paul.[1] In so doing, we are promised that we will be able to withstand every fiery dart of the adversary.

As we come to Christ, He will bring us back to the Father on this celestialized Earth. We will be there in glory with our loved ones who also chose Christ as the center of their lives — there to receive a fullness of Joy. Father's plan of happiness is perfect — being brought to full fruition because of the infinite and perfect atonement of our Lord and Savior, Jesus the Christ. The atonement is perfect and infinite in that it can take any or all of us out of our imperfections —including our addictions—and make us perfect. What an incredible promise.

There are some who say, " Earth is not our home." But it will be our home in celestial realms of glory. The Savior said, "Blessed are the meek: for they shall inherit the Earth."[2] We will look back from the celestialized Earth and know not only that our Earthly challenges were worth it but were necessary for us to enjoy a fullness of joy in those eternal realms. We saw them from a pre- mortal time perspective. In fact, we participated in choosing our trials, tribulations and sufferings as part of our mortal sojourn here on the Earth. Our premortal-existence perspective gave us celestial vision, and we rejoiced in the opportunity to have these experiences and to come to Earth and gain a body. Our body should be treated as a sacred temple because that is what it is. We saw how sacred when we were in our pre- mortal existence.

The Earth in its journey to perfection has been "baptized of water" through Noah's flood. It is now preparing to be "cleansed by fire" — born of the Spirit — in preparation for the reign of our Savior as King of kings and Lord of lords for a thousand years — as the Lord prepares a people to be delivered to the Father pure and without spot to receive the fullness that He has planned for each of us. We will rejoice greatly as we come to fully understand His marvelous and perfect plan of happiness.

As mentioned earlier, those who believe that Christ is just a great moral teacher, C. S. Lewis points out that this thought is absurd. Either Christ is

---

1) Ephesians. 6:11-18 (see also D&C 27:18-21)
2) Matt. 5:5

what he said he was, the Son of God, or he is a liar. If a liar, then He cannot be a great moral teacher. If He is the Son of God, then we best get on with true Christianity — loving God and our fellow man and looking forward to being like Him. This was Christ's great promise, as Lewis profoundly taught.

In overcoming the lusts of the flesh through Christ, Satan would have us believe that once we slip into sin that we might as well give up; if we repent, we'll just do it again. So, what's the use? The Lord's way is the opposite. There is always hope, "and His arms are stretched out still" — willing to help us out of any sin or addiction, if we will but turn to Him. His atonement is infinite, and in Him we can be made free — body, mind, and spirit. In Him we will find love, joy, and peace as we exercise faith in Him and fully repent of every sin. How thankful we should be for Him. He will help us overcome all things because He has already.

Mark B. Kastleman in his book *Healing Hearts & Mending Minds* tells his story, as a member of The Church of Jesus Christ of Latter-day Saints, and how he got hooked on pornography — and how it nearly destroyed him. In finding the Savior he found freedom from this horrible chemical — and it has been shown to be a chemical addiction. He outlines a very powerful and successful recovery program, which in summary I share below. You may remember that we have discussed Kastleman's important contribution in overcoming addictions in Chapter 4 where we see the enormous importance of wholesome loving relationships in the healing process.[3]

From his book we read the following account:

> Pornography addiction is a chemical addiction, producing a response in the brain like that of externally administered drugs.
>
> There are many well-intentioned, yet ignorant people, who advise those buried in pornography addiction to "just stop looking at it." This advice is akin to counseling an alcoholic to, "just stop drinking." This advice is irresponsible and often increases the addict's failure rate, shame, hopelessness and isolation, driving him deeper into his porn habit. Just as one would put their arm around the alcoholic and gently

---

[3] Specifically, if you wish to refer back to that, it is in Section 4.1, III.

counsel, "Brother, there are wonderful recovery programs, you're going to be fine," the attitude and response toward the pornography user should be no different.

Chris, an individual who successfully broke free from pornography, explained it this way: When I sought counsel from my religious leaders, they told me to pray and read my scriptures more, and to "just stop looking at it." When I tried and failed over and over again, I was sure that there was something wrong with me, that I was basically evil and no-good. This caused me feelings of deep shame and despair. Then when I learned that just like other drugs, this is a chemical addiction, I remember shouting, "I'm not a freak and a loser — there's a reason why I keep failing at this!" After that I started my recovery program and today I have a great life free from pornography!

The central feature of this recovery is for the person to feel the Lord's love and to be loved by those around them. In his acknowledgments he says, "To my beloved Savior, for rescuing a soul so rebellious and proud as mine. His grace and spirit have made all things possible and I am eternally in His debt." He then goes on to thank his wife and children for their great love and support. Being surrounded with this love is incredibly helpful in the recovery process — and for people to stay out of judgment. Only the Lord knows our hearts.

This love along with meaningful, procedural activities in thought and action are outlined in his book. These put the person on the path to full recovery — receiving the healing powers of the atonement, and it is Christ alone who can truly heal our bodies, minds, and spirits. Giving Christ our hearts to mold and asking Him to help train our minds so that our thoughts are continually turned to Him is fundamental.

Since Mark's book is out of print, for those interested I outline below the key steps.* What we do with our thoughts is so critical.

It's About Time

Thought:

A power which binds or makes men free.
The precedent of actions yet to be.
A guide, a builder of self-control,
Shaping the destiny of the soul.
— *Amy Baker*

David O. McKay:

> The greatest power in the world today, and the power that is needed to thwart the schemes of the adversary, is the power of the Lord Jesus Christ. That man is greatest who is Christlike, and what you think of Christ is largely what you will be. What you think about when you are not compelled to think, determines what you really are.

Alexander Pope's quatrain tells the all-too-common fate of exposure to worldly vices:

Vice is a monster,
of so frightful a mien,
As to be hated,
needs but to be seen.

Yet seen too oft',
familiar with her face,
First we endure,
then pity, then embrace.

It does not have to be like Pope poetically describes. When a person becomes renewed in Christ, that person is repulsed by vices of this world — while the world endures, pities, and embraces them.

Before the Second Coming, there will be a great division centered on two kinds of people:

Chapter 8

1. Those that are Christ-centered — serving Father's children, seeking first the Kingdom of God and His righteousness, and are motivated by their love of God and their fellow men, helping to bring about Zion, which will be glorious.

2. Those that are self-centered — serving their appetites and passions, seeking the things of this world rather than laying up treasures in Heaven, motivated by the lusts of the flesh, partaking of Babylonian (worldly) activities. We need to reach out to these folks with a perfect hope in Christ and that as the Apostle John prophesies: " Babylon the great, the mother of harlots and abominations of the earth…" shall fall.[4]

I was touched by this poem from *The Human Culture Digest:*

Let your food be plain and wholesome,
From all stimulants abstain.
Keep the body you live in,
Clean and pure from every stain.

Let each day be one of doing,
Idle moments are seeds of death.
Strong minds and strong bodies,
Require work as well as rest.

Let your mind be chaste and pure,
An evil mind makes vice and crime.
Honest thinking and noble deeds,
Build a character grand, sublime.

Let each hour be full of sunshine,
Pleasure comes from doing good.
Life is full of happy moments,
When life's plan is understood.

And Heavenly Father's Plan is perfect — bringing perfection out of imperfection.

---

[4] Revelations 17:5-18:5

The importance of being "pure in heart" is paramount as we approach the end time. "Blessed are the pure in heart; for they shall see God."[5] "Therefore, verily, thus saith the Lord, let Zion rejoice, for this is Zion—THE PURE IN HEART; therefore, let Zion rejoice, while all the wicked shall mourn."[6]

Aristotle: Tolerance [is] the last virtue of a dying society.

The world has become tolerant of a vast variety of sins and even holds them as good. Well did Isaiah say of the last days: "Woe unto them that call evil good, and good evil; that put darkness for light, and light for darkness;..."[7]

Marcus Aurelius: If it is not clean, do not think it. If it is not true, do not speak it. If it is not good, do not do it.

Ezra Taft Benson, who served as Secretary of Agriculture under Dwight D. Eisenhower said the following in 1985:

> When you choose to follow Christ, you choose to be changed. "No man," said President David O. McKay, "can sincerely resolve to apply to his daily life the teachings of Jesus of Nazareth without sensing a change in his own nature. The phrase 'born again' has a deeper significance than many people attach to it. This changed feeling may be indescribable, but it is real.
>
> The Lord works from the inside out. The world works from the outside in. The world would take people out of the slums. Christ takes the slums out of people, and then they take themselves out of the slums. The world would mold men by changing their environment. Christ changes men, who then change their environment. The world would shape human behavior, but Christ can change human nature.
>
> Human nature can be changed, here and now," said President McKay, and then he quoted the following: You can change human nature. No man who has felt in him the Spirit of Christ even for half a minute can deny this truth. You do change human

---

5) Matt. 5:8
6) D&C 97:21
7) Isa. 5:20

nature, your own human nature, if you surrender it to Christ. Human nature has been changed in the past. Human nature must be changed on an enormous scale in the future, unless the world is to be drowned in its own blood. And only Christ can change it.[8]

The following is from Beverly Nichols, in *Stepping Stones to an Abundant Life.*[9]

Finally, men captained by Christ will be consumed in Christ. To paraphrase President Harold B. Lee, they set fire in others because they are on fire.[10] Their will is swallowed up in His will.[11] They seek to do always those things that please the Lord.[12] Not only would they die for the Lord, but more important they want to live for Him. Enter their homes, and the pictures on their walls, the books on their shelves, the music in the air, their words and acts reveal them as Christians. They stand as witnesses of God at all times, and in all things, and in all places. They have Christ on their minds, as they look unto Him in every thought. They have Christ in their hearts as their affections are placed on Him forever.[13]

Highlights from Mark Kastleman's recovery program as he overcame pornography, as outlined in his book:

The Solutions

- Teach the truth about sex.

- Shut down the gateway to porn addiction.

- Don't underestimate pornographers and overestimate people - protect!

- Fight for decency.

- Create and nurture true intimacy in your family relationships.

---

8) CR April 1962, p. 7
9) pp. 23, 127
10) *Stand Ye in Holy Places*, p. 192
11) John 5:30
12) John 8:29
13) In Book of Mormon language, they "feast upon the words of Christ" (2 Nephi 32:3), "talk of Christ" (2 Nephi 25:26), "rejoice in Christ" (2 Nephi 25:26), "are made alive in Christ" (2 Nephi 25:25), and "glory in [their] Jesus" (see 2 Nephi 33:6). In short, they lose themselves in the Lord, and find eternal life (see Luke 17:33)

The Recovery Program

- Recognize and admit you have a problem.
- Understand "How you got here."
- Clearly identify your motive for recovery.
- Shed the shackles of shame, isolation and secrecy.

Four Simple Daily Commitments

- Engage in consistent daily self-care for your spiritual, physical, emotional, and mental well-being.
- Build new mental models & habits through consistent daily practice; the Lord will help you greatly to realign your thought patterns as you come to Him and ask for help with real intent.
- Nurture healthy, intimate relationships.
- Keep a journal to consistently track and assess your progress.

Learning how to love as the Lord loves is our primary purpose in life, which, guaranteed, will lead us to the greatest happiness in this life and Eternal Life in the life to come. This learning is key to overcoming all problems and all addictions.

Chapter 9

Can All Cancer Be Cured?

It's about time to know there is a cure for cancer! Dr. Leonard Coldwell has cured more cancer patients than any other doctor in the world — 35,000 cases —, and he and the doctors he trains and works with in Europe continue to do so. Dr. T. Simoncini in Italy has written a book, *Cancer is a Fungus: A Revolution in Tumor Therapy*, and a summary of his book is as follows:

> His book describes how a fungus infection always forms the basis of every neoplastic formation, and this formation tries to spread within the whole organism without stopping. At the moment the constant, uniform, and implacable growth of a tumor is in no way affected by current oncological treatments. A recovery rate for cancer that fluctuates at around 7% is mentioned in the classical books and treatises in spite of all the tricks and distortion of statistics. After making the necessary corrections, this amounts to virtually nil. The rest is propaganda for orthodox oncology. On the basis of the scientific considerations in this book which

demonstrate that cancer is caused by fungal masses (of the Candida type), sodium bicarbonate is the only useful remedy that is now available for healing the disease.

Several places throughout the book, we talk about the Big C (cancer). It is our hope and prayer that the cancer information we share herein will help you appreciate in the phrase from Dr. Lorraine Day, " Cancer doesn't scare me anymore."

Dr. Batmanghelidj (Dr. B.) has cured patients from cancer with proper hydration. We have discussed his ground-breaking work in Chapter 7.

Cancer cells cannot live at a certain level of body alkalinity. Our oldest son, Sterling, assisted a professor during his graduate work at the University of Arizona at Tucson in a literature research project. The essence of that research effort was that cancer cells will not reproduce in a sufficiently strong alkaline environment; they need a degree of acidity to reproduce. Cancer is essentially an acidic condition in the body.

A meat and dairy-centered diet, the diet for most Americans, moves the body to be too acidic. Part of Dr. Coldwell's cure for cancer is a vegan diet, which moves the body to an alkaline condition. Another part of Dr. Coldwell's cure is proper hydration — drink a gallon of pure water a day with a ½ teaspoon of sea salt added. This diet aligns with Dr. B's approach as well. Both Dr. B and Dr. Coldwell have shown that the body needs the proper amount of salt for the body's electrical system to work properly.[1] The typical recommendation for a low-salt diet is because most people are too dehydrated.

Cancer is more than just a physical disease. It ties to the emotions as well and stress is a driving force for cancer as well as diseases in general. Dr. Lorraine Day's healing from "near death's door" breast cancer using natural means is well documented.[2] We will discuss some more the importance of properly dealing with stress and emotions in the next chapter based on the inspiring and pioneering work of Dr. Lissa Rankin.

---

1) The following seven minute video is very informative: *http://drleonardcoldwell.com/tag/anti-cancer-protocol/*
2) It is an inspiring story you can find on her web site: *www.drday.com*

Chapter 10

Should We Have " Mind Over Medicine?"

Dr. Lissa Rankin's father was a doctor. She wanted to be a doctor and started practicing medicine at age seven on squirrels. She was written up in the newspaper several times and was known as the "squirrel girl." She healed and raised 21 squirrels and learned the critical lesson that love and nurturing are fundamental to healing. She worked as a candy striper at the hospital and scrubbed in on her first surgery at age 12.

She went on to become an OB-GYN and practiced with a clinic for eight years in obstetrics and gynecology. Even though she was taught to be nice to her patients in medical school, in practice, with only 12 minutes per patient, there wasn't time to extend love and nurturing that she had learned was so critical in the healing process with squirrels.

At a crisis time in her life and at great sacrifice she left the OB-GYN world to keep her integrity and to find her true mission in life. She followed her heart, and that made all the difference. Being right brained as well as left, she had in parallel followed a career of writing and painting and was very good at both. She finally found her niche and it is summarized in her excellent book, *Mind Over Medicine*.

Her remarkable life story is shared in an interview by Jonathan Fields with the Good Life Project.[1] It is inspiring and well worth listening to. I find my mindset is in close agreement with hers as it comes to healing the body, mind, and spirit.

At the end of the one-hour interview, she summarizes profoundly a great testimonial on the importance of listening to what she calls her "inner pilot light." It is the light of Christ with which all of us are born.[2]

She also calls it the voice within, the divine spark, the inner guidance, your highest self, and your inner doctor. She said that the minute she started to listen to that voice, everything changed. "It was pervasive. Everything fell into place." She said she couldn't have even imagined it before. To listen, we need to get "quieter; less busy [beware of the bareness of a busy life]; start trusting that voice."

At the conclusion of the interview, she said, "The good life is listening to that voice and align your life with your personal truth — to fully self-actualize. We often wear masks… [Conforming to what society expects us to look like]." She said, "My truth may put me at risk of criticism [as with her book]." "Take action according to that voice."

Her book flies in the face of medical tradition. She said that they may try to "burn me at the stake. The braver I get to live my life 'out of the box' the better my life gets and the better the lives of the people who are important to me get."

I was touched by her last phrases, "The good life is a life filled with love… living with an open heart. Having your life have meaning and purpose. Being aligned with what your soul is here on this Earth to do."

Recently on her web site she shared:

> There is a Universal Intelligence that is guiding everything. We have nothing to fear. Love is all around. Each one of us matters equally and is connected in oneness to a united whole that needs you, loves you, supports you, and connects you to all living beings

---
1) *http://www.goodlifeproject.com/lissa-rankin- mind- medicine/*
2) John 1:9

and all of matter. This Guiding Life Force of consciousness energizes everything, so you are never alone. Light will for sure prevail over darkness, and the time is now for all of us to wake up to these truths…

Toward the end of the interview with Jonathan Fields, Lissa shares her findings regarding the placebo effect. She found the work of Dr. Ted Kaptchuk at Harvard in looking at the placebo effect as a healing modality rather than — as is typically done — as a comparison to test some new drug or treatment modality. Her findings were life changing. Dr. Herbert Benson — also at Harvard — helped Lissa to appreciate that the placebo effect works if you move out of the " stress response" and into the "relaxation response." Then your mind can help your body move into a mode where it can heal itself. Like with the squirrels, she found from their research and her experience that this best happens when the recipient feels love from the care-giver. This helps the person move into the "relaxation response" mode and documented physiological changes — bringing about healing occur.

The good Lord designed the body that way, but in our high- stress level society, this can be a challenge. Dr. Dharma Singh Khalsa's guidelines for relaxation outlined in Chapter 13 are very helpful in this regard. Add those relaxation modalities to that of a caring-loving care-giver and you have healing as the Lord designed.

I listened to her talk on the very popular internet program *www.ted.com* (Technology, Entertainment and Design — TED; ideas worth spreading) — very inspiring as well. The interview with Jonathan Fields gives you a much better feel for her life and how she found her mission — a remarkable living story.[3]

A loving Father-in-Heaven gave us our missions in pre- mortal that would be optimum for happiness and to best use our time and talents in our mortal journey. If we listen to that "inner voice," we find that path of happiness, peace, love, and joy.

She is the doctor for the new millennium. It pulls the best of naturopathic medicine and self- healing modalities together, but adds that most important

---

3) *https://www.youtube.com/watch?v=CZ8MaLuBreQ*

dimension of love. When a care-giver is filled with love and compassion, this helps the person move out of stress-response mode, which is the major killer, and into the relaxation-response mode, where the body — designed by our infinitely loving Maker — can heal itself.

I highly recommend her book. Another remarkable dimension of her book is that she gives scientific proof that "You Can Heal Yourself." In the "flycover" of her book it says: "By the time you finish Mind Over Medicine, you'll have made your own Diagnosis, written your own Prescription, and created a clear action plan designed to help you make your body ripe for miracles." Her scientific data further prove "that loneliness, pessimism, depression, fear, and anxiety damage the body, while intimate relationships, gratitude, meditation, sex, and authentic self-expression flip on the body's self- healing processes."

As you know, cairns are used as landmarks, monuments, and to keep us on track. Dr. Rankin has developed a "Whole Health Cairn." It is built out of ten key stones — making up our "Whole Health." The "Cairn" is built on a foundation stone of your "inner pilot light." Building on it the stones are named: work and life purpose, relationships, spirituality, creativity, sexuality, environment, money, mental health, and physical health. The cairn is surrounded by four words: love is at the top; gratitude is at the bottom (like another foundation feeling), and on the left and the right are service and pleasure, respectively. I believe your happiness is your choice. Surround your cells with love; your love and your family's love. Find a care-giver who cares and loves with a pure love. You already have God's love surrounding you.

I would change the word pleasure to joy, because joy includes uplifting pleasures and much more. In my opinion, she and her colleagues are tuned into the kingdom of heaven, and are doing a great healing work on the Earth to help that kingdom come here as it is there. I also see a crescendo in the number of folks who are tuning into their "inner pilot light," and are finding their missions in their mortal journey. It is exciting to be part of this worldwide redemptive effort. And as with her cairn; it is capstoned with love, with wings of service and joy, and with a foundation of gratitude. With Dr. Rankin, let us work for a oneness — a healing of ourselves and society in body, mind, and spirit.

Chapter 11

The " Biology of Belief" and Healing Modalities

In this chapter we will feature three major contributors to healing and health modalities: Dr. Bruce Lipton, Dr. Bradley Nelson, and Michael Brook. Their contributions are most timely as you will see. I will review Dr. Nelson's and Michael Brook's important contributions at the end of this chapter.

Our youngest son, Nathan, brought to my attention the work of Bruce H. Lipton, Ph.D.[1] Though Dr. Lipton comes from a different background than Dr. Lissa Rankin's, his conclusions are remarkably similar. His work actually precedes hers. After earning his Ph.D. at the University of Virginia, he joined the Department of Anatomy at the University of Wisconsin's School of Medicine in 1973. There he did stem-cell research and developed a novel form of human genetic engineering. Frustrated with his life, he left Wisconsin·and went to the Caribbean to teach. There he had what he called "a scientific epiphany that shattered my beliefs about the nature of life." He came to the major biogical paradigm-shift that a cell's life is controlled by the physical and energetic environment and not by its genes. He realized this contradicted mainstream science. He did further research at Stanford validating this new and biology-shattering paradigm. He demonstrated pathways connecting the mind and body.

---

1) *https://www.brucelipton.com/*

Much like Dr. Rankin, Dr. Lipton's novel scientific approach transformed his personal life as well. He enthusiastically said, "I was instantly energized because I realized that there was a science-based path that would take me from my job as a perennial 'victim' to my new position as 'co-creator' of my destiny." This deepened understanding of cell biology allowed him to see how the mind controls bodily functions and "implied the existence of an immortal spirit." When he applied what he had learned "his physical well-being improved, and the quality and character of his daily life was greatly enhanced." Many others have benefited as well who have applied what he shares in his books.[2]

I could write much more of Dr. Lipton's work; it is very disruptive to the current biological model practiced by the medical and pharmaceutical communities, and many changes need to be made in college courses to accommodate the new information coming forth. If you Google his name, you will find several of his talks, and he has written books documenting much of his work as well as that of his colleagues that are in harmony with this new direction in biology. His work is a major contribution to our health and well being. I suggest you listen to some of his talks and/or read his books — especially *Biology of Belief*. I will leave this monumental work at that.

Michael Brook is an amazing individual and is a dear friend. Michael draws on a wide variety of experience: professional athlete, educator, trainer, public speaker, life-long student of holistic health, and high-performance living. As an athlete he was a trampoline champion, a world-class aerial-acrobatic-freestyle skier, and a professional high diver. In his investigation of various alternative-healing modalities, it became clear to him that many of the physical ills that affect us today are preventable. Through his study of various mind- body-spirit teachings, he learned that there were certain principles of health that, when incorporated into one's life, dramatically lessened the likelihood of disease. At the same time they enhanced one's vitality, performance, and the over-all quality of life.

Michael founded the Positive Air Team to take the message of prevention and optimal health to young people. Using a dynamic trampoline performance, this group of professional athletes presented "High Performance Living Programs" to over a million students, staff, and parents across the western United States, as well as to numerous businesses and organizations.

---

2) Bruce H. Lipton, *The Biology of Belief; Unleashing the Power of Consciousness, Matter & Miracles*

Chapter 11

His new book — just now becoming available — captures "the golden thread of truth" as he brings to light the "essence of health, healing, and wholeness. As mentioned, it is entitled, *New Dimensions in Health; Simple Secrets to Creating Optimal Health*. You will be impressed with his web site where you can watch him perform in skiing, diving, and on the trampoline. I calculated the force he encounters on the trampoline when he told me that he reaches 10 feet vertical rise, and it is three times the gravity force of the earth; yet, he has never been injured. I highly recommend his book. The link to his web site is below,[3] where you can view amazing acrobatics done safely because of Michael's "New Dimension in Health..." way of living.

Dr. Bradley Nelson's book: *FUTURE MEDICINE, The Emotion Code, How to Release Your Trapped Emotions for Abundant Health, Love and Happiness*, is a great resource. He has had great success in helping people be healed from a large variety of health challenges, and these problems can be in body, mind, or spirit. The book gives several testimonials of healing from personal experiences he has had. He is a renowned holistic physician and lecturer.

His web site is *www.healerslibrary.com*. His healing modality is easy to learn, and he has a large set of folks who have been trained and are helping share what he has developed and documented. It's about time this modality were known by the world, who critically needs this effectual-alternative for good health of body, mind, and spirit – being emotionally clear of worldly encumbrances that have been piled upon us.

---

[3] See *http://www.newdimensionsinhealth.com*

It's About Time

Chapter 12

Stress, the Killer of Body and Brain? How Can We Best Deal with It?

Dr. Dharma Singh Khalsa, M.D. has shown that excess stress causes excess cortisol release in the body, which can kill millions of neurons per day. Excess stress is deadly in many other ways as well. Dr. Coldwell has a book showing how stress is at the root cause of essentially all disease. Dr. Don Colbert's book *Stress Less* is an excellent resource for how to deal with stress as is Dr. Khalsa's book, *Brain Longevity*.

I was so impressed with Dr. Khalsa's book that I wrote a fairly comprehensive book report and placed it on our web site.[1] It is called The Abundant Life, Regeneration or Degeneration, Your Choice. The article referenced in the link below[2] pulls some of the main highlights from Dr. Khalsa's work. Having this article on the book's website allows this to be a short chapter. Don't let the size of the chapter diminish in your mind the great importance of his work.

1) http://www.allanstime.com/Health/Abundant_Life.htm
2) http://ItsAboutTimeBook.com/health-secrets-for-living-to-be-100/

I consider Dr. Khalsa to be the front pioneer in "Brain Longevity," which, appropriately, is the title of his monumental book. His book is available as an audio book as well.

I also use some of Khalsa's work in the next chapter. You will see additional aspects of the importance of his work there.

Chapter 13

Growing the Brain and Fullness of Joy

Recently, a dear friend of ours from Boulder, Delta Bement, told us of a new and fascinating book Make Your Brain Smarter by Sandra Bond Chapman, Ph.D. Dr. Chapman is founder and chief director of the Center for Brain Health and is a Distinguished Professor at The University of Texas at Dallas. Since Delta's brain is already "smarter," I paid attention! She is 10 years older than we are, and my wife and I are 77! I enjoy synthesizing and condensing down the essential kernels of a person's work that I may utilize it in a most efficient way. I will summarize Dr. Chapman's work toward the end of this article.

Many believe that we are born with a certain IQ and that we need to learn to live with that throughout our lives. Dr. Chapman proves that wrong and shows that we can grow our IQ.

Over the years, I have read several very good books on the brain, and I will tie those into who we are and the beautiful truth that " Adam fell that men might be, and men are, that they might have joy."[1]

---

1) 2 Nephi 2:25

We are three-part beings: intelligence, spirit, and body. Our intelligence was first coupled to our spirit-matter brain when we were born spirit children to our infinitely loving Heavenly Parents. Then as part of our physical birthing process, our intelligence and our spirit-matter brain were coupled to our physical brain. The intelligence coupled to our brain makes our mind. The nurturing of our mind, body, and spirit leads to the full and abundant life here and a fullness of joy in the life to come. That overall nurturing also provides the greatest peace, love, joy, and happiness in this life. It is fascinating to me that the Lord has designed His perfect plan of happiness and program of "be ye therefore perfect, even as your Father which is in heaven is perfect"[2] such that we can best nurture ourselves by nurturing others in all three areas of our beings. This nurturing brings three dimensions to the second great commandment to love our neighbors as ourselves that we may help one another be nurtured in body, mind, and spirit.

There is a fascinating section in Roy Mills' book *The Soul's Remembrance* where he describes what happens as part of our physical birthing process when our pre- mortal intelligence and spirit- brain gets, as it were, down-loaded into our physical brain — becoming the mind of man. I have included much of Roy's work already in this book In addition, a brief book report on Roy's book is available in the link at the following footnote.[3] A loving Father-in-Heaven gives us all we need in this "down-load" for our mortal sojourn, and the amount is enormous; it is much larger than we think and is there for us to tap. It makes the biggest computer seem insignificant. This transfer is done most efficiently and effectively as part of the birthing process.

The question is often asked, "When does the spirit enter the body?" The data on this interesting question indicate that the spirit can come and go, but clearly has to be there at birth. The next question one may ask is, "When does the above "down-load" occur?"

Our intelligence is coupled to the universe through the diallel-field line structure described in the research we have shared on our web site; it is not confined to our physical brain.[4] We share experiments we have conducted in this regard in Chapters 21 and 22. A beautiful example of this coupling is how

---

2) Matthew 5:48
3) For a brief book report on Roy's book see the following link:
http://www.allanstime.com/Spiritual/BookReports/souls_remembrance.htm
4) *http://www.allanstime.com/Spiritual/In_Touch_with_Eternity.htm*

we connect with Heavenly Father through prayer. Science has recently proven that the " mind," as they call it, is not confined to the brain.[5] One very exciting aspect to our intelligence is that it is coupled to the Spirit of Christ — given us at our physical birth[6]. Our intelligence can come to know all truth, if we choose to seek it.[7] This is such an exciting doctrine and promise.

After the above-mentioned "download" occurs, our spirit can leave our body and is coupled, independent of time and space, to our intelligence. So our spirit can leave our body both before and after our birth, giving us remote spiritual eyes for instruction and edification.

Inherent in our intelligence is agency — the right to choose. This doctrine is explained profoundly in Doctrine and Covenants 93:29-36. We will spend more time on these very important principles toward the end of this article. We know that agency — our choices — makes all the difference, and that there is an eternal law that there must needs be opposition in all things so that choice is meaningful and operative in time and eternity.[8]

This doctrine gives full meaning to the Savior's command: "Be ye therefore perfect, even as our Father which is in heaven is perfect." (Matt. 5:48) In other words, we are spirit children of perfect parents and are hence gods in embryo; we have the potential to be like them as we use our intelligence in following the Spirit of Christ. The Lord states it extremely poignantly where He describes those who receive celestial glory: "These are they who are just men made perfect through Jesus the mediator of the new covenant, who wrought out this perfect atonement through the shedding of his own blood."[9]

As mentioned before, Dr. Dharma Singh Khalsa, M.D. is another pioneer in the field of how the brain works. I highly recommend his book be on your shelf: *Brain Longevity; Regenerate Your Concentration, Energy, and Learning Ability for a Lifetime of Peak Mental Performance*. I have a review of his book available at a link in the following footnote.[10] Here, I share the most important highlights.

---

5) See and hear the talk of Dr. Rupert Sheldrake, for example: *http://www.youtube.com/watch?v=0waMBY3qEA4* (part 1) and *http://www.youtube.com/watch?v=VRKvvxku5So* (part 2).
6) John 1:9
7) D&C 93:26-28
8) 2 Nephi 2:10-11
9) D&C 76:69
10) *http://www.allanstime.com/Health/Abundant_Life.htm*

He has also unveiled a very important cause of mind and memory degeneration, and gives ways to avoid that cause. His claims have been clinically validated, and help us to:

- think better,

- become more productive and optimistic,

- overcome age-associated memory loss,

- increase our brain's strength and vitality, and

- meet the world with energy and joy.

Stress Management and the Cortisol Connection

"Cortisol is one of the hormones secreted by the adrenal glands. It's secreted in response to stress. In moderate amounts, cortisol is not harmful. But when produced in excess, day after day — as a result of chronic, unrelenting stress — this hormone is so toxic to the brain that it kills and injures brain cells by the billions," reports Dr. Khalsa. Hence, we see the need for being at peace in the midst of the storms of life. Dr. Khalsa further reveals that the secretion of excess cortisol is a major cause of memory loss problems. He also believes that this "is one of the primary causes of Alzheimer's disease." The article in the footnote below lists dr. Khalsa's four complementary means for dealing with stress and for brain longevity.[11] These allow us to correct much of the mental degeneration problem, which is very serious and wide-spread problem in our society.

Dr. Khalsa points out that "Wisdom comes naturally to older people. It is the natural product of experience. Our neurons are biologically built for wisdom. As the years of our lives. . . unfold. . ., each of our neurons. . . sprout. . . new dendrites, and. . .[make] new dendritic connections with other neurons. . . Living long is the only way to achieve this abundance of multi-branched dendrites, with billions of connections." Hence, we see that wisdom can come with age if we haven't suffered serious degeneration and if we have used our brains to process worthwhile information.

Dr. Khalsa's meditation exercises were so successful that he offered a

---

11) *http://ItsAboutTimeBook.com/health-secrets-for-living-to-be-100/*

course for doctors getting ready to take their board exams. He says, "I showed all of them how to meditate, and how to apply the mental training techniques of peak performance. . . .the doctors taking my course had over a 90-percent pass rate, compared to the national average of about 50-percent.."

His guidelines for how to meditate are very simple:

- Find a quiet place with no distractions;
- Allow 10 to 20 minutes; don't use an alarm;
- Sit comfortably and relax all of your muscles from bottom of toes to the top of your head;
- Stop all internal dialog and mind chatter;
- To help you stop thinking and calm down, silently repeat a pleasant word or phrase — like peace, love, that has religious significance;
- Don't be concerned if thoughts intrude; inhale and bring yourself back, relax, and meditate;
- When you finish, sit quietly for a minute or two, and try to merge your calm state of mind with your normal outlook.

"That's all there is to it. If it sounds simple, that's because it is."

Exercising the Brain as a Muscle

In the ground-breaking work of Dr. Marian Diamond of UC Berkeley in the 1960s, she was the first one to show that we are able to actually grow the brain at any age. It is much like a muscle; use it or lose it. In her famous book *Magic Trees of the Mind*, she discusses how that growth process can be augmented. Her work was so successful, that they allowed her to do the autopsy on Einstein's brain.[12]

Dr. Diamond discusses the seven different areas of brain development and how they can complement each other. When they do, then brain growth is significantly enhanced. These areas are: 1) language, 2) math and logic, 3) spacial representation, 4) music, 5) physical skills, 6) interpersonal and intra-personal skills, and 7) knowledge of nature. We have, for example, all heard

---

12) For a very brief book report on her work see: *http://www.allanstime.com/Spiritual/BookReports/magic_trees.htm*

of the Mozart effect — how listening to Mozart helps us retain and learn the material we are studying. The best dendrite growth occurs when the environment is pleasant and positive. The best retention occurs when we have cross-fertilization among these seven areas.

Our oldest son, Sterling, in wanting to memorize Isaiah 52, composed it to music. He did the first half of it back in 1993 — 20 years ago. Without practicing, he recently shared that with us; it was very inspiring and impressive. It was a beautiful example of how we can enhance mental retention as we harmonize these different areas of our brains.

Following up on Dr. Diamond's work, Dr. Khalsa recommends "at least a couple of hours each day doing some form of mental exercise." This is especially valuable for older people. He says that, "Studies have shown that when people do engage in moderate pleasant forms of mental exercise, it increases not just their knowledge, but also the efficiency and power of their brains." My little personal joke is that I try to learn faster than I forget!

There are some excellent on-line brain exercise programs, and just playing games — alone or with others; the latter gives good social interaction as well, which is very healthful for the brain. My wife and I have some favorite games we enjoy. There are also CDs and DVDs that can help a lot.

Prayer — a Fundamental Brain Exercise

I was so happy to read of Dr. Khalsa's great support of the power of prayer. He has a whole section on The Healing Science of Prayer. He points out some excellent documented sources of the power of prayer in healing and in moving one into a healthy meditative state. He prays for his patients, and believes that health care is just that — truly caring for his patients.

In the previous chapter, we discussed how Dr. Lissa Rankin has documented scientifically the value of love and caring for her patients. As I mentioned before her remarkable life story is well told when Jonathan Fields interviews her.[13] Her book and web site are additionally very helpful tools in this regard.

In discussing the four kinds of prayer Dr. Khalsa categorizes them so:

---

13) *http://www.youtube.com/watch?v=CZ8MaLuBreQ*

(1) colloquial conversation with God; (2) petitioning God; (3) ritualistic prayers (employing a prepared script); and (4) meditative prayer. He says that all four can "elicit the meditative state, and the relaxation response.

"It appears, however, as if the formal meditative prayer — which involves listening to God more than talking to Him — is the most effective style of prayer for creating a subjective perception of closeness to God. Survey respondents were twice as likely to feel a 'strong relationship with God' if they used the meditative style. . . The style that least elicited a subjective sense of connection to God was the petitionary style."

I derive a significant level of peace and enjoyment by early morning meditation, prayer, and study of the scriptures. I find that when I am in the low-frequency (α-state) while waking — which is usually around 5:00 a.m. for me — I receive some of my most profound insights and creative thoughts. This state coupled with the beautiful colors associated with the sunrise — refracting through the clouds with spectrum of red hues and reflecting off snow-covered Mt. Nebo in shades of pink — is an inspirational sight to behold.

For me, the Christian way provides the opportunity for the abundant life and for perfect balance. Ironically, as we "lose" our lives for the sake of the work of the Lord; then we "find" our lives at peace with Him and fulfilled to the fullest extent — body, mind and spirit. Then we are "free indeed."

The new book (2013) that I mentioned at first in this chapter, *Make your Brain Smarter*, by Dr. Sandra Bond Chapman adds some interesting dimensions to the use of this incredible organ (the brain) that consumes 20% of your blood flow to run your CPU (Central Processing Unit). In summary the key message of her book fits within the phrase: "Win the PRIZE," which stands for Prioritize, Reason, Innovate, and do it with Zeal moving you towards the joys of Eternity. This phrase ties her work into God's work in some fascinating ways.

In *prioritization*, Dr. Chapman calls it "Strengthen Your Strategic Brain Habits." Don't let things that matter least get in the way of those that matter most. A common problem is that people have several tasks they want to get done. With all of them on their minds they often get none of them done efficiently and/or effectively. Listen to your heart and the Spirit to strive to know the Lord's will as to what is most important from an eternal perspective.

Then pick the most important, and if it is timely, "do it." Then take satisfaction in its completion. This process is very satisfying and healthful for the brain. Focus and finish! Then go to the next task that the Lord has helped you prioritize if it is the right time to do it. Timing is often very important. Only the Lord knows the heart of another, and having us reach out to them at the right time is a wonderful and fulfilling experience as we let the heavens guide our lives. "Thy will, not mine be done," was the perfect example Jesus set before us.

A very touching example in scripture of the Lord being at the right place at the right time is when he walked from Capernaum to Nain in one day, so that He could be there for the widow of Nain to raise her son from the dead. Google maps shows this to be a 50.6 km trip and He needed to cross a mountain pass. I estimate that He walked over 30 miles — hiking up some 1,700 feet to cross over the mountains — to get there on time to meet the widow and those carrying the bier with the body of the widow's son. This is a most touching scene with perfect timing.[14]

As we prioritize, it is always good to remember that people are more important than things. It is best to align our work with the work of the Father: to bring to pass the immortality and eternal life of His precious sons and daughters. If we will listen to the Spirit, He will best use our time and talents to bless the lives of others and it will bring the greatest fulfillment to us. Listen to your heart.

In *Reasoning*, Dr. Sandra Bond Chapman calls it "Enhanced integrated reasoning to accelerate performance." She has a whole chapter on this important subject as she also does on prioritization and innovation. She says that "integrated reasoning is your brain's platinum cognitive asset," and "there is nothing that we cannot get better at doing." Think of new ideas on how to do things in all areas of our lives as we reason out the whys and wherefores. Synthesize things to their essence — thriving on the "meat" of ideas. Have the attitude that every problem has a solution. Through reasoning, pondering, and praying, answers will come. Rejoice in the light and truth that come to us as we seek, knock, and ask. The Lord's promises are sure and He says, "Ask, and it shall be given you; seek, and ye shall find; knock and it shall be opened unto you: For every one that asketh receiveth; and he that seeketh, findeth; and to

---

14) Luke 7:1-17

him that knocketh it shall be opened."[15]

Dr. Chapman further says to, "stretch and challenge your mind to construct deeper-level, thought-filled ideas when presented with any type of information (magazine articles, movies, books, television shows, lectures, sermons, songs, political speeches, physician reports, comic strips, jokes, emails, etc.)." "A… brain habit to adopt is to practice synthesizing ideas into one or two abstracted statements when presented with information. Ask… questions that push… [to] deeper thinking and action."

As an example, Edna, my wife, and I recently attended some motivational seminars conducted by Kirk and Kim Duncan, who founded 3-Key Elements. Currently, they are one of the most successful and fastest growing companies in Utah and recently received the Inc. 500 award. I synthesized their 3-key elements as: Visualize, Verbalize, and Actualize, and pointed out to Kirk that this is how Heavenly Father does it. God visualized the Earth, verbalized its creation, and then His Beloved Son actualized it into being so that Father had a place for His precious children (you and me) to come and dwell, as a major step in our path to Heaven. Continually synthesize positive meanings from life's experiences, whether it is personal interaction, reading a book, watching a movie, or whatever. This tendency brings a real richness to life.

She adds that there are "three strategies that will help you enhance your integrated reasoning, thus improving your frontal lobe function and brain efficiency. They are:

- Zoom In

- Zoom Out

- Zoom Deep and Wide

The brainpower of zoom in requires attending to facts, content, and the situation at hand. Gathering facts and using them to support a novel approach is essential to enhancing integrated reasoning and deeper level thinking. However, it's a delicate balance of knowing when to gather more information and knowing when to stop… It is not enough to understand all the facts; it is highly critical to fit them into a larger schema…

---

15) Matt. 7:7-8

"…the brainpower of zoom out" allows you to see "the big picture… merging them into the major themes, core concepts, and broad principles…

The brainpower of zoom deep and wide requires the deepest level of thinking where you apply novel developments from one area to other issues, other problems."

Innovate to inspire your thinking. She defines innovation as "the ability to generate and exploit new ideas to solve problems; to see, devise, and employ improved ways of dealing with unknown and unfamiliar contexts; or to create something that is original and valuable."

She goes on to say that:

- Innovation is about improving upon and changing old ways of doing things through novel thinking.
- Innovative thinkers practice mental flexibility — stretching their creativity and imagination.
- Ingenious thinkers are open to experimentation to rethink practices and are at ease with ambiguity.
- Innovators are not beaten down by failure; rather, they constantly ask what they can learn from their mistakes.

Einstein said, "Imagination is more important than knowledge. Knowledge is limited. Imagination encircles the world." Dr. Chapman further states that "You can spark creativity in your brain and stoke the smoldering fire of your imaginative capacity at any age… if you practice and unlock its immense capacity to create and innovate."

Robert Sternberg said, "Creativity is not just a matter of thinking in a certain way, but rather is an attitude toward life." Our innovative capacity is limited by ourselves. Exercise it continually and it will grow. Examine traditions in the light of truth. Break with past bad habits. Ask challenging questions to get others to thinking creatively. We can all be creative and it is very rewarding to create something that helps the world be a better place to live.

Chapter 13

Laughter helps the brain. I quote the following from www.helpguide.org[16]

> Humor is infectious. The sound of roaring laughter is far more contagious than any cough, sniffle, or sneeze. When laughter is shared, it binds people together and increases happiness and intimacy. Laughter also triggers healthy physical changes in the body. Humor and laughter strengthen our immune systems, boost our energy, diminish pain, and protect you from the damaging effects of stress. Best of all, this priceless medicine is fun, free, and easy to use.
>
> Laughter is a powerful antidote to stress, pain, and conflict. Nothing works faster or more dependably to bring our minds and bodies back into balance than a good laugh. Humor lightens our burdens, inspires hopes, connects us to others, and keeps us grounded, focused, and alert.
>
> With so much power to heal and renew, the ability to laugh easily and frequently is a tremendous resource for surmounting problems, enhancing our relationships, and supporting both physical and emotional health.

Laughter Is Good for Our Health

- Laughter relaxes the whole body. A good, hearty laugh relieves physical tension and stress, leaving our muscles relaxed for up to 45 minutes after.

- Laughter boosts the immune system. Laughter decreases stress hormones and increases immune cells and infection-fighting antibodies, thus improving our resistance to disease.

- Laughter triggers the release of endorphins, the body's natural feel-good chemicals. Endorphins promote an overall sense of well-being and can even temporarily relieve pain.

- Laughter protects the heart. Laughter improves the function of blood vessels and increases blood flow, which can help protect you against a heart attack and other cardiovascular problems.

---

16) *http://www.helpguide.org/life/ humor_laughter_health.htm*

Here is a good joke for you: Two elderly couples were walking down the street, the women a couple of metres [obviously an English joke] ahead of the men. One man told the other that they'd had a wonderful meal the night before — great food, and reasonably priced. His friend asked for the name of the restaurant. "Well, I'll need your help on this. Let's see, there's a flower that smells great and has thorns on the stem?" "That would be a rose," his friend responded." "That's it!" the man exclaimed. Then he shouted to his wife, "Hey, Rose! What's the name of the restaurant we ate at last night?"

In 1977, I was invited to give a paper in Tokyo on my work with atomic clocks. Professor Musha was my host. He asked me a searching question, which in summary was, "Why have most of the major inventions in the world been in the United States?" Environment is important, and the " land of the free" fostering liberty to think, to act, and to realize one's dreams is a great motivator to creativity. This land of promise and melting pot for all nations has brought forth many marvelous inventions that have contributed greatly to the world.

In Summary

In the Lord's divine design, there are seven parts of the brain working which complement the seven different areas of brain development. As we go forward with faith and Zeal on our path to Eternal Life we will both find the abundant life along with great joy and a fullness of joy in the life to come. Our intelligence, mind, brain integration is given us by our loving Father to this end. He wants us back and all the tools are there for us to receive these enormous blessings if we but choose to do so.

Chapter 14

About Energy

In the new Unified Field Theory we are researching, we have suggested a diallel-field line structure connecting matter. Inherent in it is an energy field not yet appreciated. In her book The Field, Lynne McTaggert reports on several experiments showing the existence of a field which may correspond to what we discovered. Others have reported over-unity experiments — possibly coupling energy from this field. Our oldest son, Sterling has covered in his news service for fourteen years, the development of several exotic/free-energy devices[1] by various parties across the globe. Those who follow these burgeoning developments in the free energy arena will observe the intense activity; inventors, humanitarian and commercial interests all vie for that world-changing potential trophy.

The allure to meet our energy needs is understandably appealing. Just as easily understood is the range of personalities that are rising to the occasion. Garage tinkerers, PhDs, machinists, electricians, philosophers, crackpots,

---
1) *http://FreeEnergyNews.com*

professors, armchair and laboratory experts all form this fantastic melting pot of thinkers, who could help bring about an extremely helpful energy revolution. It hasn't been so very long that farmers without engineer's degrees developed agricultural machinery that optimized production, or that computers used to save their data on paper punch cards.

The point here is this: formally credentialed or not, people are driven to solve this energy problem because there's so much to gain. Early solutions are likely to be, or at least appear awkward. They'll be clunky, like the first mobile phones. In 1908 professor Albert Jahnke claimed to have a wireless telephone. He and his group were accused of fraud, and didn't proceed with production, and perhaps they were fraudulent, yet it's possible that he was merely ahead of his time. You may be reading this book in its electronic format between phone calls and videos in the palm of your hand. The vision is the seed, and in exotic energy there appear to be seeds aplenty.

The world's primary energy sources are presently petroleum-based. The greed for oil is one of the drivers for the current wars in the world. Evidence suggests that we are on the verge of having some exciting exotic clean energy sources coming forth that will not only be clean but affordable, making obsolete most of the mainstream, polluting energy sources to which society is presently addicted.

A few of the more promising technologies currently in development:

- U-Plug 2kW magnet motor. It is 24 inches long and weighs 20 pounds.

- Brilliant Light Power, Inc., in New Jersey, is developing a product they call SunCell®. They claim 200 kW from a unit that weighs 250 pounds and hope for commercial availability in 2017.

- Mike Waters has designed a windmill that claims twenty times greater efficiency and less bird-kill than conventional bladed windmills.

You can find these and many other promising technologies on Sterling's web site listed above.

I believe with Sterling that free energy is about freedom. It will not emerge in full to bless the human family as long as we deserve tyranny because of the

Chapter 14

seeds of corruption society has sown. The key is to repent and turn to God so that these new technologies can be used for good — not war and tyrannical control. Could it be that those managing our divine destiny are holding these exotic solutions from emerging in the market until we deserve them? Sterling thinks so, and he says as much when interviewed in the alternative media, such as Coast to Coast AM.

Conventional clean energy sources, such as solar, wind, geothermal, and tidal are available, and fortunately, more and more countries are using them. While these are a move in a positive direction, being clean, and fostering independence, they usually tend to be quite a bit more expensive than grid-based power, which are often predominated by polluting modalities. But with the glut of companies pursuing sustainable energy worldwide, the price has been coming down significantly, making off-grid solutions much more attainable by the thrifty and resourceful.

When I retired from the lab in Boulder in 1992, we built a home in Fountain Green, Utah, without a furnace. We are at 6,000 feet elevation (1,828 meters), and we have seen the temperature drop to -20 degrees Fahrenheit (-29° C) in the winter. We've integrated around ten different solar gain principles, including the obvious: photovoltaic, hot water, green house; as well as some more unknown: trombe wall, phase-change of eutectic salts, passive air circulation, passive-solar air conditioning, wood chips in growing spaces as a water diode against evaporation by the sun.[2] For years, when a person Googled "solar home," ours was the first on the list. We have had a lot of people reference it and visit our home, so we feel we have done some good in helping move away from petroleum-based heating systems for our homes.

Here are two fascinating and remarkable things about sunlight: If you study the spectrum from the sun, the peak energy wave length matches the human eye's maximum sensitivity — Divine design. And the photosynthesis process generates chlorophyll molecules, which are quite complex and have magnesium at the center. The hemoglobin molecule — making up our blood — is identical to the chlorophyll molecule except the magnesium is replaced by iron.

In other words, the Lord designed greens to be the best blood builders and are one of the best things we can eat on a regular basis — moving our bodies closer to alkaline with all the health benefits from a strong immune system.

---

2) *http://www.allanstime.com/SolarHome/*; see also: *BackToEden.com*

The DVD and movie *Thrive*, by Foster Gamble, does a great job of giving hope for alternative energy sources in the midst of a world that is "oil hungry" and drowning in energy problems.[3] It is amazingly insightful, and for the most part consistent with the UFT work and research we have done as described in Chapter 21.

Technology exists for flying cars, hovering cities, novel farming techniques, and expanded space travel but these technologies are all limited by the current methods of harvesting or generating energy. With new forms of energy generation and storage, many of the amazing contraptions seen in science fiction could be realized now.

We live in a most interesting time, when the world is saying there are too many people, while the Lord says "…there is enough and to spare." The mainstream teachings about population control don't take into consideration the responsible abundance that will be ours when these clean energy technologies become available to lift humanity, empowering them both literally, figuratively, and most importantly spiritually as we "Trust in God" — like it says on our coins, and like it says in the center chapter and center verse of the Bible. The world is looking in the wrong place, and until they turn back to God and live the commandments, they will not find the right place to look.

---

3) *http://www.thrivemovement.com/the_movie*

Chapter 15

Is Our Defense for Our Defenses?
If Not, What Can We Do?

I am most grateful for those who have given their lives for the liberty we have enjoyed in this land of the free. Our US Constitution provides "for the common defense. . . and secure the blessings of Liberty for ourselves and our Posterity. . ." However, a violation of our Constitution of enormous magnitude was perpetrated on the American people by the elite, which included some government leaders and the media, with the events of 9/11. Scientific evidence clearly points to controlled demolition of the Twin Towers and of Building 7, which had no airplane impact — all three of them coming down in "near free-fall motion," as happens in controlled demolition. Super-thermite has been documented in the dust, which explains how they were destroyed. Falling as they did in "near free-fall motion" is physically impossible from just the impact of the airplanes and from the fire in Building 7. Several peer-reviewed science papers and books have been written documenting this demolition.[1]

---
1) See for example a large quantity of information at the following link: http://www.journalof911studies.com

In addition, as you look at the data, you find some very unusual evidence. Most people don't know that buildings 3, 4, 5, and 6 were also seriously damaged in highly unusual ways, and not from falling debris from the Twin Towers. Building 6, for example, was an 8-story building and had a massive hole in it going from the top all the way to the ground. A lot of the destruction reminded several people, who were familiar with it, of the John Hutchison effect, and John got numerous phone calls after 9/11. From our work with the new unified field theory, we know that energy densities are very important. The magnetometer readings and whether systems on that day were highly unusual. Our oldest son, Sterling has documented this in an article he wrote last September 11th.[2] In addition, the world has been duped into believing a lie of enormous proportions.

Additionally, the Constitution provides for the "calling forth of the Militia. . . [to] suppress Insurrections and repel Invasions." Our AF fighters were kept on the ground, when they could have been launched to protect us during the 9/11 disasters. Why? Vice President Dick Cheney is recorded monitoring the airplane (missile?) hitting the Pentagon. With plenty of time to launch counter measures he took no action to avoid that disaster. President George W. Bush lied on several occasions about why we were attacking Iraq.[3]

Nick Rockefeller told Aaron Russo 11 months before 9/11 happened that an event was being planned that would make Americans support the "war on terrorists" — justifying the attack on Iraq and Afghanistan — making it look like we are defending our nation.[4] See also his excellent movie: Freedom to Fascism, in which he documents that the IRS is operating outside the bounds of the Constitution. When people ask him, "Should we pay our taxes?" He says, "Yes; otherwise, they will hurt you."

President Dwight D. Eisenhower prophetically spoke of our military getting out of control and out of the bounds set by our Constitution. Excerpts from his closing talk (17 January 1961) just before exiting as president are shared here; you will find this talk to be amazingly insightful:

---

2) http://pesn.com/2013/09/11/9602370_Magnetometer-and-Hurricane-Correlations_with_9-11-2001/
3) See the documentary DVD *Why We Fight* (2005 PG-13 98 minutes) — filmed during the Iraq War and directed by Eugene Jarecki's Sundance Grand Jury Award-winning film documentary. For additional information, watch on-line *Loose Change and Invisible Empire*.
4) See this insightful one-hour interview with Aaron:
http://video.google.com/videoplay?docid=5420753830426590918

Chapter 15

Good evening my fellow Americans...

Three days from now, after half-century in the service of our country, I shall lay down the responsibilities of office as, in traditional and solemn ceremony, the authority of the Presidency is vested in my successor. This evening, I come to you with a message of leave-taking and farewell, and to share a few final thoughts with you, my countrymen...

We now stand ten years past the mid-point of a century that has witnessed four major wars among great nations.

Throughout America's adventure in free government, our basic purposes have been to keep the peace, to foster progress in human achievement, and to enhance liberty, dignity, and integrity among peoples and among nations. To strive for less would be unworthy of a free and religious people...

Progress toward these noble goals is persistently threatened by the conflict now engulfing the world. It commands our whole attention, absorbs our very beings. We face a hostile ideology [communism] global in scope, atheistic in character, ruthless in purpose, and [insidious] in method. Unhappily, the danger it poses promises to be of indefinite duration. To meet it successfully, there is called for,... those [who] enable us to carry forward steadily, surely, and without complaint the burdens of a prolonged and complex struggle with liberty the stake.

A vital element in keeping the peace is our military establishment. Our arms must be mighty, ready for instant action, so that no potential aggressor may be tempted to risk his own destruction...

Until the latest of our world conflicts, the United States had no armaments industry. American makers of plowshares could, with time and as required, make swords as well. But we can no longer risk emergency improvisation of national defense. We have been compelled to create a permanent armaments industry of vast proportions. Added to this, three and a half million men and women are directly engaged in the defense establishment.

Now this conjunction of an immense military establishment and a large arms industry is new in the American experience. The total influence — economic, political, even spiritual — is felt in every city, every statehouse, every office of the federal government. We recognize the imperative need for this development. Yet, we must not fail to comprehend its grave implications. Our toil, resources, and livelihood are all involved. So is the very structure of our society.

In the councils of government, we must guard against the acquisition of unwarranted influence, whether sought or unsought, by the military-industrial complex. The potential for the disastrous rise of misplaced power exists and will persist. We must never let the weight of this combination endanger our liberties or democratic processes. We should take nothing for granted. Only an alert and knowledgeable citizenry can compel the proper meshing of the huge industrial and military machinery of defense with our peaceful methods and goals, so that security and liberty may prosper together.

You and I, my fellow citizens, need to be strong in our faith that all nations, under God, will reach the goal of peace with justice. May we be ever unswerving in devotion to principle, confident but humble with power, diligent in pursuit of the Nation's great goals.

... Those who have freedom will understand, also, its heavy responsibility; that all who are insensitive to the needs of others will learn charity; and that the sources — scourges of poverty, disease, and ignorance will be made [to] disappear from the Earth; and that in the goodness of time, all peoples will come to live together in a peace guaranteed by the binding force of mutual respect and love.

Thank you, and good night.

He also said, "God help this country when somebody sits at this desk [as

President in the Oval Office] who doesn't know as much about the military as I do." No one since him knows as much and we now critically need God's help. One of the hardest things for me right now is when I hear people say that our military is protecting our freedoms. It breaks my heart that our men and women go off to war with good intentions that they are fighting for our liberty when, in fact, I know the New World Order has effectively taken over our " military industrial complex" with power and gain motives in their hearts, as President Eisenhower warned against. They make it appear that they are for our defense, but in fact they could declare martial law and our military would be our enemy.

Our Constitution also provides that only congress can "declare war." Now, through unconstitutional executive orders, the President declares war — coordinating with the UN — further destroying our sovereign nation status. An excellent historian, Trevor Loudon, has shown in his book, *Barack Obama and the Enemies Within*, that President Obama has been groomed over the last decades to move the (far left) communist agenda forward in this country. His marching orders are from them. Even though he appears to be serving the middle-class American workers using the typical communist disinformation strategy, in fact they are the target of the conspiracy designed by the elitists. Several have used Loudon's book as reference material — including Glenn Beck — not disputing the points of fact therein. A talk he gave in March 2012 in Colorado, summarizing his book, may be found at the bottom of the following link in two mp3 files on our web site. He not only tells why we are where we are, but what we can do. [5]

In all of this unfolding of events, I do not fear, because I know who is really in charge and to trust in Him is the greatest opportunity of our existence. I see all of this as the signs of the times and fulfillment of prophecy. For me, it is the most exciting time in the history of the Earth. We have every reason to hope if we understand the work of the Lord. If we move into fear, doubt, hate, anxiety, etc., these are tools of the adversary. If we send love to all of these people, as the Savior has asked us to do, then we neutralize Satan's tools against us. At the same time it is extremely important for us to be aware of what is going on, so that we are not deceived by the great deceiver who is behind it all and to do all we can to stand for the "Perfect Law of Liberty," which is defined in the Bible as the gospel of Jesus Christ and which our founding documents were

---

5) *http://www.allanstime.com/Government/ Barack_Obama_and_the_enemies_within/index.html*

designed to protect in its promulgation based on "free will." If we do nothing and let the state end up sovereign instead of the people, Satan smiles. If we join the Lord in taking His word and truth — which includes the "perfect law of liberty" — wherever people will receive it, we will have peace, love, and joy in our hearts. It is Liberty versus captivity, Our Constitutional principles versus Communism, to know God versus no God, freedom versus slavery, and having the individual be sovereign versus the state being sovereign. In Trevor Loudon's talk, he outlines some important things the average person can do politically. From my perspective, if we will pray and hearken to the voice of the Son of God, we will do our part and have joy in it — in the midst of the storms of life.

Chapter 16

New World Order Infiltration of the Internet;
the Anti- Christ is Proactively Saying: "There Is No Devil."

George A. Smith well said, "The mind that can assume there is no God — no good — can as consistently claim there is no devil — no evil. As well they say there is no right, no wrong, and consequently no morality for man. Man's own nature, his own consciousness teaches him other than this; that there is right and wrong, and his happiness depends upon doing the one, and not doing the other." The modern phrase for "There is no devil" is that true scientists do not believe in God. Therefore, anything speaking of God or spirituality is pseudoscience. Thus, recently, the outstanding and excellent talk by Dr. Rupert Sheldrake given on the internet talk show: TEDx, was pulled from being viewed because of complaints from a few scientists — confirmed atheists — pushing their weight in objecting to anything with spiritual overtones.

Dr. Sheldrake received a standing ovation for a talk he gave at the Electric Universe Conference last January (2013) in Albuquerque, New Mexico, which was even more detailed than his TEDx talk. My youngest son, Nathan, and I were privileged to hear that talk. I was so enthralled, I could not take notes.

This talk was based on his latest book, *Science Set Free*. He is a very well-credentialed scientist and has written several books. That his talk would be pulled from TED.COM is an injustice.[1]

TED.COM is one of the most popular web sites and shows on the internet — now at a billion hits with the theme: "Ideas worth spreading." And, "Riveting talks by remarkable people, free to the world" is their subtitle, and there are some 1,500 talks freely available. I have heard several, and they are "riveting." I felt to investigate the origin and nature of this most successful internet source of information. I learned TED stands for Technology, Education, and Design. It was founded in 1984 by Richard Saul Wurman. Born in March 26, 1935, he is an American architect and graphic designer who coined the phrase 'Information Architecture' and is considered to be a pioneer in the practice of making information easily understandable. In November 2001, Chris Anderson's non-profit The Sapling Foundation (motto: "fostering the spread of great ideas") became the owner of TED. In due process, he created a web site, where the talks are readily available and now receives over a million views per day. The speakers are given a maximum of 18 minutes to present their ideas in the most innovative and engaging ways they can. Past presenters include Bill Clinton, Jane Goodall, Malcolm Gladwell, Al Gore, Gordon Brown, Richard Dawkins, Bill Gates, Google founders Larry Page and Sergey Brin, and many Nobel Prize winners.

The ideas from the New World Order elitists have infiltrated big time and God is basically excluded from any of the ideas shared; hence, we see the reason Dr. Sheldrake's talk was pulled. Because of the outrage on the part of thousands, when this happened, they later regretted doing it, but they had to cover their mistake. The few true "pseudo-scientists" (atheists) have won the day in keeping Rupert's talk off the TED web sites — calling his work "pseudo-science." We know they will not win in the end, but to be awake and aware of the subtle deceit going on among the powerful few elitists and atheists is very important, so that we can help the world wake up to the truths of God. Christ is the way, the truth, and the life, and no man cometh unto the Father but by Him.[2]

You can find lots of truths on the internet tucked away in different corners.

---

1) You can hear his excellent response to this sham at the following link: *http://www.youtube.com/watch?v=_SGzu8TJsyo*

2) John 14:6

Chapter 16

Today you can find Rupert's banned talk online, and it is basically a condensed version of the talk he gave in Albuquerque last January.[3] His talk given in Albuquerque at the Electric Universe Conference can be heard as well as a two-part presentation.[4] Be sure to listen to both parts and you will see why he received a standing ovation.

As we may listen to tantalizing TED talk internet presentations, let us make sure that they, along with any other information, line up with "ye shall know the truth, and the truth shall make you free."[5] Only the truths of God will do that. Trust in Him and not the arm of flesh. Truth is truth, no matter where it comes from.

Wikipedia has become a very valuable internet encyclopedia, and I was sorry to learn that these pro-active elitist atheists are doing their best to remove God from it. In a similar vein, Snopes has set themselves up as the internet truth validator, and some say that they are funded by George Soros. I don't know, but I do know that I have found many examples when what they have said is "true" or "false" is to bring consistency with the NWO agenda and doesn't line up with the facts. I love the song, *Let the Holy Spirit Guide*:

Let the Holy Spirit guide;
Let him teach us what is true.
He will testify of Christ,
Light our minds with heaven's view.

Let the Holy Spirit guard;
Let his whisper govern choice.
He will lead us safely home
If we listen to his voice.

Let the Spirit heal our hearts
Thru his quiet, gentle pow'r.
May we purify our lives
To receive him hour by hour.[6]

---

3) http://www.collective-evolution.com/2013/04/10/banned-ted-talk-rupert-sheldrake-the-science-delusion/
4) http://www.youtube.com/watch?v=0waMBY3qEA4
5) John 8:32
6) Text: Penelope Moody Allen, b. 1939. (c) 1985 IRI. Music: Martin Shaw, 1875-1958

I added a few more verses to the same music by Martin Shaw and titled it *Zion*:

Zion

Zion is the pure in heart;
A society Divine.
Christ is King — we'll ne'r depart;
We are branches — He. . . the vine.

Bab'lon falls, and He will come
With the saints, and in glory —
Singing grand millennial songs.
Saved by love — His face to see.

Christ the Way — our heavenly door.
Total trust and faith in Him.
His love fills our hearts with joy.
Free indeed — cleansed from all sin.

What rapture we will enjoy,
His light — our light, we are One.
His love — our love, we employ.
Back to Father — through the Son.

Even the scriptures are not left to "private interpretation."[7] They were given by the Holy Ghost and can be properly understood only by that same power. We have the sweet assurance that whatever is spoken by the power of the Holy Ghost is scripture,[8] and that angels will speak to us by the power of the Holy Ghost the words of Christ.[9] That should be our guide and our stay for our path back to the Father and a fullness of JOY.

My prayer is that God's love be known and internalized by all His children. We know that that love will prevail.

---

7) 2 Peter 1:20, 21
8) D&C 68:4
9) 2 Nephi 32:2-5

Chapter 17

Are We a Christian Nation?

Most Americans would say without question that we are a Christian Nation. Christmas is our biggest celebration. Examining the evidence, however, leads to some surprises. The basis of Christianity is, of course, the Bible. The Bible was the moral standard used by the founding fathers of America, which was the fundamental reason why we became the greatest nation in history. The Bible had more influence on the writings of the Declaration of Independence and the Constitution of the United States of America than any other book.

David W. Bercot was seeking to know what true Christians believed anciently. Along with the Bible, he decided to go back to the records contemporary to the epoch when Christianity had its birth. He found some unexpected and significant surprises, which led him to write his insightful book, *Will the Real Heretics Please Stand Up*.

As he compared current Christianity to the characteristics and beliefs of the early Christians over the first two or three centuries, he found some significant differences between their beliefs and practices. He shows, for example, that fourth century St. Augustine has had more influence on current Christianity's doctrines and teachings than have any of the Biblical writers — including the Christ.

In summary, Bercot characterizes the beliefs, practices, and teachings of the members of the primitive church as follows:

I. They Were Not of This World; for Jesus Taught:

- to the Jews, "Ye are from beneath; I am from above: ye are of this world; I am not of this world."[1]

- to the Twelve, "These things I have spoken unto you, that in me ye might have peace. In the world ye shall have tribulation: but be of good cheer; I have overcome the world."[2]

- to the Father He prayed, "I pray not that thou shouldest take them out of the world, but that thou shouldest keep them from the evil. They are not of the world, even as I am not of the world. Sanctify them through thy truth: thy word is truth."[3]

- to Pilate, "My kingdom is not of this world: if my kingdom were of this world, then would my servants fight, that I should not be delivered to the Jews: but now is my kingdom not from hence. Pilate therefore said unto him, Art thou a king then? Jesus answered, Thou sayest that I am a king. To this end was I born, and for this cause came I into the world, that I should bear witness unto the truth. Every one that is of the truth heareth my voice."[4]

A large percentage of the early Christians lived up to these admonitions from the Savior as are clearly taught in the Bible. A Roman commented, ". . .they despise honors and purple robes. . . They are not afraid of present torments. . . . . .they do not fear to die. . . . . .you abstain from. . . pleasures. You do not attend sporting events. You have no interest in public amusements. You reject the public banquets, and abhor the. . . games."

---

1) John 8:23
2) John 16:33
3) Jn 17:15-17
4) John 18:36-37

Chapter 17

A second-century elder exhorted his congregation, "Brothers, let us willingly... do the will of Him who called us. And let us not fear to depart out of this world,... deeming the things of this world as not belonging to us, and not fixing our desires upon them... The Lord declares, 'No servant can serve two masters,' If we desire to serve both God and Money, it will be unprofitable for us, 'For what will it profit if a man gains the whole world, and loses his own soul?' This world and the next are two enemies... We cannot therefore be the friends of both."

Cyprian, the respected overseer of the church in Carthage, wrote in a letter to a Christian friend, "The one peaceful and trustworthy tranquility... is this: for man to withdraw from the distractions of this world, anchor himself to the firm ground of salvation, and lift his eyes from Earth to heaven... He who is actually greater than the world can crave nothing, can desire nothing, from this world. How... unshakable is that safeguard, how heavenly is the protection... — to be free from the snares of this entangling world, to be purged from the dregs of Earth, and fitted for the light of eternal immortality."

II. They Loved With Pure Christ-Like Love (Their Friends and Their Enemies)

Jesus taught:

In the sermon on the mount, "... Love your enemies, bless them that curse you, do good to them that hate you, and pray for them which despitefully use you, and persecute you; That ye may be the children of your Father which is in heaven: for he maketh his sun to rise on the evil and on the good, and sendeth rain on the just and on the unjust. For if ye love them which love you, what reward have ye? do not even the publicans the same? And if ye salute your brethren only, what do ye more than others? Do not even the publicans so? Be ye therefore perfect, even as your Father which is in heaven is perfect." [5]

We are perfected in Christ. We cannot perfect ourselves but Christ can. In the sermon on the plain, "But love ye your enemies, and do good, and lend, hoping for nothing again; and your reward shall be great, and ye shall be the children of the Highest: for he is kind unto the unthankful and to the evil. Be ye therefore merciful, as your Father also is merciful. Judge not, and ye shall

---

5) Matthew 5:44-48

not be judged: condemn not, and ye shall not be condemned: forgive, and ye shall be forgiven: Give, and it shall be given unto you; good measure, pressed down, and shaken together, and running over, shall men give into your bosom. For with the same measure that ye mete withal it shall be measured to you again."[6]

- To the Jews regarding which is the greatest commandment, "...Thou shalt love the Lord thy God with all thy heart, and with all thy soul, and with all thy mind. This is the first and great commandment. And the second is like unto it, Thou shalt love thy neighbour as thyself. On these two commandments hang all the law and the prophets."[7]

- To the Twelve, "A new commandment I give unto you, That ye love one another; as I have loved you, that ye also love one another. By this shall all men know that ye are my disciples, if ye have love one to another."[8] "If ye love me, keep my commandments. He that hath my commandments, and keepeth them, he it is that loveth me..."[9]

Hence, they obeyed because they loved; their obedience was bathed in love, not just out of duty, which is the lesser law and which is so common in our time. They obeyed because they wanted to and because they loved the Lord and all His children. (the good and the bad — enemies and friends) A primary characteristic of the early Christians was their love for one another and for all mankind: Tertullian reported that the Romans would exclaim, "See how they love one another!" Justin Martyr sketched Christian love this way, "We who used to value the acquisition of wealth and possessions more than anything else now bring what we have into a common fund and share it with anyone who needs it. We used to hate and destroy one another and refused to associate with people of another race or country. Now, because of Christ, we live together with such people and pray for our enemies."

Clement wrote, describing the person who has come to know God, "He impoverishes himself out of love, so that he is certain he may never overlook a brother in need, especially if he knows he can bear poverty better than his brother. He likewise considers the pain of another as his own pain. And if he suffers any hardship because of having given out of his own poverty, he does not complain."

---

6) Luke 6:35-38
7) Matt. 22:37-40
8) John 13:34-35
9) John 14:15, 21

Chapter 17

When a devastating plague swept the ancient world during the third century, "Christians were the only ones who cared for the sick, which they did at the risk of contracting the plague themselves."

III. They Had Great Joy in Sharing the Gospel and Rejoiced in the Lord Regardless of Tribulation

Jesus taught:

- In the sermon on the mount, "Blessed are they which are persecuted for righteousness' sake: for theirs is the kingdom of heaven. Blessed are ye, when men shall revile you, and persecute you, and shall say all manner of evil against you falsely, for my sake. Rejoice, and be exceeding glad: for great is your reward in heaven: for so persecuted they the prophets which were before you."[10]

- In the sermon on the plains, "Blessed are ye, when men shall hate you, and when they shall separate you from their company, and shall reproach you, and cast out your name as evil, for the Son of man's sake. Rejoice ye in that day, and leap for joy: for, behold, your reward is great in heaven: for in the like manner did their fathers unto the prophets.[11]

- The Seventy returned with joy, and the Savior said, ". . . rejoice not, that the spirits are subject unto you; but rather rejoice, because your names [and those you have brought into the fold] are written in heaven.[12]

- To the Twelve, "If ye keep my commandments, ye shall abide in my love; even as I have kept my Father's commandments, and abide in his love. These things have I spoken unto you, that my joy might remain in you, and that your joy might be full. . . Ye have not chosen me, but I have chosen you, and ordained you, that ye should go and bring forth fruit, and that your fruit should remain: . . . If the world hate you, ye know that it hated me before it hated you. If ye were of the world, the world would love his own: but because ye are not of the world, but I have chosen you out of the world, therefore the world hateth you.

---

10) Matt. 5:10-12
11) Luke 6:22-23
12) Luke 10:17-20

Remember the word that I said unto you, The servant is not greater than his lord. If they have persecuted me, they will also persecute you;"[13]

Origen told the Romans, "When God gives the Tempter permission to persecute us, we suffer persecution. And when God wishes us to be free from suffering, even though surrounded by a world that hates us, we enjoy a wonderful peace. We trust in the protection of the One who said, 'Be of good cheer, for I have overcome the world.' And truly He has overcome the world. Therefore, the world prevails only as long as it is permitted to by Him who received power from the Father to overcome the world. From His victory we take courage. Even if He should again wish us to suffer and contend for our faith, let the enemy come against us. We will say to them, 'I can do all things through Christ Jesus our Lord who strengthens me.'"

Ignatius, bishop of Antioch, wrote, "Bring on the fire and the cross. Bring on the packs of wild beasts. Let there be the breaking and dislocating of my bones and the severing of my limbs. Bring on the mutilation of my whole body. In fact, bring on all the diabolical tortures of Satan. Only let me attain to Jesus Christ!. . .I would rather die for Jesus Christ than to reign over the ends of the entire Earth." Shortly after, he was brought before a screaming mob in the Coliseum and he was torn in pieces by wild animals.

Tertullian reminded the Romans that, "The more you cut us down, the more in number we grow. The blood of Christians is seed. . . For after thinking about it, who among you is not eager to find out what is really at the bottom of it all? [JOY in Christ and the promise of Eternal Life] And after inquiring, who does not end up embracing our teachings? And when he has embraced them, who does not also willingly suffer so that he may partake fully of God's grace?"

Well did Paul say:[14]

Great is my boldness of speech toward you, great is my glorying of you: I am filled with comfort, I am exceeding joyful in all our tribulation.[15]. . .we glory in tribulations:. . . knowing that tribulation worketh patience; And patience, experience; and experience, hope: And hope maketh not ashamed; because the love of God is shed abroad in our hearts by the Holy Ghost. . .

---

13) John 15:10-20
14) Rom. 5:3-5
15) 2 Cor. 7:4

Chapter 17

Now, as we measure current Christianity against these early Christian Church standards, categorically, we come up way short. We see that the fundamental problem that we have as Americans is that we are not true Christians.

In this regard a book that is tell-tale is, The Ranking of the (100) Most Influential Persons in History, the author, Michael Hart, has Muhammad first, Isaac Newton second, and Jesus Christ third. For most people, that Jesus is not first seems absurd, but when you listen to Hart's reasoning, it gives you significant pause. His ranking is more on how their teachings affected the course of history. For example, to love God and to love your neighbor are not attributed to Jesus as they are Old Testament teachings. Michael Hart ascribes "loving your enemy" as unique to Jesus, and because the Christian world is not living that doctrine for the most part, he gives Jesus third place. We know Hart is wrong in his assessment, because Christ is the great Jehovah — the God of the Old Testament. But now, we are all greatly concerned because America has turned from the God of this land (Jesus Christ), who gave us our freedom in the first place. The scriptures are clear that if we keep the commandments — the greatest of which is to love — we will be blessed and prosper in the land.

Now, we see the signs of the times telling of His Second Coming. These signs are rapidly increasing. Again, the scriptures are clear, the Bridegroom will come when the Bride is ready; He will come to a pure people. While the Bride is being purged and purified and prepared for His coming, the wicked will be destroying the wicked; Babylon the great (the wickedness of the world) will fall and great will be the fall thereof. The righteous will be pulled out to build a Zion society (the Bride).

Christ will then reign as King of kings and Lord of lords during the 7th (Sabbath) millennium of the Earth's sojourn, at the end of which the Savior will present to the Father those who have chosen to follow His great plan of happiness and to partake of the fullness which the Father will give to those who will come unto His Beloved Son, and partake of a full measure of the promises and blessings offered through His infinite and perfect atonement.

When we consider the awesome size and rapid growth of the forces of evil, we may move into fear and to doubt that we can do anything, which is exactly what Satan wants. We should never lose hope, as the servant of Elisha asked, when the Syrian hosts were upon them, "How shall we do?" Elisha answered, "Fear not: for they that be with us are more than they that be with them. And Elisha prayed, and said, LORD, I pray thee, open his eyes, that he may see. And the LORD opened the eyes of the young man; and he saw: and, behold, the mountain was full of horses and chariots of fire round about Elisha."[16] We can always trust the Lord will be there for us.

In fact, we live in the most exciting time in the history of the Earth. All the prophets looked forward to our day as they anticipated the Messiah's glorious return. We will see, after all the purging and purifying is over, how blessed we are to live in this day and time.

A great Christian, Immaculée Ilibagiza learned the correct perspective, as shared in her incredible book, *Left to Tell*. I was so impressed with her book that I prepared a report for our web site,[17] and a summary of this book report is available at the link in the following footnote because of its importance.[18] She was a survivor of the Rwandan holocaust. She learned to truly love her enemies, allowing her not only to survive but to be a great blessing to her nation and to the world as she shares her life-changing story.

As we approach His Coming, the most important thing we can do is to come unto Him, hearken to His voice, and help as many of our family and friends to do the same — to be true Christians. Then we will find peace, love, and joy in the midst of the storms as the world is purged. He will prepare the Bride (Zion) for His glorious return (the wedding feast). Though He may take us through great trials to get there, we can be part of it if we choose Him. I love the French translation of Matthew 7:24. . . 'les met en pratique,' which literally means, "Put these teachings I have given to you into practice." Then we would be true Christians and become a Christian Nation!

---

16) 2 Kings 6:16-17
17) *www.allanstime.com/Spiritual/BookReports/Left_to_Tell.htm*
18) *http://ItsAboutTimeBook.com/left-to-tell-immaculee-ilibagiza/*

Chapter 18

How Are We Saved?

What does the Bible say? Are we saved by grace? Yes! Are we saved by works? Yes! How can this be? Which is it, grace or works? Can it be both?

As mentioned in the previous chapter, David W. Bercot has researched the writings of the early Christians (the first three centuries of the Church's existence) and has documented in his excellent book, *Will the Real Heretics Please Stand Up*, that they believed in both. These writings are from people who were taught by Peter, James & John, and by Paul or by people who knew them. He says:

> The early Christian universally believed that works or obedience play an essential role in our salvation. This is probably quite a shocking revelation to most Evangelicals. But that there's no room for doubt concerning this matter, I have quoted below. . . from early Christian writers of virtually every generation. . . to the inauguration of Constantine.

In a section of his book, entitled, *Are Faith and Works Mutually Exclusive?* He shares a very enlightening historical perspective:

> You may be saying to yourself, "I'm confused. Out of one side of their mouths they say we are saved because of our works, and out of the other side they say we are saved by faith or grace. They don't seem to know what they believed!"
>
> Oh, but they did. Our problem is that Augustine, Luther, and other Western theologians have convinced us that there's an irreconcilable conflict between salvation based on grace and salvation conditioned on works or obedience. They have used a fallacious form of argumentation known as the "false dilemma," by asserting that there are only two possibilities regarding salvation: it's either (1) a gift from God or (2) something we earn by our works.
>
> The early Christians would have replied that a gift is no less a gift simply because it's conditioned on obedience. Suppose a king asked his son to go to the royal orchard and bring back a basket full of the king's favorite apples. After the son had complied, suppose the king gave his son half of his kingdom. Was the reward a gift, or was it something the son had earned? The answer is that it was a gift. The son obviously didn't earn half of his father's kingdom by performing such a small task. The fact that the gift was conditioned on the son's obedience doesn't change the fact that it was still a gift.
>
> The early Christians believed that salvation is a gift from God but that God gives His gift to whomever He chooses, And He chooses to give it to those who love and obey Him.

Interestingly, Bercot shares his own struggles and paradigm shift as he read these early writings:

> When I first began studying the early Christian writings, I was surprised by what I read. In fact, after a few days of reading, I put their writings back on the shelf and decided to scrap my research altogether. After analyzing the situation, I realized the problem was that their writings contradicted many of my own theological views.

Chapter 18

He had the integrity to continue his research, and he came to the profound realization that the reformation actually introduced some heretical teachings; hence, the title of his book. Saved by grace alone is one of them. This doctrine was unknown in Christian teachings before Augustine and Luther. It came in primarily through the translations and interpretations of Paul's writings as part of St. Augustine's works and those of the reformers — especially Luther. There is no denying that the reformation did a great amount of good in helping Christianity correct many of the false teachings introduced after the apostles were killed and during the Dark Ages.

If we are to believe that the early Christians had it correct, who were taught personally by Paul, then we must conclude that salvation is by grace and works. Bercot clearly documents this conclusion. Paul was confronting those who believed that the works of the Mosaic Law were necessary. Christ's atonement fulfilled the law, and those works were no longer applicable. This truth is what Paul meant "For by grace are ye saved through faith; and that not of yourselves: it is the gift of God: Not of works, lest any man should boast."[1]

Because of these doctrinal distortions that occurred during the dark ages and during the reformation, the Lord came to Joseph Smith in the spring of 1820 to commence the restoration in its purity of the fullness of the gospel and called him as a modern prophet. That restoration included the coming forth of the Book of Mormon — a record of the fullness of the gospel as given to the ancient inhabitants of America. It beautifully teaches grace and works.[2]

Ironically, with the discovery of the Dead Sea scrolls, as they were being translated, a real conflict arose because 200 years before Christ, there were Christian teachings. Most Christians believe that Christ and the Apostles introduced Christian doctrines. This Qumran society was a break-off from the Hebrews, and the Jews do not want to accept that these Christian teachings were part of their tradition. The Book of Mormon peoples had the fullness of the gospel before the first advent of the Savior with significant similarities to the Qumran society. The gospel taught in the Book of Mormon is totally consistent with the Bible and with those teachings of the early Christian Church as documented by Bercot.

---

1) Ephesians 2:8-9
2) e.g. 2 Nephi 25:23

Another profound message of the Book of Mormon is that this land of the free (America) was raised up by the Lord through our founding fathers and mothers to provide a place for this restoration of the fullness of the gospel and for it to go forth to all nations in preparation for the great and glorious Second Advent of our Lord and Savior, Jesus the Christ. Somewhat like the finding of the Dead Sea scrolls, this idea is profoundly documented in a recent book by Dr. Bruce H. Porter and Rod L. Meldrum, *Prophecies & Promises; The Book of Mormon & The United States of America* and in another book by Porter, *The Everlasting Decree*.

Coming out of the Ancient American Archaeology Foundation, Edwin G. Goble and Wayne N. May have documented in a five-volume set, entitled, *This Land*, a prolific number of photos and archaeological artifacts that are consistent with the Book of Mormon religious record of these peoples. The above books are similar to the large number of paleontological and archaeological evidences that we now have authenticating the Bible.

If we can set aside the paradigm bias coming out of the reformation that we are saved by grace — not of works — we find the Bible is clear in teaching that we are saved by grace (faith) and works. There aren't different gospels for Paul, or James, or John, or Old Testament prophets. As Peter was profoundly taught by the Lord, "Of a truth I perceive that God is no respecter of persons: But in every nation he that feareth him, and worketh righteousness, is accepted with him."[3] We have the following scriptures as examples:

- Even so faith, if it hath not works, is dead, being alone. Yea, a man may say, Thou hast faith, and I have works: shew me thy faith without thy works, and I will shew thee my faith by my works. Thou believest that there is one God; thou doest well: the devils also believe, and tremble. But wilt thou know, O vain man, that faith without works is dead? Was not Abraham our father justified by works, when he had offered Isaac his son upon the altar? Seest thou how faith wrought with his works, and by works was faith made perfect? And the scripture was fulfilled which saith, Abraham believed God, and it was imputed unto him for righteousness: and he was called the Friend of God. Ye see then how that by works a man is justified, and not by faith only. Likewise also was not Rahab the harlot justified by works, when she had received

---

3) Acts 10:34-35

Chapter 18

the messengers, and had sent them out another way? For as the body without the spirit is dead, so faith without works is dead also.[4]

- Therefore whosoever heareth these sayings of mine, and doeth them, I will liken him unto a wise man, which built his house upon a rock . . .[5]

- For God so loved the world that he gave his only begotten Son, that whosoever believeth in him should not perish, but have everlasting life.[6]

- . . . I am the light of the world: he that followeth me [believes in me and abides by my teachings — doing works of righteousness] shall not walk in darkness, but shall have the light of life.[7]

- If ye love me, keep my commandments.[8]

- He that hath my commandments, and keepeth them, he it is that loveth me: and he that loveth me shall be loved of my Father, and I will love him, and will manifest myself to him.[9]

The logical conclusion from these most meaningful scriptures is that to love is to obey and to obey is to love; they are equivalent, a concept which raises obedience to a celestial level and out of the duty (do it or else) level.

Other relevant scriptures include:

- I call heaven and Earth to record this day... that I have set before you life and death, blessing and cursing: therefore choose life, that both thou and thy seed may live: That thou mayest love the Lord thy God, and that thou mayest obey his voice, and that thou mayest cleave unto him: for he is thy life, and the length of thy days...[10]

- . . . he considereth all their works.[11]

- Also unto thee, O Lord, belongeth mercy: for thou renderest to every

---

4) James 2:17-26
5) Matthew 7:24-29
6) John 3:16
7) John 8:12
8) John 14:15
9) John 14:21
10) Deut. 30:19-20
11) Psalm 15:33

man according to his work.[12]

- ... and he that keepeth thy soul, doth not he know it? and shall not he render to every man according to his works?[13]

- Unto the pure all things are pure: but unto them that are defiled and unbelieving is nothing pure; but even their mind and conscience is defiled. They profess that they know God; but in works they deny him, being abominable, and disobedient, and unto every good work reprobate.[14]

- Wherefore, my beloved, as ye have always obeyed, not as in my presence only, but now much more in my absence, work out your own salvation with fear and trembling.[15]

Finally, as the Lord — speaking apocalyptically through John — "And, behold, I come quickly; and my reward is with me, to give every man according as his work shall be."[16]

As we look at the trend in the world — and in our own great nation — away from those most important teachings of our Savior to both have faith in Him and walk the Christian walk, we can see the need for a full return unto Him and for massive repentance. In Bercot's perspective, modern Christians are — for the most part — in the camp of the heretics. Are we then, as most would believe, truly a Christian nation? Do we worship the works of men's hands or the God of Israel (Jehovah and Jesus). As He taught, do we believe Him and do His will? Do we love all including our enemies?

We live in the most exciting time in the history of the Earth in anticipation of the glorious return of our Lord and Savior. Now is our opportunity to truly come to Christ — believe and hearken; then we have the promise that we need not fear (worry, doubt, hate, envy, fear, discouragement, and anxiety are tools of the adversary and he is very real — knowing it is his last hour). Christ promises love, joy, and peace as we fill our lives with faith and trust in Him. He knows what is coming, and He will help us through it all to receive

---

12) Psalm 62:12
13) Proverbs 24:12
14) Titus 1:15-16
15) Philippians 2:12
16) Rev. 22:12

Him, if we will but totally trust in Him and love Him with all of our hearts, might, minds, and strength. This purging time coming has its Divine purpose and will purify His people and the Earth. We can look forward to the most glorious day this Earth has ever seen when He comes again and reigns as King of kings and Lord of lords. May we exercise pure faith in Him and hearken to His voice I pray.

It's About Time

Chapter 19

The Ideal Society and the First Law of Heaven

In 1909 the Jews began their kibbutzim projects. They are collective communities in Israel working together around a common goal. They were first based on agriculture, but later branched into other economic opportunities — including some high-tech ideas. The early Christians — both in the old world and in the new world — had "all things in common."[1] Enoch and Melchizedek established ideal societies — called Zion societies — and were taken up into heaven.[2]

Prophetically we know that Jerusalem will become such a Zion society as will the New Jerusalem, and those taken up will join those here on Earth at the Second Coming of our Lord.[3] We anticipate a very exciting time in this regard — greater than anything the world has ever seen.

---

1) Acts 2:44; 4:32; 4 Nephi 1:3
2) Moses 7:67-69; JST- Genesis 14:25-40
3) Revelations 21:2; Doctrine and Covenants 84:96-102

These societies learn how to internalize and live the first two great commandments in all that they do, and that makes all the difference; for it is on these two hang all the laws and the prophets.[4] Many books have been written about this ultimate society and how we can each be part of it. If we will truly come unto Christ, He will teach us how to become so. This is a society whose members are motivated by their pure love of Christ, and their characteristics are pure in heart of one mind and one heart and they dwell together in righteousness with no poor among them and with "all things in common." They are filled with love, joy, and peace. Words are inadequate to describe such, but this you can know, "You want to be there." That society takes us to the greatest of all the gifts of God — Eternal Life and a fullness of joy.

It is difficult for most to envision such a society as we grovel around in this fallen world, which often looks a lot like the survival of the fittest. Our lack of vision is because we have forgotten who we really are as the literal spirit offspring of an infinitely loving Heavenly Father. Also, many don't realize that our struggles here are a necessary part of our growth in reaching the goal our Savior gave to us of being perfect like our Father-in- heaven is perfect. In the wisdom of Solomon we learn, "Where there is no vision, the people perish: but he that keepeth the law, happy is he."[5] And the law is the Law of Love: When we learn to love like our Heavenly Father and Jesus love, we will be there and we will rejoice.

Over the last two years, I have been invited to Russia twice to give invited papers at an international symposium on time and space in Suzdal, and at an international navigation conference in St. Petersburg. I was fascinated to learn that there are now more Christians per capita in Russia than in the USA and that they have increased in 17 years from 31% to 72%. I took as many copies of my 2014 addition of my book as I could, and they sold very quickly. I probably could have sold 100 more if had had them. I was asked to write a paper for the navigation conference and chair a panel on using the Allan variance for assessing navigation errors. I felt to include in that paper an equation on how to be one with God as we look forward to a Zion society here on earth. They kindly asked me to remove the religion part, which I did, but the full paper can be found as the last publication on my other web site[6] and the equation showing how to become one with God is available as an article

---

4) Matt. 22:36-40
5) Proverbs 29:18
6) *http://allanstime.com/Publications/DWA/index.html*

on the book's web site.[7]

So much of life is about competition and not about loving cooperation — sharing and caring. Well did the Apostle Paul say:

> We must not be conceited, challenging one another to rivalry, jealous of one another. If a man should do something wrong, my brothers, on a sudden impulse, you who are endowed with the Spirit must set him right again very gently. Look to yourself, each one of you: you may be tempted too. Help one another to carry these heavy loads, and in this way you will fulfill the law of Christ. For if a man imagines himself to be somebody, when he is nothing, he is deluding himself. Each man should examine his own conduct for himself; then he can measure his achievement by comparing himself with himself and not with anyone else. For everyone has his own proper burden to bear.[8]

Of ourselves, we are nothing. But with God, we are of infinite worth for He has designed His perfect plan of happiness so that — if we so choose — we can "be like Him."[9] Let us pray "with all the energy of heart" to be filled with His love.[10] That will make all the difference! The society of heaven is so designed.

For those desiring to be part of this great society and to help in its preparation, the Lord promises:

> And blessed are they who shall seek to bring forth my Zion at that day, for they shall have the gift and the power of the Holy Ghost; and if they endure unto the end they shall be lifted up at the last day, and shall be saved in the everlasting kingdom of the Lamb; and whoso shall publish peace, yea, tidings of great joy, how beautiful upon the mountains shall they be.[11]

Now that we have dealt with several of these false traditions that have crept into society, let us turn to understanding man's time. We will find some interesting misunderstandings there as well. Correcting these could give us a much better world methodology for both keeping and using man's time.

---

7) http://ItsAboutTimeBook.com/spiritual-science-one-with-god-equation/
8) Galatians 6:1-5 *New English Bible*
9) 1 John 3:1-3; Philippians 3:21
10) Moroni 7:48
11) 1st Nephi 13:37

Man's time and space are mortal limitations as we will see in the next chapter. We have the interesting scripture in Revelations, "time is no longer." Since it used to be my job back in the lab in Boulder, Colorado, to generate time for this nation, when I would read that scripture I humorously thought to myself, "I'll be out of a job when the Savior comes!" What it actually means is that then will be the end of worldliness as Babylon the great falls and wickedness is destroyed. We will have a "new heaven and a new Earth" as we move into the millennial era when Christ reigns as King of kings and Lord of lords for a thousand years of peace — during most of which epoch Satan is bound. Man's time as described below will be "no longer" as it relates to astronomical phenomena — sun rise, sun set, the full moon overhead at night, the sun overhead during the day, etc. I wonder if atomic clocks will be useful during the millennium. Maybe I will have a job.

In addition, we will share some ideas on how GPS could be improved significantly.

Section III.

Man's Time and GPS, God's Time, Transcendental Time, Where We Are in God's Time, and the Wheel of Time

Chapter 20

It's About Man's Time and How GPS Works Using Precise Time, and How to Make Both Better

In our world, precise timing is used much more than most would imagine. As you will learn in this chapter, in the high-tech society we live in, precise timing is vital. My efforts in this chapter are to help you understand and appreciate some important aspects historically, where we are now, and where we can go with precise timing.

You may think to set your clock or watch to the " correct time," but even if you succeed, it will immediately start to depart from " correct time" because of the random and systematic variations in the clock, and all clocks have these variations. Both the above random variations as well as systematic variations are the cause of the time deviations in clocks, and they naturally exist and have an interesting character.

So when we ask the question, "What is the " correct time?" it is not a simple question. Correct time is a manmade construct and official time is what society defines it to be. My 32 years with NBS and NIST dealt largely with helping to provide the best standard for time and frequency for the USA and for international timing. A common misunderstanding is that the atomic clocks are radioactive; they are not. Instead, they utilize the electromagnetic-frequencies of photons given off or absorbed by quantum transitions outside the nucleus in atoms and molecules.

For 24 of our 32 years in Boulder my group — working with atomic clocks — had the responsibility to generate official time for the civil sector of the USA, and we coordinated that time very closely with the official time for the military sector, which included GPS. That coordination continues to this day, and military time is maintained by the United States Naval Observatory (USNO) at 34th and Mass. Ave. in Washington D.C., which property more recently is where the Vice President's house is, and the security has been increased dramatically, as you can well imagine. I visited there often as I also did the Air Force Space Division in El Segundo, CA, as GPS was being developed.

Historically, man's time has been kept by using physical-periodic events, such as the rotation of the Earth — solar time; sun up, sun down and the seasons, the orbit period of the moon, the orbit period of the Earth going around the sun, the period of a pendulum; etc. Any of these timekeeping devices can be thought of as a two-part system: 1) an oscillator with some well-behaved period of oscillation; and 2) a means of counting those oscillations to keep time using that stable or repeatable period, such as a calendar, an escapement mechanism, some counting electronics, etc. We don't often think of a calendar as a counter, but that is what it is doing for us. The best-behaved clock in the world has the most stable-repeating period of oscillation and an adequate counter. We will see that this is a very far-reaching statement.

For clarity, let us realize that a clock with a periodic device, having a period of length T, has a frequency of oscillation which is the reciprocal of its period, $\nu = 1/T$. Physicists use the Greek letter nu "ν" to designate frequency. The unit for frequency is the Hertz (Hz), which is one cycle per second, and so the units for T are seconds per cycle.

When it was discovered by astronomers that the Earth's spin rate was both slowing down and not as stable as its period of one year as it goes around the sun, scientists redefined the second in 1956. We think of the period of the Earth's spin with respect to the sun, T_{Earth}, is 86,400 seconds (24 hours per day x 60 minutes per hour x 60 seconds per minute) in a day, and prior to 1956 the definition of the second was simply 1/86,400 of what is known as a mean solar day. We see then that the frequency of the Earth's spin rate is $\nu = 1/86,400$ Hz $= 1.1574074074...10^{-5}$ Hz. However, the Earth is stable to only nine significant digits. In other words, it speeds up and slows down each day about 0.1 millisecond. A millisecond is one thousandth of a second. This amount seems small, but for atomic clocks this is enormous. Not only that, the long-term (year-to-year) variations of the Earth's spin rate are more divergent than almost any other kind of clock, and in general, the Earth's spin-rate is slowing down.

The next definition of the second, after 1956, was based on the one-cycle-per-year frequency of the Earth going around the sun and was called Ephemeris Time (ET). The Ephemeris Second was defined as "the fraction 1/31 556 925.9747 of the tropical year for 1900 January 0 at 12 hours ephemeris time." It was very difficult to measure and took years of averaging, so when the stable period of atomic clocks came along, this was exceedingly attractive and exciting — a major science breakthrough for time keeping.

The idea of using periodic phenomenon in atoms for a clock was first thought of in the 1940s by Nobel Prize winner, I. I. Rabi. He based his thoughts on the idea that quantum transitions in atoms and molecules absorb or emit photons with energy $E = h\nu$, where h is Planck's constant and ν is the frequency of the electromagnetic wave associated with that photon. So with Rabi's idea we move from a physical oscillation — pendulum, Earth spin, Earth orbit — as was historically used for timing, to a photon with an electromagnetic oscillation. However, building an atomic-clock frequency standard is a big job. I got to work closely with those who did this fundamental work and used these frequency standards to calibrate a secondary set of atomic clocks for keeping time — the second part of counting cycles! I didn't have to do it; I can't count that fast! The electronics and computer software we developed did all the hard work.

In 1948-49 Harold Lyons and his colleagues at NBS, Washington D.C., built the first atomic clock frequency standard based on a hyperfine resonance transition in ammonia, which had a nominal frequency of 23 GHz, where giga (G) is the official designator for 10^9. This is a microwave-frequency transition. Soon after, in the early 1950s, Lyons, et al, succeeded in making a cesium-beam frequency standard using the ground-state, nuclear-magnetic-resonance transition in the Cs_{133} atom (non-radioactive). The cesium-transition oscillation was much more accurate than that in ammonia. However, they never did step two — using it to keep time. It took Essen and Perry in June 1955 to turn a cesium frequency standard into an atomic clock at the National Physical Laboratory in Teddington, England. So atomic-time keeping began with Essen and Perry.

The classical picture for this cesium quantum transition from one energy state to the other is that the spin direction of the magnetic moment for the valence electron in the cesium atom can be either parallel or anti-parallel to that of the magnetic moment spin direction of its nucleus. The atom can be moved from one energy state to the other by absorbing or emitting a photon of energy $E = h\nu$. Because this transition's frequency was shown to be greatly more stable than that of any astronomical measurements, the International Bureau of Weights and Measures (BIPM) redefined the second in 1967: "The second is the duration of 9 192 631 770 periods of the radiation corresponding to the transition between the two hyperfine levels of the ground state of the cesium 133 atom."[1]

The precision and accuracy with which time and frequency can be measured using atomic clocks has improved by more than a factor of a 100 million over the last half-century; the progress has been amazing, and it has been exciting to be part of it. Systems like GPS became possible with such precision and accuracy. Atomic clocks are at the heart of GPS; it would not work without them.

We will see that our temporal clocks — including atomic clocks — are somewhat like people; each one has a different personality and all are imperfect, except one — our Lord and Savior. An ideal society is where everyone's time and talents are optimally utilized. Just as it is with an ensemble of clocks. Their time readings can be used together optimally if each of their abilities to keep time is properly utilized. Like as our Lord and His word are the standards of truth to which an ideal society can look, similarly, the time and frequency community has standards to which they can refer and approach optimal timing performance. The ability to do precise timing in the world has improved by a factor of a billion in my life-time — most remarkable achievements having been made with a very large impact on society — GPS (the Global Positioning System) being one example.

Every clock's time differs from every other clock's if measured with enough precision, and every clock is wrong except the one defined to be

---

1) See *http://www.physics.nist.gov/cuu/Units/current.html*

correct. This is the message I shared as I taught seminars for the Time and Frequency Division of the National Bureau of Standards (NBS and later the National Institute of Standards and Technology). Just like people, every clock is different. We are what we are because of our intelligence, our pre-mortal spirit birth, our mortal birth, our environment, and the choices we make — the use of our agency. Similarly, clocks perform as they do as a result of how they are manufactured, their environment, and what we call random-noise perturbations. All these things come together — determining a clock's performance and its timing errors.

In the 1960s, Dr. James A. Barnes (Jim was my mentor at NBS) and I along with other colleagues investigated and characterized these random-noise aspects in detail and we found and characterized different kinds of noise including what is called flicker-noise; these various noise processes are direct contributors to timing errors in clocks. It was very satisfying to me to see the Allan variance, which came out of my master's thesis, be a useful tool for characterizing these noise processes.

Because of our work, I was invited to consult for the Apollo program regarding the time dispersion of the clocks that they planned to use on their space vehicles. This dispersion was important as the vehicle would go into the shadow of the moon, and when it came out the other side they needed to pick up communications as quickly as possible and this flicker-noise time-dispersion component was significant.

Edna, my wife of 57 years, asked me to explain flicker-noise in laymen's terms; well, here goes! For those familiar with mathematics, I have that written up in the article referenced in the footnote as an explanation of flicker noise[2]

When Jim and I were characterizing clock noise back in the 1960s, we and others found that we could model clock noise and time deviations with six different kinds of noise processes. This is like a rainbow of noise colors that cover much of the variability's that we also see in nature — not just in clocks. The more one studies these, the richer one's observations become. Once you know about them, then you can see them in nature in abundance. These various spectral noise colors are around us and they have an effect on much of what we see. As a business example, stock market pricing often behaves in a flicker-noise fashion. Flicker noise is sometimes called pink noise.

---

2) http://ItsAboutTimeBook.com/natures-natural-noise-processesss-science-and-religion/

To do this explanation in laymen's terms is a bit of a challenge. There is a whole international symposium dedicated to understand flicker noise it is so ubiquitous in nature, and its origin is not that well understood.

One of these six noise processes is called white noise. It can be thought of as like time variations going as the flip of a coin. If you flip a coin — not a biased Las Vegas one — then this series of flips creates a random and uncorrelated sequence of heads and tails. In other words, the next flip of the coin remembers nothing of all the past flips and has equal probability of heads or tails. We can think of it as if an atomic clock's rate is one nanosecond per day fast if the coin flips heads and one nanosecond per day slow if the coin flips tails; a nanosecond (ns) is one billionth of a second. This thought experiment is very close to reality for white-noise frequency modulation (FM) is the noise for an ideal atomic clock. Whether the clock moves fast or slow a nanosecond depends on the flip of the coin each day in this thought experiment. Since the flip of a coin is a random and uncorrelated process from one flip to the next, all clock rate variations are equally probable.

As our next thought experiment, we will generate what is called random-walk or drunkard-walk noise. One can think of it as taking a step forward if the flip of the coin is heads and backwards if the flip of the coin is tails. If you line up a row of people side-by-side (shoulder-to-shoulder), each having a coin, and you ask them all to flip their coin a 100 times and they walk forward or backward if they flip heads or tails, respectively, you can show statistically that as you look down that row of people you will see what is called a normal distribution, and the width of the distribution will be square-root of 100 or 10 steps away from the origin. Every person's position in that distribution is a perfect memory of their individual 100 flips of his or her coin. In contrast, our previous white-noise coin toss thought experiment, each flip of the coin remembers nothing of all previous flips — as if it had no memory of the past.

In a very similar sense, theoretically, the time-error of a clock is a perfect memory of its entire past clock rate (frequency) variations. Similarly, the last ideal atomic clock's frequency variation has no memory of all the previous variations (white-noise, frequency modulation, FM). In this case, measuring from when the clock was synchronized the time-error will be a random-walk noise with a perfect memory of past frequency variations after the point of synchronization.

Chapter 20

This kind of clock rate or frequency noise happens in atomic clocks because they are designed to always look for the intrinsic accurate and stable frequency of some quantum transition, which has a specific electromagnetic frequency associated with the photons being absorbed by or being emitted by this desired quantum transition. As the electronics of this atomic clock hunts for this ideal atomic resonance frequency, at one moment the frequency may be too high and the clock runs too fast during that moment. The next moment the frequency may be too low and the clock runs too slow during that moment. The accumulated time error is a "perfect memory" — like the random-walk process above — of every white-noise frequency deviation of the atomic clock's electronic system seeking the true resonance of the atom. So we see a random-walk of the time error, while the frequency deviation spectral color is a white-noise process.

Flicker noise has a spectral behavior half way between white noise and random-walk noise. It appears to have some correlation from step to step, but seems to lose memory of the past as time goes on. Jim and I and others observed that the long-term frequency deviations of almost all clocks, atomic or otherwise, have a flicker noise character.

If you average measurement values from a white-noise process, it has a well behaved average value. The excellent accuracy of this average value is one reason why atomic clocks are so good — the longer you average the better you can know its ideal or natural resonance frequency. Your confidence on the estimate of its average value will improve as the square root of the number of measurements.

For flicker-noise, you find, mathematically, that its average value no longer exists; it is unbounded. The statistical tools that we have developed over the years allow us to deal with this unbounded nature and to characterize these different processes in nature. The mathematics of flicker noise are challenging with some interesting overtones; it reminds you of agency or free choice when looked at a certain way as you approach its unbounded nature.

At the same time Dr. Barnes and I were doing our work, Dr. Benoît B. Mandelbrot was doing his work at IBM on fractals, which word he and Dr. Richard F. Voss coined as the title of their famous book, *Fractals*. Fractals are like flicker-noise; these are self-similar processes and are ubiquitous in nature — like lake shorelines, mountain range profiles, the atmospherics causing the twinkling of a star, etc. It is these kinds of deviations that occur in the error signals of clocks. This is fundamentally why all clocks read different times.

The idea of how GPS works can be thought of in the following way. The speed of light is about 30 centimeters per nanosecond (about one foot per billionth of a second). Let us suppose that the 32 GPS satellites circumnavigating the globe have atomic clocks on board each satellite, and all of them are synchronized to an accuracy of a nanosecond. They exist in six orbital planes tilted 55° to the ecliptic, and there are four or more satellites in each plane — giving good coverage of the Earth. They continually broadcast their times and positions.

Suppose, further, that you have a GPS receiver with an atomic clock in it also synchronized to GPS time to a nanosecond, and you listen to the signal from three GPS satellites in very different parts of the sky. Since all the clocks are synchronized, you can measure how long it takes for each satellite's signal to get to you. Knowing how long the GPS signals take to come from each of the three satellites allows you to calculate the length of each leg to a precision of one foot (discounting for the moment atmospheric and ionospheric delay issues, etc.). Since the satellites also continually broadcast their positions, you effectively have an upside-down tripod with each of the three legs going to one of the three satellites, which allows you to calculate your position with a precision of about one foot in this thought experiment.

Having an atomic clock in your GPS receiver is too costly and impractical for size, power consumption, cost, and other reasons. So, what is done in practice is to measure the delay from four or more satellites. Measuring four satellites gives you four equations with four unknowns — longitude, latitude, altitude, and time (x, y, z, and t in mathematical terms).

What is used in practice in GPS receivers is a frequency stable quartz-crystal oscillator. Because it is stable, the measurements can be made over some reasonable integration interval and the quartz-clock can be effectively synchronized to GPS time. Hence, it acts as if it were a GPS synchronized atomic clock. So, as long as you have four or more GPS satellites in view, your receiver can continuously calculate and know x, y, z, and t (which translates to longitude, latitude, altitude, and GPS time).

Chapter 20

Also, in practice, delays for antennas, ionospheric delays, tropospheric delays, multipath reflections, receiver processing software, the quality of the quartz-clock in the receiver, and many other considerations enter into the final precision and accuracy achieved for the time and position solution from a GPS receiver. A military receiver can now achieve sub-meter accuracies in real-time. A few meters of accuracy are now obtained for civilian receivers in real-time.

The applications for the use of GPS have grown exponentially over the last three decades, and now the civilian applications vastly outnumber the military even though it is a military system. In the beginning, the military applications were paramount and significantly outnumbered the civilian applications. GPS receivers were very expensive. The first GPS timing receivers used at USNO cost over $100,000. With the large number of civilian applications, that has changed dramatically.

In many of the civilian applications of GPS, it is used in what is called the differential mode. This allows one to cancel or reduce significantly a lot of the systematic delay errors. In this mode, centimeter accuracies are obtained. For example, this mode is used in surveying, road construction, farming, earthquake monitoring, etc. This requires two GPS receivers: one at a known and fixed position and which communicates its information to the other receiver, which in turn observes the difference between the two readings — bringing about the cancellation of most of the systematic errors that otherwise would be present.

Hewlett Packard asked me to write a general tutorial on timekeeping and GPS.[3] I asked Professor Neil Ashby, who did the relativity calculations for GPS, and Dr. Cliff Hodge, a space-clock expert, to help me with it. With permission from Symmetricom (now MicroSemi), who now owns the copyright, I have it on our web site, and it is referenced at the link in the footnote.[4] This booklet is called *The Science of Timekeeping*.

At the time I was writing this booklet, I was privileged to meet Dava Sobel, the author of the classic book, *Longitude*, which is the incredible story of John Harrison. Harrison solved the problem of determining longitude at sea by building a sea-worthy chronometer. As the banquet speaker at an annual Institute of Navigation (ION) meeting in Santa Monica, California,

---

3) Application Note 1289
4) http://ItsAboutTimeBook.com/the-science-of-keeping-time/

in 1997, Dava shared Harrison's solution of how to make a clock that keeps accurate time at sea in all conditions. She received a standing ovation from the best of navigation experts in the world.

I gave her a draft-copy of our 88 page booklet, and that began our friendship. As you can see from her web site,[5] she has published several interesting books. I value my autographed copies of *Longitude* and *Galileo's Daughter*. She honored me greatly when she wrote, "…keeper of Harrison's Legacy" in my copy of *Longitude*. She is the science writer for the Pluto-encounter space-program.

The Science of Timekeeping contains details on timekeeping and Einstein's relativity equations, which, in part, gives GPS its high accuracy, and without such considerations GPS would not work. The following is written to update this Application Note, to complement its contents, and with a minimum of equations. Perhaps, most importantly, the following contains some suggestions on how to improve GPS dramatically as well as international time and frequency comparisons.

The Application Note contains the history of timekeeping in much more detail – including the famous John Harrison chronometers, how atomic clocks work, the relativity equations for GPS, and how time is communicated and kept around the world. It was published the year I meet Dava in 1997 and two years after the copyright on her book – providing some slight corrections. I am pleased the Science of Timekeeping has had some usefulness. It certainly complements this book well – and hence the reason for its inclusion as an article on the book's web site.[6]

The following is written to update this Application Note, to complement its contents, and with a minimum of equations. Perhaps, most importantly, the following contains some suggestions on how to improve GPS dramatically as well as international time and frequency comparisons.

The Application Note contains the history of timekeeping in much more detail — including the famous John Harrison chronometer, how atomic clocks work, the relativity equations for GPS, and how time is communicated and kept around the world. It was published in 1997, and I am pleased that it has had some usefulness.

---

5) *www.davasobel.com*
6) *http://ItsAboutTimeBook.com/the-science-of-keeping-time/*

Chapter 20

There have been a lot of improvements in GPS since we wrote this article in 1997, and this is not the place to enumerate them all. Most can be found on the web. It is worth mentioning a couple of very significant improvements in terms of the title of this book — for timing accuracy. Prior to 1997 only the military had access to two frequency signals from GPS, allowing an estimate of the ionospheric delay. This delay can amount to several tens of nanoseconds across the globe, and the civil sector was obliged to use the eight parameters of the Klobuchar model broadcast by the satellites for an estimate of this ionospheric delay.

The physics of the ionospheric dynamics is extremely complex and Dr. John A. (Jack) Klobuchar, who is a good friend, has done a great job of dealing with these complexities; regardless, this model is only about 50% accurate. So if the delay were 50 ns then you may be off by about 25 ns in your estimate of GPS time. With the new improvements, the civil sector will be able to receive two GPS frequencies and estimate the ionospheric delay to an accuracy of about a nanosecond. Other delay effects may prevent the overall accuracy from being this good, however. Another major improvement for the civil sector GPS user community occurred in 1999 as the Selective/Availability (SA) signal was turned off. This action removed the degradation of that signal for the civil sector.

Over the years in my work with GPS I have studied and been involved with the research on ways that one may improve its performance and utility. When GPS was first deployed, short life-time failures of the rubidium gas-cell atomic clocks then being used nearly caused congress to discontinue funding the program. Had it not been realized that GPS could be used for nuclear-event detection during the cold-war, it might well have died. My group was involved with sorting out why these clocks were failing and we found some significant problems. Rockwell found the fundamental problem of buffer-gas leakage out of the rubidium gas-cell, and the problems were remedied — giving much greater reliability and improved performance for future clocks.

My group developed the GPS common-view technique in the 1980s, which became the primary means for communicating the times of atomic clocks around the globe to the International Bureau of Weights and Measures (BIPM) for the generation of International Atomic Time and UTC. I also suggested going from a hardware-reference clock to a software clock as we had done at NBS, making a significant improvement in GPS timing stability and this change was implemented. While at Hewlett Packard in the 1990s, I taught them how to overcome the effects of the SA degraded signal on the GPS civilian available signal (L1) and helped them develop a GPS smart-clock receiver for synchronizing cell sites using this technique. A large number of these cell-tower synchronization GPS receivers were successfully deployed around the globe — helping to give seamless hand-off as a cell phone moves through different cell sites.

Recently, as part of this book, I felt to document the best of ideas that I know of for additional improvements in GPS. In documenting these, I prepared a letter for those folks in charge of GPS for them to consider these opportunities. These improvements could amount to as much as a 100-fold increase in GPS accuracy. The letter contains technical details and the essential equations for the suggested improvements.[7] Specifically, in the letter, I listed four improvement areas:

- GPS satellite positions could be determined with accuracies of under a centimeter; this approach represents about a 100-fold improvement over the current Kalman-filter estimates;

- A different architecture could result in an annual savings of many millions of dollars for the GPS control segment while adding robustness, redundancy, and increased system security;

- With cooperation between the people responsible for controlling and making improvements in GPS and the international timing community, greatly improved international time and frequency comparisons could be brought about, which could also significantly enhance the timing at the GPS tracking stations;

- A more efficient method of dealing with and estimating the frequency drift for the current rubidium atomic clocks on board the SVs (GPS space vehicles) could be implemented, which could significantly help the long-term performance of GPS timing in case of a crisis.

---

7) A copy of the letter is included at the following link for those desiring to know more details. *http://ItsAboutTimeBook.com/letter-to-coordinators-of-gps-program-science/*

Chapter 20

Atomic Timekeeping and Earth (Solar) Time

As mentioned earlier, there are two parts to a clock: oscillator (frequency standard) and a counter (for accumulating the periods of the oscillator and generating time from that clock). Over the years, under congressional mandate, it is the job of NBS/ NIST to provide for the United States the primary frequency standard for determining the length of the second — the first part of the "two parts." In other words, when the folks at the Time and Frequency Division in Boulder, Colorado, determine the frequency of the Primary Frequency Standard at NIST — removing all biases as best as possible — per the definition of the second as given above, then that information is used to calibrate the frequencies for anyone needing a calibration and for an ensemble of secondary atomic clocks used for the second part of a clock (time keeping). For reliability and redundancy reasons this continuously calibrated ensemble is used to keep time for the NBS/ NIST laboratory and for the nation. These calibrations are also communicated to the BIPM for international timekeeping purposes.

In 1965, I was responsible for using this ensemble in a proper way for keeping time at NBS. With the help of colleagues and inspiration I felt from above, I wrote what is called the AT1 time-scale algorithm for generating official time for the USA; with many significant improvements made by my colleagues, this software clock is still being used today and generates official time, UTC(NIST).

The theory of this algorithm has some interesting features:

- The software clock has better performance than the best clock in the ensemble

- The best clock cannot take over and is optimized

- The worst clock is optimally used as well and enhances the output

- The algorithm is robust and rejects clocks with mal-performance

- It optimizes both the short-term and long-term stability performance of each contributing clock by giving proper weights to each clock in the ensemble

- The algorithm dynamically tracks the performance of each clock, which can change over time — especially when a cesium-beam atomic clock runs out of cesium and dies!

Since clocks are like people, all different, one can think of clock weighting for optimum performance in the following way. One moves forward in life the best by giving the greatest weight to people's words who have the greatest wisdom and integrity. This idea is why trusting the Lord and believing His words, who has perfect integrity and all wisdom, are fundamental to gaining light and truth. Similarly, clocks get weighted according to their stability — their predictability in timekeeping — to give optimum performance for the ensemble time (software clock).

Other countries best use their resources and do similar things.

For International Atomic Time (TAI), the BIPM collects data from about 400 atomic clocks from more than 69 timing institutes around the world. The following is taken from their web site. Notice that the above numbers are significantly larger than what they share. They have not updated this web site for some time, but the other information is very useful and informative:

> A practical scale of time for world-wide use has two essential elements: a realization of the unit of time and a continuous temporal reference. The reference used is International Atomic Time (TAI), a time scale calculated at the BIPM using data from some two hundred atomic clocks in over fifty national laboratories.
>
> The long-term stability of TAI is assured by a judicious way of weighting the participating clocks. The scale unit of TAI is kept as close as possible to the SI second by using data from those national laboratories which maintain the best primary caesium standards.
>
> TAI is a uniform and stable scale which does not, therefore, keep in step with the slightly irregular rotation of the Earth. For public and practical purposes it is necessary to have a scale that, in the long term, does. Such a scale is Coordinated Universal Time (UTC), which is identical with TAI except that from time to time

a leap second is added to ensure that, when averaged over a year, the Sun crosses the Greenwich meridian at noon UTC to within 0.9 s. The dates of application of the leap second are decided by the International Earth Rotation Service (IERS).

In the 1970s the calculation of atomic time replaced the calculation of time based on the irregular rotation of the Earth. Studies of the Earth's dynamics show that the velocity of the Earth's rotation is decreasing, and in consequence a rotational day is longer than a day of 86,400 atomic seconds.

When atomic time was adopted, some communities of users — in particular those using celestial navigation — requested that atomic time be synchronized with the rotation of the Earth. To compensate for the Earth's irregular velocity of rotation, the International Telecommunication Union (ITU) defined in 1972 a procedure for adding (or suppressing) a second as necessary, to ensure that the difference between the international time reference and rotational time remained less than 0.9 s. The resulting time scale is Coordinated Universal Time (UTC), the atomic time scale maintained at the BIPM with the contribution of 69 national institutes that operate about 400 atomic clocks.

The International Earth Rotation and Reference Systems Service (IERS) is responsible for monitoring the Earth's rotation and announces the dates of application of any leap seconds required, usually timed for the end of 30 June or 31 December. The IERS announced that a leap second be added to UTC ON 30 June 2012 at 23 h 59 m 59 s UTC. This was the 26th occasion since 1972 that a leap second adjustment has been made. The previous leap second was introduced on 31 December 2008.

The application of the leap second will be made at the same time (UTC hour) on all clocks maintaining local representations of UTC in the world.

These leap seconds are a real nuisance to many users. GPS time ignores them. Serious consideration is being given for ways of eliminating them.

But if one desires to have UTC track Earth spin-rate (solar) time, which historically the world is used to, then there are challenges. The Earth has lost about 58 seconds since the year 1900, which is the epoch the international community decided to use in switching from ephemeris time (ET) to atomic time, which in turn led to the definition for the second cited above.

There is an effort on the part of each nation (designated k) to keep their UTC(k) as well synchronized as they can to UTC as generated by the BIPM. This is a bit of a challenge since the times of UTC — UTC(k) are calculated and published on only a monthly basis. In other words, when this Circular-T is published by the BIPM, it tells you what time it "was" a month ago! We published a paper in 1994 showing how to circumvent this problem.[8]

The international community has taken a slightly different path which works well. Starting 1 July 2013 they publish an interim weekly report which they have been able to keep within 2 ns of the final official published times for UTC, and each nation can use this interim report to make it easier to synchronize their nation's UTC(k). NIST and USNO keep their UTC(k) time scales well synchronized within a few nanoseconds. Since USNO controls GPS time, one can simply use a GPS timing receiver to keep time very close to UTC after accounting for the leap-second difference which is broadcast as part of the GPS data word.[9]

UTC as generated by the BIPM is the official time for the world; all countries synchronize to it. The folks in each area of the Earth account for their own time-zone corrections and for each country's daylight-savings-time changes. As an interesting aside, if everyone were born and raised using TAI, we would not have to worry about leap-seconds, time zones, and daylight savings issues. We are victims of traditions — some of them good and some not so good. We are in an internet global community; why not everyone use the same clock? It would avoid the confusion of meeting times for people in different time zones, and would eliminate the need for time zones and congressionally mandated day- light savings time changes. It would take some getting used to for those accustomed to our past traditions, but like with the metric system, it would have some significant advantages. If you were born and raised with it, you would look back at the old system, and wonder why civilized people didn't make the change earlier. The military effectively used this type

8) *http://tf.boulder.nist.gov/general/pdf/217.pdf*
9) In the following link you can see an example of the Circular-T for December 2013: *ftp://ftp2.bipm.org/pub/tai/publication/cirt/cirt.312*

of timing starting with WW II — using GMT as their timing reference to avoid confusion across time zones. As another example of confusion with the current timekeeping traditions, whether 12:00 A.M. is noon or midnight has never been resolved, and the same for 12:00 P.M. A 24-hour clock solves that problem, and using TAI would do that. The international community agreed to change the name of GMT to UTC in 1972. People still use GMT because of tradition and not knowing of the 1972 name change.

Having two clock rates (the Earth and atomic) for official time, UTC will never work well. The second used in generating UTC is based on atomic clocks, while the accumulated time for UTC keeps step with Earth solar time. So UTC is a contaminated time-scale — having to chase the time of the Earth with a long-term variability having a spectral density going as f^5 — the most divergent of all the noise processes discussed below, and at a level that is a billion times worse in variability than the best atomic clocks in the world today.

An interesting case of using two clock rates occurred between 1967, when the second was redefined in terms the Cs_{133} atom, and 1972, when " leap seconds" were introduced for chasing the Earth's time variability with official time, UTC. During that five-year interval, the international community decided to chase the Earth with both time steps — of 0.1 second size — and frequency steps — of 5×10^{-9} size. The public paid no attention when we inserted 0.1 second steps, which change happened frequently in chasing Earth time. When frequency steps were inserted, this was as my friend and colleague, Judah Levine, in Boulder would say, "A four-ply pain!" The manufacturers of commercial atomic clocks had to add extra electronics — using a synthesizer to change the output frequency, since you can't change the natural resonance of an atom's frequency. This change added confusion and another point of possible failure in a clock where you want total reliability in keeping time with it. You want a time-keeping device never to fail.

The adding of "Leap-seconds" in 1972 allowed the rate of UTC to agree with the definition based on the cesium transition of 9,192,631,770 Hz. Moving from 0.1 second steps to one second steps caught the public's attention, and the TV news cameras would often roll up from Denver to our lab in Boulder to take pictures of the minute having 61 seconds. Lots of funny stories behind that one. One of the funniest is that not uncommonly people keeping time elsewhere would put it in the wrong way and be two seconds off UTC until they realized their mistake; very embarrassing for a timekeeper. The last leap-second was 30 June 2015 – making 26 leap-seconds in all since 1972. Between 1900 and 1972 the Earth had slowed down about 32 seconds. The year 1900 is the reference year for the Earth's rate of rotation for the last two definitions of the second. So currently, the Earth is about 26 + 32 = 58 seconds slow with respect to the year 1900.

As mentioned earlier, the accuracy of frequency standards has advanced dramatically since atomic clocks were first thought of and were invented in the 1940s. They have improved by an astounding factor of about a billion. There is a very important equation in relationship to how accurate and stable the rate of an atomic clock can be. It is written as follows, where we can think of "Fractional Frequency Instability" (FFI) as the error in a clock's rate averaged over some time interval, τ.

$$\text{Factional Frequency Instability} = \sigma_y(\tau) = \frac{\Delta \nu}{\nu\, \text{Signal/Noise}}$$

where $\Delta \nu$ is the quantum transition line-width at a transition frequency ν. The Signal to Noise ratio is improved as the square-root of $N\tau$, where N is the number of atoms or molecules available for the measurement per second and τ is the averaging time for the measurement in seconds. One can clearly see from this equation that the higher the frequency the smaller is the instability and the greater is the potential accuracy. The stability also improves for longer averaging times and as more atoms or molecules are available for the measurement.

For example, suppose a clock has an error of plus or minus a second in a day (86,400 s), then its FFI would be 1 second per 86,400 seconds = 1.1574 x 10^{-5}. One can see why FFI is useful. It is simply designated as $\sigma_y(\tau)$ in the time and frequency literature, where y is the fractional frequency deviation of the clock averaged over some data set, and τ is the averaging time over which the clock is measured. This measure is the square root of what is commonly called the Allan variance and is called the Allan deviation.

Using this measure one observes that atomic clocks have advanced dramatically since they came into being — moving from 10^{-9} to 10^{-18} — the factor of a billion improvement mentioned. The Earth is stable to about 10^{-9} — about where the first ammonia-based atomic clock started in the late 1940s.

In practice, the FFI (square-root of the Allan variance or Allan deviation) can be measured using the time or frequency deviations of a clock for different averaging times, τ. The following equation is symbolic of how to do that: $\sigma_y(\tau) = \frac{1}{\sqrt{2}} <(\Delta y)^2>^{1/2}$. This equation may look formidable, but it is simple to implement in software; don't let all the Greek symbols scare you. Δy is simply the frequency change from one measurement to the next, each of which has been averaged over an interval τ and divided by the carrier frequency, ν_o, to normalize it to a simple dimensionless number. This measure of instability is not dependent on the frequency of the clock. So you can compare the performance of your wristwatch with the best atomic clock. The brackets < > simply mean to average all the possible squared values of the frequency changes for that value of averaging time, τ; then take the square root and divide by root 2. 'Voila,' you have an instability number for that clock. Then you can change τ in the software and ascertain how the instability changes with τ; that will tell you the kind of noise you have. So in a $\sigma_y(\tau)$ diagram, you can ascertain both the kind of noise as well as the level. We sometimes call it a super-fast Fourier transform. The electrical engineers appreciate "super-fast" since they have developed fast Fourier transforms because Fourier transforms are tedious and time consuming to perform on long data sets.

Dividing by the square-root of 2 makes $\sigma_y(\tau)$ equal to the standard deviation for classical white noise, which is random and uncorrelated. The typical symbol for the standard deviation is σ (the Greek letter sigma). For many of the noise models we use in characterizing precision clocks the standard deviation does not converge to a useful value as the data length increases — making it not a useful measure. In contrast, $\sigma_y(\tau)$ is convergent, easy to compute, and is well behaved mathematically. It also turns out to be the optimum instability estimator for the classical noise in atomic clocks as deduced from the fundamentals of physics and statistics. It is also mathematically convergent for all of the interesting, noise-process models used for clocks as well as being easy to compute.

How to Characterize the Performance of Your Own Watch or Clock

One can show from fundamental physics that every clock or watch almost always disagrees with every other if measured precisely enough, because of systematic and random errors in each. So we have as many time readings from these physical devices as there are clocks or watches in the world. The only time that is correct is the one society agrees upon; it is an artifact of man's creation. Hence, characterizing the performance of a timing device is very important.

Commercially, you can buy what is called an "atomic clock." It is not an atomic clock, but is a quartz-crystal-oscillator clock with a radio receiver that listens to the broadcast signal from radio-station WWVB in Ft. Collins, Colorado. This station's clock is synchronized with the official- USA-time standard in Boulder, Colorado, at the NIST laboratory using the GPS common-view technique. The net is that your "atomic clock" is synchronized with a continually accuracy of about a tenth of a second. Otherwise, your clock will gain or lose time in an unbounded way. The following allows you to know the level of unboundedness of your timing device. Similar systems are available elsewhere in the world. In fact, the train stations in Europe used this idea for synchronizing train-station time long before it was developed in the USA.

To show you how easy it is to obtain a computation of the Allan-deviation, you may conduct the following experiment to characterize the performance of your wristwatch or wall clock. The United States official time standard can be accessed by telephone: 1(303) 499-7111. This phone call gives you access to the voice announcement on the NIST- WWV, official-time, radio transmissions. It announces the time (Universal Time Coordinated, UTC) on the minute. Don't worry about the hour, because your time zone will most likely be different than that of the Greenwich Meridian. Decide on some convenient minute each day that you can call the above telephone number. If you happen to miss the time you decided on a few minutes, don't worry about that; just compensate for the number of minutes off, which will normalize your reading back to the time you decided upon. Note the reading of your wristwatch or clock and write it down for 31 days; 31 days is enough to gain some statistical significance to your calculations. If you can note the reading to better than a second, write it down; if you can estimate your reading to a tenth of a second that would be good. At the end of your analysis, you will know your measurement noise as well as the performance of your time piece. At the end of 31 days you will have 31 values of what we will call $x(n)$ — the time difference between your time piece and official time, where n ranges from 1 to 31.

Chapter 20

Now calculate the value of $x(31) - 2\,x(16) + x(1)$; this is called the maximum second difference from your data, $\Delta^2 x(\tau=15$ days$)$. Now divide $\Delta^2 x(\tau=15$ days$)$ by the square-root of $2=1.414$ and by the number of seconds in 15 days, which means that you will divide by 1,832,820.777 seconds; you can round this divisor off to 1.83×10^6 if you wish. If you have measured well, this will be an approximate estimate of what in the literature is called the "Allan deviation" for an averaging time of 15 days for your time piece with one degree of freedom.

Now let's calculate the rate error of your time piece in seconds per day. The rate error is given approximately by $R = (x(31) - x(1))/30$, if you have measured well. If you have a "quartz-crystal" in your time piece, these numbers should be very small. The optimum estimate of R is obtained by a linear regression fit to your 31 values, but we will let the regression experts calculate that rate.

Now let's estimate your measurement noise, MN. To do this we need to subtract from your data the systematic errors in your time piece. This will give you 31 new values of just measurement noise. These systematics are the synchronization error and the rate error, R. To calculate these new 31 values — let's call them $x'(n)$ — simply calculate the following for each of the 31 values: $x'(n) = x(n) - x(1) - (n-1)R$. The standard deviation of these 31 numbers is your measurement noise, MN.

The standard deviation can be calculated as follows. Add up the squares of each of these 31 numbers and divide that by 30. The square-root of that result is the standard deviation. If you have measured well and your time piece is reasonably well behaved, this number should be less than a second. If your data are written down with 1/10th of a second precision, then the smallest your measurement noise can be is 29 milliseconds (0.029 seconds).

If you divide your measurement noise, MN, by 15 days x 86400 seconds/ per day = 1,296,000 seconds, and that result is smaller than your Allan deviation value for 15 days average for your time peace, you have measured how good your time piece is, because your measurement noise is less than the clock noise at a 15 days averaging time. However, if that result is larger than your Allan deviation value for 15 days average for your time peace, you have not measured how good your time piece is, and you are limited by your measurement noise. If that is the case, the modified Allan variance approach deals with this problem in an optimum way. You can get an approximate estimate of what is called in the literature the modified Allan deviation by taking the difference between the average of the first 15 values of x'(n) and the last 15 values; then divide this difference by 1,832,820.777 or by 1.83 x 10^6, which ever you chose, but be consistent.

I have cheated here a little in your behalf for simplicity and to satisfy the definition of the modified Allan deviation, but don't tell anyone! You would need 46 values instead of 31 to get a true estimate, but this short cut works. If this value is larger than your measurement noise divided by the averaging time to the 3/2s power: $MN/1.83 \times 10^9$, then you have measured the performance of your time piece. Otherwise, you are still limited by measurement noise. The way around that is to take more data, because the measurement noise is most likely white-noise PM and the modified Allan deviation for the measurement noise reduces as the averaging time to the 3/2s power. Since the performance of your time piece will likely not improve with averaging time, if you take enough data, you will see the performance of your time piece — guaranteed.

Once you have determined the performance of your time piece for an averaging time of 15 days or for how many days it takes, then you can estimate its unbounded nature by multiplying that performance value by whatever interval is of interest to you. For example, suppose you want to know the time dispersion of your time piece for a year after you have synchronized it with the Boulder-official-time standard, then you need to calculate the accumulated errors arising from the rate offset and from the random component.

The rate offset you calculated above as R is in units of seconds per day. Suppose you measured R to be 0.1 s/day. Since there are 365.25 days in a sidereal year, 365.25 x 0.1 = 36.5 seconds is the estimated-accumulated error of your time piece over a year. This value can be, of course, plus or minus depending on whether your time piece is running fast or slow.

The random component is given by the Allan deviation. Suppose you calculated a value for the Allan deviation for an averaging time of 15 days of 1×10^{-7}. This will be probably the clock's flicker FM component of noise, and this noise doesn't change with averaging time. So the random-component of

Chapter 20

error for your clock for a year is 1 x 10^{-7} x 3.14 x 10^7 seconds = 3.4 seconds, since there are approximately π x 10^7 seconds in a year. This error is random, and hence it can be plus or minus. Our example is typical. The systematic errors are usually much larger than the random errors.

If you want to see a big rate error in your time piece, put it in the refrigerator. Pull it out once a day to make the measurements, and then put it back. You will see a much larger rate error, R, because the quartz-crystal oscillators in most clocks are made to run near the correct rate at room temperature. Or if it is a wristwatch, they are made to run at the correct rate at body temperatures. Quartz-crystal oscillators have a significant temperature coefficient.

Because the systematic errors are almost always much larger than the random errors, the official-time-scale algorithm, which generates official time for the US, subtracts the systematic components from the data by calibration and we are only left with the random components. When the ensemble of atomic clocks at NIST in Boulder are optimally combined according to their performance, we have official time, UTC(NIST), for the United States. Easy?

How to Define a Time Deviation Measure (TDEV)

In the late 1980s the telecom industry came to us to develop a time-dispersion measure for their network-synchronization challenges. The above Allan-deviation equation gives an estimate of the frequency stability or instability of a clock. Out of our work for the telecom community came a time-stability measure called TVAR for time variance. It is symbolized as $\sigma_x^2(\tau)$ where x denotes the time deviations of a clock and τ is the time interval over which the time deviations are averaged for the variance calculation. There was a need, and this measure was readily accepted by the telecommunication community both nationally and internationally. They gave me the "Time Lord Award" in 2011 at an international telecom conference in Edinburgh, Scotland, in appreciation for and in recognition of this work. Since they paid for my wife and me to go there, I was happy to accept as we had a great time after the conference touring Scotland, which is the place of origin of the Allan part of my ancestry. In the literature, TVAR is defined as $\sigma_x^2(\tau)$ = τ^2 MVAR/3, where MVAR is the modified Allan variance, and TDEV is the square-root of TVAR. Dividing by 3 makes TVAR equal to the classical variance if the noise is classical white-noise phase-modulation, PM, as in our above experiment. TVAR is mathematically convergent and well behaved for all the different models of clock and network noise that we typically encounter in practice.

As an interesting characterization of the above-mentioned six different kinds of noise processes, TVAR allows you to determine the level of noise and the kind for each of these six. The kind of noise is indicated by how TVAR changes with averaging time, τ^n, where n = -1, 0, 1, 2, 3, and 4 for the six different kinds of noise. They each have a name. White-noise PM (phase or time modulation) corresponds to n = -1. Flicker-noise PM corresponds to n = 0. White-noise FM, that we mentioned before, corresponds to n = 1; this is the same as random-walk PM — like keeping track of how many steps forward or backward you move as the clock rate ends up fast or slow with each step you take (early or late for a clock). Flicker-noise FM corresponds to n = 2. Random-walk FM corresponds to n = 3, and Flicker-run-noise FM corresponds to n = 4.

The first five noise processes are useful in characterizing clocks — whether atomic, quartz, or mechanical like an escapement mechanism or a pendulum — and are useful in characterizing telecommunication networks and clocks as well. The random-time variations of the Earth's spin are characterized with n = 2 and n = 4 — the latter of which is the most dispersive of the six noise models. The spectral density of this noise, as mentioned before, is proportional to f^5, where f is the Fourier frequency variations in the noise. Why the Earth behaves this way would take another book to tell the story. While this story is interesting, it is not well understood. The level of the Earth's variations for the n = 4 noise type is such that the Earth could end up spinning the other way with the sun coming up in the west for the ages of the Earth given by traditional science. For some time in the past, there is some evidence that there were 460 days in the year from some coral-ring data analysis, but the uncertainty on that estimate is large.

The n = -1 white-noise PM is an interesting concept and observance. It implies that your knowledge of time (phase) improves with longer averaging times, τ. This kind of noise is observed for the short-term stability of masers, which were the first atomic clocks. Maser stands for microwave amplification by stimulated emission of radiation. Charles Townes, et. al., were the inventors, and later Townes invented the laser (Light Amplification by Stimulated Emission of Radiation).

The reason for this kind of noise in masers is that atoms or molecules (ammonia and hydrogen were the first to be used) in an excited quantum state are directed into a resonant cavity tuned to the frequency of its quantum clock transition. Natural noise in the cavity stimulates the atom or molecule to emit its photon at its natural frequency. When it does, it increases the energy of the electromagnetic field in the cavity and stimulates another atom or molecule to emit its photon in phase (synchronized) with the phase or time of the previous ones. More and more join the crowd as atoms or molecules enter the cavity and the field builds up in synchronicity — all have the same time! So as you average the phase or time readings, you arrive closer to what the ensemble of atoms or molecules are trying to tell you is their true phase or time. If this white-noise PM process were the dominant noise process for all averaging times, then your time or phase estimate would continue to improve indefinitely. Nature gives us many much more divergent processes that become more dominant for longer averaging times. Thus this unusual phenomenon does not persist.

The prediction time-error of a clock is very important in GPS. As a GPS satellite goes over a tracking station, the tracking station uploads its best estimates — coming from Colorado Springs, Colorado, — of the correct time and position for that space vehicle's (SV) atomic clock. Going forward, that SV has to predict its time and position as it continuously transmits that information to the GPS users on the Earth until it gets another upload some hours later. An estimate of the prediction time-error can be obtained by multiplying the prediction time τ times the frequency instability for that clock, $\tau\,\sigma_y(\tau)$. The clocks have gotten better and better for their prediction time error and the end is not in sight. These errors are typically of the order of a nanosecond, but there is the potential for them to be much smaller, as mentioned above.

Great progress has been made with Cs_{133} atomic clocks by dramatically decreasing $\Delta\nu$. This was accomplished by going from cesium-beam frequency standards to cesium-fountain frequency standards. When one interrogates the cesium resonance to determine the length of the second, the observation time of the atoms determines $\Delta\nu$ through the Heisenberg uncertainty principle. For a beam one observes it for less than a millisecond, while for a fountain the observation time increases to nearly a second — giving beautiful narrow line widths, $\Delta\nu$, and from the above equation one can see the benefits. The definition of the "second" today is determined by cesium-fountain, frequency standards because of this line-width advantage. There are several such standards at the various Standards Laboratories around the world, and their values are communicated to the BIPM for determining the length of the second for official time for the world.

Going from the cesium microwave frequency (~10^{10} Hz), where the definition of the second is determined, to optical frequencies (10^{15} Hz) would give an advantage of about 100,000 from the above equation. Optical frequency standards have been around for decades, but there are two fundamental problems in turning them into clocks: 1) having a stable reference oscillator at those frequencies and 2) being able to continuously and reliably count at 10^{15} cycles per second (Hz).

Jim Bergquist and colleagues at NIST originally solved the first problem — developing extremely stable laser signals that were needed as local oscillators for interrogating these optical clocks. The second problem was solved by John L. Hall, Theodor W. Hänsch, and Roy J. Glauber with their pioneering work on laser-based, precision spectroscopy, and the optical frequency comb technique. They were awarded the Nobel Prize in physics in 2005 for this very significant accomplishment.

Now, optical clocks are measured and counted and compared with microwave clocks with great finesse and are now part of every major timing laboratory. Accuracies and stabilities of parts in 10^{18} have been achieved, which is astounding. These are the most accurate measurements ever made by man or woman! To try to grasp this degree of accuracy, 10^{-18} is like 1-second error over 32 billion years, or like measuring the distance to the sun (93 million miles) to 3/1000 the thickness of my human hair. I sacrificed one of mine to make the measurement comparison, and I don't have many to spare! Comparing such clocks remotely located is currently one of the biggest metrology challenges. The following footnote contains a link to the letter to GPS headquarters where you can read a proposed solution to this problem.[10]

---

10) *http://ItsAboutTimeBook.com/letter-to-coordinators-of-gps-program-science/*

Chapter 20

The following list gives the frequencies of some of the more prominent atomic clocks:

H maser	1.4 GHz	most stable microwave clock
Rb gas-cell	6.8 GHz	stable and small; used in GPS
Cs (defines the second)	9.2 GHz	most accurate microwave clock
Al+ trapped ion	1,121,000 GHz	most accurate clock in the world (UV)
Yb optical lattice clock	519,032 GHz	most stable clock in the world (optical)

Again, a GHz is 10^9 (a billion) cycles per second. Last year, the most accurate (7×10^{-18}) and the most stable ($\sigma_y (\tau=7 \text{ hours}) = 2 \times 10^{-18}$) clocks in the world have been developed by my incredible colleagues in Boulder.[11] This year, I went to Boulder to lecture at their annual Time and Frequency Seminar and learned that significant improvements have already been made. I learned that JILA — using strontium — set some new records for precision and stability using some very clever optical-lattice-clock technology.[12] JILA is the Joint Institute for Laboratory Astrophysics — working both with NIST and the University of Colorado. We see the exciting trend continue of about a factor of ten improvement every ten years.

A sense of the importance of atomic-clock physics is had when noticing that a significant number of the Nobel Prize in Physics winners are associated with this discipline. It has been my privilege to know several of them and to work with some of them. My friend and colleague, Dave Wineland, just won the Nobel Prize in 2012. Out of the work of his group in the Time and Frequency Division in Boulder, CO, has come some of the best atomic clocks in the world along with quantum computing and quantum information processing. There seems to be no limit to the progress and the benefits. The biggest limit may be our imaginations!

As we have shared, GPS employs the four dimensions of space and time using Einstein's theories of relativity. Hence, a GPS receiver can give us our

---

11) *http://tf.boulder.nist.gov/general/pdf/2438.pdf http://tf.boulder.nist.gov/general/pdf/2688.pdf*
12) *http://www.nist.gov/pml/div689/20140122_strontium.cfm*

longitude, latitude, altitude and time. Because precise time is part of the solution, the receiver can also give us our velocity vector — how fast we are going and which direction. In the next chapter we will share experimental evidence for a fifth dimension, which we call the Eternity-Domain. The implications and applications of this fifth dimension in our lives are both exciting and life changing.

Chapter 21

It's About the Unified Field Theory Leading to the Fifth-Dimension Beyond Man's Time and Our Mortal Sphere of Existence, Where God and Angels Dwell

Albert Einstein's theories are built around four dimensions. We can think of them as longitude, latitude, altitude, and time. While we were in West Africa serving a mission in Cote d'Ivoire, John Wiley and Sons asked if I would write a book about GPS and Precise Timing. I suggested that we wait until I got back to the States to evaluate the need of such an effort. Upon returning, I could see that there was a need, so I started on the book. Brad Parkinson, the father of GPS, agreed to write the forward; we had worked on GPS together for years. I was excited about writing this book, and I was making good progress when we felt impressed to look at the Unified Field Theory (UFT) work that Einstein didn't finish before he passed over. Following those impressions has led us into some remarkable truths including scientific data giving evidence of a fifth dimension and a new direction in physics called diallel-field lines.

Those impressions came as a direct result of a dear friend and neighbor at the time — Ranae Lee — and her promptings and inspiration. You will find this a unique and highly unusual story. One day she was visiting us in the living room of our solar-heated home. I remember this conversation well, even though it occurred over 14 years ago. She asked how the book was coming, and then asked, "Has Albert been helping you?" I told her I was not aware of that. She said, "He wants to help."

If you were to know this sweet lady and her spiritual gifts as we do, she not only has a close tie to the Savior, but has been given the opportunity to communicate with some of the folks who have passed over, in particular those who have a mission to help us here. She went on to explain that Albert — as she called Einstein — had learned some important physics after passing over that were not known here on Earth, and that it was the Lord's will that they be made known. If we wanted to help in that process, it was my choice, but there would be a lot of opposition and misunderstanding.

My wife and I discussed it and had our own prayer time with the Lord. We felt impressed that since the Lord wanted us to do it, then we would trust in Him and give it our best shot. I can tell you we have learned a lot. Indeed the opposition has been there, but it has been a great experience and is not over yet. The Lord has much more to tell us. This chapter in this book is written to encourage others to listen to the Spirit of the Lord, and if you feel so prompted to contribute in this area or any other, please follow your heart. The Lord wants to use our talents in the cause of truth.

As explained in the beginning of the book, the scientific method becomes much richer when one adds the sixth sense, which includes listening to the Spirit and to the heart. That has been indispensable in this new approach to the UFT. When I learned of these diallel-field lines, I asked the Lord for experiments to know of their existence and characteristics that I might convince friends and colleagues, and a flood of impressions and experiments came in my mind.

Diallel-Field Lines

To date, we have conducted seven experiments all validating both the existence of diallel-field lines and some of their significant characteristics. These are described in part in articles and papers that you may find on our web site.[1] I was pleased to have a good friend and colleague, Jeff Lorbeck of Qualcomm, help with one of these papers. Jeff is a brilliant, Christian compatriot.

---

1) *http://www.allanstime.com/UFT_private/*

Chapter 21

Here, as simply as I can, I will explain and give the characteristics of diallel-field lines from what we have learned, and then I will share some highlights from some of the experiments that have been done showing their existence and some characteristics. These diallel-field lines are extremely important and our understanding is minute. They permeate the universe and basically connect everything to everything. Independently, Lynne McTaggart dedicates her whole book, *The Field; the Quest for the Secret Force of the Universe*, to documenting the experiments of others, which provide evidence that directly relate to these diallel-field lines — though she does not name them nor does she know of our work, as far as I know.

Targ and Katra in their book *Miracles of Mind* provide consistent and repeatable data, using the empirical- scientific method and drawing upon their own and several others' experiences. They couple into this Eternity-Domain in their research and experiments. Their book is the first to report some of the recently declassified research findings of over more than twenty years and coming from the files of the CIA, the Defense Intelligence Agency, the Army, the Navy, and NASA.

Their data base includes documented cases of remote viewing, mind-to-mind communication, mind-to- body, mind-to-cell, object-to- mind, mind-to-object, and most importantly mind-to-God and God-to- mind. They often refer to God as the "universal higher consciousness." Dr. Jane Katra shares some very personal experiences in how she came to know and feel the excitement from the " light" emanating from this Universal Consciousness, which is her name for God, and which she believes is available to all.

Peter Tompkins and Christopher Bird document some remarkable examples of mind-to-plant and plant-to- mind interactions. For example, a plant attached to a lie detector reacted violently when an expert thought about burning one of its leaves. A philodendron, activated by a thought impulse from an electronics technician, started a car two-and-a-half miles away. From their introduction we read,

> Evidence now supports the vision of the poet and the philosopher that plants are living, breathing, communicating creatures, endowed with personality and the attributes of soul. It is only we, in our blindness, who have insisted on considering them

automata. Most extraordinary, it now appears that plants may be ready, willing, and able to cooperate with humanity in the Herculean job of turning this planet back into a garden...

As we will see, both Tompkins and Bird's data as well as Targ and Katra's data are consistent with this new UFT and model of how all these communications and interactions work together. As Dr. Dossey writes in the Foreword to their book, "Russell Targ and Jane Katra provide compelling reasons why the limited view of consciousness must give way to an expanded one in which the mind knows no bounds... This means that in some sense our consciousness is infinite — soul-like and boundless, limitless and immortal." As Dossey further states, "In the modern era, the belief in the infinite nature of consciousness has been considered illusion at worst or a matter of faith at best. But as Targ and Katra show, it is now a matter of data. And in our culture, in which science is so highly valued, this four-letter word makes an immense difference... We thus stand at a landmark period in human history."

In this same spirit, the January 2013 talks at the Electric Universe Conference in Albuquerque, New Mexico, were titled *The Tipping Point*. During the conference both Dr. Dean Radin and Dr. Rupert Sheldrake shared some additional and exciting results showing the mind is not constrained to the brain. Radin not only shared solid data showing that ESP works — using a Faraday cage to try to block any electromagnetic signal — but that it also had no-time-delay using a global set of folks to participate and find that result. As I listened to his talk, I predicted the no-time-delay result from our UFT work.

Einstein said, "There are only two ways to live your life: One is as though nothing is a miracle. The other is as if everything is. I believe in the latter." And from David Bohm, "The great strength of science is that it is rooted in actual experience. The great weakness of contemporary science is that it admits only certain types of experience as legitimate." Clearly, we need to think "outside the box" as we investigate this new UFT and these diallel-field lines.

In summary, these diallel-field lines emanate from any entity — affording its connection and communication channel with any other entity. Our oldest son, Sterling, is a better writer than I, and he has on our web site a nice tutorial on diallel-field lines.[1] I extract some of his language here and add some of my own to aid your understanding.

On a more practical or approachable level, diallel-field lines are like a naturally built-in wireless internet that has always existed and which will always exist, whether or not it is acknowledged. But our awareness of its existence can facilitate our use of its full capacities, which are only partially tapped in the case of most mortals — just as the brain's capacity is far, far beyond what is typically put to use. By knowing it exists and pondering its function, we can augment its capacity and increase our ability to communicate and learn truth. Using Internet vernacular, we can boost our connection speed, increasing the amount of information that can flow, and thus enhance our efficiency and effectiveness.

In near death experiences, people describe the mode of conversation between individuals in that sphere, they say that while verbal speech is possible, that most communication takes place telepathically and is instantaneous and perfectly accurate. This is via diallel-field lines. Likewise, when the scriptures describe the process of receiving revelation from the divine, the lines of that communication as well are through these diallel-field lines. These lines are the conduit of light and truth (in a temporal and spiritual sense).

Most practically, when we pray or meditate, our communications with the divine and God's angels, who are always with us, are conveyed in this way.

By the same token, just as the internet is not discriminating in what sort of information is passed through its lines, likewise, the demonic elements also convey their messages of deception in this way.

But also, like the internet, 'you get what you want' is true of diallel-field lines. We consciously and subconsciously choose our connections. Also, much like the brain, or muscles, where we 'use it or lose it,' these diallel-field-line connections can be fine-tuned.

Nevertheless, one does not have to be aware of their existence to put them to use. Whether a person believes in these lines or not, they use them as naturally as the synapses that convey electrical impulses through their central nervous system.

---

1) *http://www.allanstime.com/UnifiedFieldTheory/Learning/Spiritual_Internet.htm*

The inventor who comes up with a 'new' idea, has most likely received that idea from unseen angels who have conveyed it to them by inspiration via the diallel-field lines of communication. The mother who suddenly receives a feeling of warning about one of her children — or even on a more seemingly mundane matter such as suddenly 'remembering' that a child needs to be picked up from a friend's house — may have received that impulse from a departed loved one who communicated that feeling of warning or the 'remembrance' to her thoughts via the diallel-field lines. The pet, who comes to the rescue of his owner in distress, knowing what to do to save the person's life, most likely received that outside knowledge through the diallel-field lines.

So how might we augment our diallel-field-line connection to the inexhaustible well of infinite divine wisdom? Fundamentally, our connection is dependent on our relationship with God, which is typically a function of our love of God and our love and service rendered to God's children. One does not need to know about the existence of diallel-field lines in order to commune with the divine any more than a person needs to know how the body works physiologically in order to climb a hill. But for the professional mountain climber or racer, a knowledge of the body's physiology is going to come into play in being able to optimize his or her performance. In like manner, the intent seeker of things spiritual will be benefited by an awareness of the existence of diallel-field lines and a study of their function in the process of that communication.

First, on a simple level, by realizing these connections exist we can increase our ' faith' in things that previously might have been more difficult for us to believe, such as communicating with God and angels, telepathy with other individuals, with animals and plants, and the value of our intuition. The person who knows about the Internet and its possibilities is more likely to want to use it. These diallel-field lines are like the phone lines through which the Internet is connected, but in this case the lines are 'invisible,' — made of spirit matter — and definitely more reliable. There is no such thing as 'down time' for diallel-field lines.

Second, in realizing that these lines exist, we can begin to conceptualize ways to 'speed up the connection' so to speak. Here is where the various religious modalities come into play. Each presents their model of how to 'connect' to the divine. Some methods are more effective than others. What works for one person given where they are at in life, may not necessarily work for another person. God's communication with His children is optimized for each individual's situation.

Because some information could be injurious if dispensed prematurely, God places certain protections (like 'passwords' or filters on the Internet) so that we will only receive that for which we are prepared.

Some of the universal principles to increase receptivity that can be found in just about any religious approach include such factors as: (1) a desire to know God; (2) willingness to serve as an instrument for God; (3) concentration — freeing oneself from distraction; (4) humility — preparing one's heart and mind to receive new truth and not reject it because it might seem at first to contradict something in our belief system; (5) a conducive environment, such as uplifting music, fragrances, loving, supportive family and friends; (6) belief in the ability to make such a connection with the divine; (7) genuine and virtuous living — aligning one's life with the truths that he/she has learned, also called ' obedience' to the teachings one has received from deity; (8) purity of heart — through the sanctifying power of the divine; (9) forgiving those who have wronged oneself; (10) repentance for wrongs one has committed; (11) asking a question with the belief that an answer will come, seeking that one might find what they are looking for; and (12) deep and heartfelt gratitude for the many blessings from the divine. There are undoubtedly other factors as well.

The more these factors are in place, the more these diallel-field-line connections will be facilitated. In the Internet parlance, having these factors in place is like increasing one's connection speed and bandwidth.

One may imagine in the ultimate society, where the pure love of God and of one another abounds, that communication would often be at the highest level using these diallel-field-lines telepathically. Picture in your mind a world connected by this spiritual internet as it is now connected by a corporeal internet. Everyone's talents could be readily accessible for the good of the whole. Everyone would be growing in light and truth in optimum diallel-field-line coupled communion with each other and with the source of all light and truth. A society of pure-in- heart people of one mind and one heart — would fulfill the prayer of our Savior:

> [I pray] for them also which shall believe on me through their word; That they all may be one; as thou, Father, art in me, and I in thee, that they also may be one in us: that the world may believe that thou hast sent me. And the glory which thou gavest me I have given them; that they may be one, even as we are one: I in them, and thou in me, that they may be made perfect in one; and that the world may know that thou hast sent me, and hast loved them, as thou hast loved me.[2]

Interestingly, seven is an important number in the Lord's arithmetic; it stands for completion or perfection. As is known, there are seven basic electron shells that provide the fundamental quantum states for the electrons making up all of the known elements. These, nominally, have spherical symmetry. The diallel-field lines, nominally, have cylindrical symmetry and seven communication channels and carry all four of the known force fields.

In completeness of this diallel structure, there are seven spectral bands in these communication channels. They are increasing in energy and may be divided up in generic terms as follows: 1) the communication band (TV, radio, satellite, etc.); 2) the molecular band (coming from quantum transitions outside the nucleus); most of the visible light from the sun is in this band; 3) the nuclear band (coming from quantum transitions inside the nucleus) — a nuclear blast is a manifestation of this band; 4) the creation and/or annihilation band (where $E = mc^2$ as matter and light convert with the emission and/or absorption of cosmic rays); 5) this is nominally the gravitational-field band; 6) this is the band used for mind-to- mind, mind-to-object, mind-to and from-God, ESP, etc.; 7) this is the band used by our great-loving Creator to bring about the grand harmony we see in the heavens and the Earth. Bands five, six and seven are not limited by space and time and the velocity of light, for example, consistent with that suggested in the book of Targ and Katra and by the mind-to- mind experiments conducted by Dean Radin mentioned above.

As discussed in our papers, even though these diallel-field lines are made of spirit matter, there is actually a lot of evidence for their existence. Some simple examples are the aurora-borealis for the Earth, the aura of an individual, which some can see and as may be scientifically observed using

---

2) John 17:20-23

Chapter 21

Kirlian photography. The data in Targ and Katra's book and in Thomphin's and Bird's book also provide direct evidence of the existence of a heretofore unknown communication channel that can be well explained with these diallel-field lines.

Because of its relevance to this new theory, I have made a study of several peoples' near-death experiences. These experiences are of particular interest because they move into this fifth dimension and the Eternity-Domain during the NDE. I have become very good friends with some of the authors of books who have had NDEs, and they have greatly appreciated some scientific explanation for their experiences.

This fifth dimension opens up some extremely important understandings about how the universe works — validating not only the existence of God, but also our relationship to God. This new theory has come as a result of opening ourselves up to a very different paradigm — providing a harmony between all of the force fields in nature as well as with this new, fifth dimension beyond relativity. This new UFT has provided consistencies to heretofore unexplained inconsistencies as well as providing potential solutions to previously-unsolvable problems.

I have an article on our web site called *We are in Touch with Eternity*, which gives additional explanations of this new UFT. Some of the information above is taken from it. That article evolved from an invited talk that I gave in Colorado Springs at the 50th anniversary meeting of the Institute of Navigation. Eight experts were invited to give talks on various aspects of navigation. We were asked to look at the last 50 years progress and then look into the future 50 years. It was a fun and exciting conference. I had more compliments on that talk than any other scientific talk I have given, and that number is not small. I was later invited by the Wisdom Society in San Diego to write an article from a creationist's point of view. I modified the Colorado Springs talk to make it into this article. It is one I have referenced a lot in this book and explains the spectral structure of these field lines.[3] The following are some experimental examples of diallel-field lines. When a nuclear blast goes off, it makes the familiar mushroom cloud. Why doesn't the energy go out in all directions following paths of least resistance? It is because the energy goes up the Earth's diallel-field lines in a column following the cylindrical

---

3) : *http://www.allanstime.com/Spiritual/In_Touch_with_Eternity.htm*

symmetry of these field lines. The mushrooming effect is like pushing a wet toothpick against a wall; it flares out in all directions as the energy from the blast pushes against the atmosphere. If one observes a hydrogen bomb with energy massively greater than an atomic (nuclear fission) bomb, the fusion energy is so much greater that observing beyond the mushroom flaring out against the atmosphere one sees the column continue up into the stratosphere — making the whole thing look like a giant spindle as that energy continues to follow the diallel-field lines of the Earth. One can observe this data at a facility in Las Vegas where records have been kept of all nuclear testing activities. I visited there when I was first gathering data for the evidence of these diallel-field lines.

Another fascinating diallel-field line effect one observes in these blasts is that there is a massive amount of energy going down the Earth's diallel-field lines, and some of it reflects off of the solid core of the Earth. When that energy-density comes back up it builds up the energy-density already present, and one can observe on some occasions a quantum relaxation in the mushroom cloud column. It is initially just white, but with this buildup of energy-density and quantum relaxation, it will collapse inward some and turn black. This was predicted by the new UFT and observed in several of the blasts. The delay for the collapsing is just the time for the energy-density to reflect off the Earth's core and come back up to the column seen above the blast site.

The Foucault pendulum gives direct evidence that the Earth is spinning. The equation for the motion of the pendulum has three terms: one for the local gravitational field, one for the latitude of its location, and one for the spin-rate of the Earth. When carefully measured this motion is modified during a lunar and solar eclipse. Because the three bodies (sun, moon, and Earth) have their diallel-field lines in parallel, the pendulum speeds up and slows down very slightly during this time. Classical theory cannot explain this phenomenon. This effect was first observed back in the 1960s in India and has been observed by others since. With currently available technology, refined eclipse experiments could be done today that could shed some interesting light on this subject, and it is a beautiful example of diallel-field-line coupling as these three heavenly bodies line up.

The theory of diallel-field lines predicts that one can stand an egg on end and it will be barely stable. On the web you can find people saying

Chapter 21

that it can only be done during an equinox and not knowing why. I did several experiments over the course of three months to show — as the theory predicts — that it can be done anytime. The diallel-field lines of the Earth set up a slight resonance between the shell at the top and bottom of the egg and the principle of least action causes a slight potential well making it slightly stable in its vertical position. I stood up several eggs over our dish washer at home in Fountain Green so there would be vibrations and disturbances present, and I had them standing for weeks.[4] It is a very sensitive experiment. The eggs have to be raw and even some raw eggs won't stand. I have never been able to stand one up on a piano. I believe this is a disruption of the energy-density of the Earth's diallel-field lines by the piano wires being under very high tension. More experiments are needed here.

We felt to look at the energy-density of the planets — considering both their masses and their magnetic fields — to see if that correlated with the sun-spot activity. I obtained the last 100 years of sun-spot data for our study. I am a member of the International Astronomical Union (IAU) and have some very bright and knowledgeable friends there, but I could not find anyone who knew the summer and winter solstice times for Saturn. I had a picture with data taken by Voyager 2; so I used my telescope to get a second view of its rings and orientation. From that data I could determine its solstices.

I was able to get up to about a 90% correlation coefficient over the 100 years of data between the UFT energy-density prediction and the actual data. The Allan variance turns out to be a very good low-frequency spectrum analyzer; so I used it and could see in the actual 100 years of sunspot data the periods of the planets. We were able to see resonance frequency nulls in the Allan-variance plot that were five times the period of Mars, ten times the period of the Earth, the period of Jupiter, the period of Saturn, and of course boldly the 11 year period of the sunspot activity. Most people think of the sun-spot cycle being an 11 year period, but 11 years is actually its approximate half period. The north and south magnetic poles of the sun reverse every 11 years. The reason we could not see the single period of Mars and of the Earth was that those planetary frequency components were swamped by the 11 year (half-cycle) signal coming from the sun-spots. This data is reported in our paper on our web site.[5]

Since this new UFT depends on energy-density, we set up the following

---

4) These experiments are described in quite a bit of detail on our web site *www.allanstime.com*
5) *http://www.allanstime.com/UFT_private/final2.htm*

experiment. The classical period of a pendulum depends on the local gravitational acceleration, g, which is nominally 32.2 feet per second per second (9.8 m/s^2). This is nominally the rate that the two Twin Towers fell and Building 7 on 11 September 2001, which can happen neither from airplane impact and explosion nor from fires within. Controlled demolition can cause such a free-fall situation. Any physicist studying the data can know this. The big question, of course, is who set up the controlled demolition. If you go to 9/11 studies on the web, you can find a lot of extremely important information on this critically-relevant topic for our day and time.[6] As mentioned before, see also the write up that our oldest son, Sterling, placed on his web site.[7]

The following figure shows the observed periods of the planets showing up in the Allan deviation plot, which is the square-root of the Allan variance. The period of Uranus could not be observed because the 100 years of data were insufficient to give a good confidence on the 84.3 year period of this planet. One can show that when the averaging times, τ, is equal to the period of a planet or an integer multiple thereof, then one obtains a null in the value of $\sigma_y(\tau)$ – leaving only the background noise. This is exactly what we observe in this log-log plot of the sunspot data.

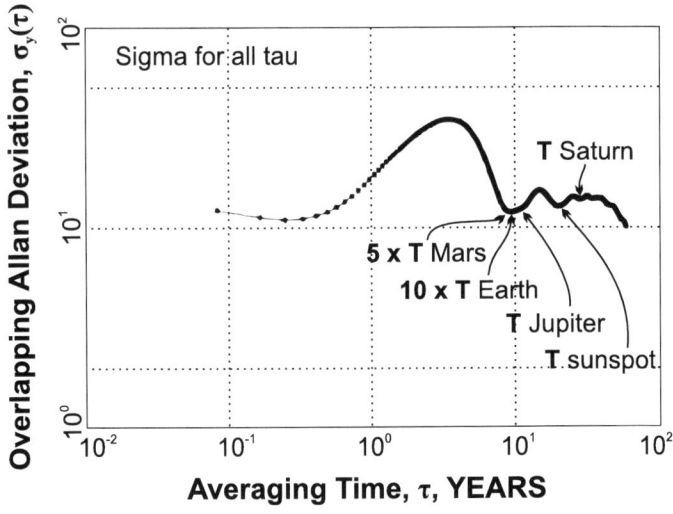

---

6) http://www.journalof911studies.com/
7) http://pesn.com/2013/09/11/9602370_Magnetometer-and-Hurricane-Correlations_with_9-11-2001

Chapter 21

If I introduce a mass underneath a pendulum, classical theory predicts that the pendulum would speed up a few parts in a billion — depending upon the size of the mass. The diallel-field line theory predicts that the energy-density of the introduced object under the pendulum bends the diallel-field lines of the Earth slightly inward — like a lens bends light — modifying the local vertical so that it slows the pendulum down a few parts in a million — depending on the energy-density introduced. You will note that this is the opposite direction of that predicted by classical theory, so we felt this would be a good experiment to do.

I drove to Provo and went to a very nice clock shop and asked if I could borrow for a week two of their best pendulum clocks. When I explained our experiment, they kindly accommodated me. I know how to measure clocks. Timing metrology has been my field of expertise for 56 years.

I built a low table to set the pendulum clocks and separated them a goodly amount to minimize sympathetic coupling (clock on the wall phenomena). As far back as Huygens, it was shown that if you have two similar clocks in proximity, they can tend to communicate with each other through the mounting structure or electrically, or in other ways as well, and that communication tends to make them synchronize.[8] I used three kinds of energy-density: electric field, magnetic field, and chemical. I bought and charged up six one-farad capacitors for the electric field. This pack of six (not a six-pack) was about the size of a car battery. I obtained from an electrician friend a transformer, which was also about the size of a car battery, for our magnetic-field energy-density. Then I used a car battery for the chemical energy-density source.

An uncompensated pendulum will be temperature sensitive as well, since its period depends on its length as well as on g, and most materials expand with increased temperature. So, I had to also worry about the laboratory temperature in our home, since the effect being measured is small, and we live in a solar-heated home with no thermostat. To deal with this problem, I did 24 hour experiments to average out the diurnal effect that may be caused by temperature. I still have the video tapes of all these experiments.

I measured the periods of the two pendulums without any introduced energy-density under them to assure they were adequately stable, that their

---

8) Christiaan Huygens, *Mututal Space-Time Synchronization Between Clocks*

periods were different and they were not coupled. I introduced one of the energy-density objects under one pendulum, and then observed the periods of both pendulums. Then I moved that object under the other pendulum to assure that the measured effect reversed. I did that for each of the three energy-density sources.

Happily, the diallel-field line theory was confirmed. Each of the pendulum periods slowed down according to the energy-density under them, and the slowing was independent of the kind of energy-density, as the theory predicted. Furthermore, as is explained in detail in the paper, this experiment gave direct evidence of a fifth dimension, which we call the Eternity-Domain. This finding has enormous implications in this new UFT. You will also find in the paper the general UFT attraction equation, and we show how the classical theory of gravity — with the attractive force being proportional to the product of the masses and varying inversely as the square of the distance between them with the proportionality constant being G (the universal gravitational constant) derived from our general UFT equation as a special case.

Another spectacular case of diallel-field line emissions and bending can be observed with the emissions above heavily charged clouds. We often see the lightening between the cloud and the ground, but as the energy-density builds up in a cloud — from time-to-time that may also be released upwards in the form of what are called blue- jets and red- sprites.

They are, indeed, spectacular in appearance and pilots have observed them for decades, but were reluctant to report these sightings for fear it would upset passengers and make people not want to fly. The University of Alaska provided some of the first documentation and later the University of Oklahoma has done some excellent work in this regard. You can now find some nice visuals of them on Youtube if you Google "blue- jets and red- sprites."

We use the U of Alaska documentation in our paper showing how with a significant energy density build up in the cloud one sees the blue-jet emission first going up rapidly. With enough energy-density they follow the Earth's diallel-field lines, which are bent inward toward a center local vertical. So it looks like the shape of the bottom section of a teepee. The energy-density of the cloud acts like a lens to bend these diallel-field lines inward in this teepee shape. This is like the bending that occurred as we affected the gravity in the previously described two pendulum experiment.

Then with the focused energy-density of these blue- jets increasing as they go upward, they go through a beautiful quantum transition — releasing energy horizontally outward in the ultraviolet region of the spectrum going out like a rapidly expanding donut at the speed of light. The new UFT theory predicts these, and to my knowledge they have not been observed yet. These horizontal emissions will also be polarized and one would need a UV photo-detector to observe them.

After this quantum release of energy, the blue- jets turn into what are called red- sprites. With this release of energy the base of the red- sprites is relaxed horizontally outward from the center line of the emissions and diallel-field line lens bending effect. As shown in the University of Alaska's documentation, the red- sprites retain the teepee like shape as these red emissions continue upward at horrendous velocities. They reach altitudes of around 80 km (262 thousand feet) in a fraction of a second; this is about eight times higher than commercial airplanes typically fly.

Because of the Earth's diallel-field lines are bent inward toward the center local vertical of the cloud, these red- sprites converge like the shape of teepee — similar to the blue- jets, as their energy-density builds up they also go through a magnificent quantum transition. This time their emissions are red and they have been observed. They are called elves and again are in the shape of a rapidly expanding donut.

Continuing upward, after this quantum release of energy, there will be a third teepee shape with emissions in the infrared. Because of their frequency, to my knowledge they have not been observed to date. With modern infrared photography technology, they could be observed.

Observation of Diallel-Field Line Quantum Transitions at BYU

One of the most important experiments I have ever done in my life was performed after the papers on our web site were written. It is similar in concept to the above description of quantum transitions due to the focused energy-density of the Earth's diallel-field lines. The big difference is that the following was documented in a laboratory environment.

Since I am an alumnus of Brigham Young University, I got permission to use their laser physics lab for some of the experiments, and they kindly let me use a couple of grad students to help in setting up the experiments and in writing some of the software we needed. It was fun working with those very bright and helpful students.

The theory for these diallel-field lines predicted that we should get ultraviolet quantum emissions if we set up the experiment a certain way. BYU did not have a UV spectrometer, so I called my good friend and colleague, the late Len Cutler of Hewlett Packard, at their Palo Alto research facility. Back in the 1980s, Dr. Cutler and his team designed and developed the best commercial cesium-beam atomic clock ever made. That design proved to be so successful that most of the timing centers in the world have acquired one or more of these 5071-A cesium-beam atomic clocks. Of the some 400 clocks making up international time today, TAI and UTC, probably more than 80% of those clocks are Len's and his team's design.

I worked with Len when I consulted for HP in the 1990s, and over the years we have attended many symposia and meetings together — have gone jogging together — gone out to eat together and visited on many occasions about common interests. Len visited our lab in Boulder many times as well. I have to say that Len was an incredible scientist and one of my best friends. We were all saddened with his passing. He was most accommodating of my request when I explained what I was doing. They just happened to have a UV spectrometer, which he kindly and carefully packaged up and shipped to me.

I asked one of the professors if they had any UV lenses. UV rays will not pass through glass; these lenses need to be made of quartz and are very expensive. He happily exclaimed that the year before BYU also happened to have acquired a box of quartz lenses, which had never been opened and which turned out to be exactly what we needed for this experiment as well.

BYU had also given me access to their science shop in the basement of the Eyring Science Center — named fafter Henry Eyring — and so I could build whatever I needed. It was a great research environment. They also gave me 24/7 access to this laboratory.

Chapter 21

We had already done some experiments in the BYU lab when we started on this diallel-field line UV quantum emissions experiment. This experiment was the most complicated to set up. I had never used a UV spectrometer before, and one of the professors helped me get it aligned and set up properly — the same professor who told me of the box of quartz lenses. He knew of my work in Boulder and was fully supportive of our experiment.

The goal for this experiment was two-fold: 1) to see the broadband UV emissions from these diallel-field lines and 2) to see a specific UV quantum transition line — a specific frequency in the UV band. Our understanding of the theory was not sufficient to know the frequency, but we knew it was in the UV range. Neither of these results had ever been observed, and we were anxious to test the new UFT theory.

By the way, this experiment has never been written up before; I just have my lab notes. This is the first publication of this most significant experiment that I have ever done. It is also very unusual. Because of its unusual nature, I chose not to fight the peer review process, which has recently been shown to be seriously flawed in terms of making sure published papers don't have significant errors.[9] The experimental setup in simple terms was as follows. Our goal was to focus the diallel-field lines coming up out of the Earth — at right angles to the surface of the Earth — through a special energy-density medium, which I will describe later — onto a vertical mirror — which we will call the focal point of the experiment. At right angles to the mirror's surface we focused a helium-neon laser beam onto the face of the mirror as well. The laser beam — traveling horizontally — was also at right angles to the quantum-mechanically-exited, diallel-field lines coming up out of the Earth.

The theory predicted UV emissions coming out horizontally and at right angles to both the diallel-field lines of the earth and to the laser beam. We placed another mirror to reflect the UV emissions from one side of the focal point of the experimental setup to reflect them into the UV spectrometer, which we had placed on the opposite side of the focal point and where we could measure the UV intensity and frequency.

To focus and excite the Earth's diallel-field lines, I used several different

---

9) Paul Rimasch, *Scientific Validation of Friction Ridge Analysis: A Case for Empiricism*, Journal of Forensic Identification, 2014 64

forms of energy-density. These were stacked up vertically underneath the mirror (the focal point of the experiment). Included were a car battery providing chemical energy-density, six one-farad capacitors for electric field energy-density, a transformer for magnetic field energy-density, and a quartz-crystal oscillator at 16.7 MHz — loaned to me by Jeff Lorbeck of Qualcomm.

I worked all one week on this experiment and was happy to see the UV broadband emissions, but I could never reach the second and most important goal of our experimental setup — to see a specific UV quantum transition line. I swept the spectrometer across the UV spectrum many, many times changing different parameters in the experiment, but never could see a unique line. It was the week of the BYU homecoming, and I stayed up all of that Friday night without success. At about 4:00 a.m., I left to go home and get some sleep, and to see if we could figure out why the theory was not working as predicted. Fountain Green is about an hour south of Provo, Utah — giving me time to think and pray.

After getting some sleep, I felt impressed to invite John and Ranae Lee up to the house to visit with my wife, Edna, and me about why I did not succeed in this second and most important part of the experiment. After counseling together I expressed my feelings that if the four of us went up to Provo to the lab using some of the ideas we had talked about, we would succeed. Edna said she had other things she had planned to do as did John. Ranae's youngest child, Shinoahè, was about ½ year old at the time. This was the fall of 2000.

Ranae loves science, and her spiritual gifts as it relates to science are pretty amazing — even though she has had no formal training in the sciences. We decided that she would bundle up Shinoahè and the three of us would head for the lab. On the way, she asked if I would educate her on terminology: what was a nanometer (one billionth of a meter), an Angstrom (1/10th of a nanometer), etc. As we were going up the interstate, she said, she had been told that the laser light focused on the mirror was too intense and needed to be defused. She said that she could see in the box of quartz lenses a lens that would give us the diffusion we needed. It turned out to be a double concave lens.

This was an excellent day to work in the lab. It was totally quiet since everyone was at homecoming. When we arrived, we had prayer, and then

Chapter 21

Ranae walked over to the box of lenses — reached inside and pulled out the lens she had seen in her " mind's eye." I made a mount for it, and we placed it in the laser beam giving us a beautiful red diffused laser signal on the face of the mirror.

I gave Ranae my lab notebook and had her write down what was happening. I began the frequency sweeping process with the UV spectrometer. At 421 nanometers the intensity meter jumped way up giving us the most beautiful signal I have ever seen in my life, as we observed this specific diallel-line UV quantum emission predicted by the theory. I nearly jumped through the ceiling for joy.

We also saw something unexpected. As we achieved this quantum resonance condition, we both saw these red beds — as Ranae called them — dancing on the ruby-red laser light line between the laser and the mirror — the focal point of the experiment. The red laser light was apparently shifted in frequency as it interacted with the diallel-field lines coming up from the Earth. The diallel-field lines had been focused on the mirror through the different kinds of energy-density mentioned above and the interaction at the mirror caused a harmonic shift in the laser light — so that as it was reflected back on itself it caused a beat frequency during the time we were exciting a quantum transition.

There were about a dozen of them along the laser light line of about two meters. From that we can estimate the size of the shift. Since $\nu = c / \lambda$, where c is the velocity of light and λ is the wave length, the shift would be about 2 GHz. We did not document this as well as we should have. Definitely, more work needs to be done here.

In many ways this BYU experiment was very much like that which occurs with the blue jets and red sprites observed above high energy-density thunder clouds. Emissions are given off at right angles to the focused diallel-field lines coming up from the Earth.

The Chair of the Physics Department told me the next week that they needed the laser table I was using; so I had to conclude my experiments at BYU. He was not aware of the significance of the results we had obtained. They being so unusual, I didn't try to explain them to him. They had been so accommodating for our experiments, I was most anxious to not be in their way at all.

That part of the experimentation ended at a good time, and we moved on to an experiment that is very important where remote clock synchronization is needed. It is called synchronistic modulation detection (SMD). The SMD is a spin-off prediction of the new UFT, and theoretically goes as follows: given two remote clocks (No. 1 and No. 2) whose times are not known, we send a timing signal from 1 to 2 and measure that time difference. That measured time difference will be the time of clock 1 minus the time of clock 2 minus the delay for clock 1's signal to get to clock 2 ($t_1 - t_2 - \tau_{12}$). At nearly the same time, we can send a timing signal from clock 2 to clock 1 and measure the reverse scenario, and we obtain clock 2 minus clock 1 minus the delay for clock 2's signal to get to clock1 ($t_2 - t_1 - \tau_{21}$). We of course can measure the round trip delay, which will be the sum of the two delays ($\tau_{12} + \tau_{21}$). We cannot assume reciprocity of these paths, but under the assumption of reciprocity, we can get an estimate of the time difference between the two clocks, which is adequate to get our measurement sequence started.

So you see, we have two measurements and three unknowns: the time difference between the two clocks and the forward and return path delays. The round-trip measurement is not an independent measurement. Classically, this problem cannot be solved, but using the SMD under the new UFT, it is solvable.

In November 2000 I invited Charles Wheatley III, who was at that time Vice President of Science and Technology at Qualcomm, to our home in Fountain Green to discuss this exciting direction. Chuck is a good friend and colleague, whom I had known for many years while working on GPS issues.

Chuck, while working for Rockwell-Collins had both designed the frequency synthesizer and the random sequence to drive the synthesizer, which was used during most of the 1990s to degrade the GPS L1 civil signal so that the civilians could not have the same accuracy as the military signal. This process was called selective availability (SA).

A military receiver had access to the secure key to remove this SA degradation. The SA degradation caused up to 100 nanoseconds of timing error on a civil GPS

Chapter 21

receiver. The Lord gave me the idea for how to filter away this degradation for a fixed position GPS timing receiver. Hewlett Packard hired me as a consultant to use this idea for using GPS to synchronize cell-phone towers. We demonstrated that we could track the timing from a military receiver over the course of a month to within a root-mean-square error of 1.5 nanoseconds.

It was fun to have as one of the first customers for this HP product to be Qualcomm. A group of them came to HP Santa Clara to see what we had done. Who would be in the audience but Chuck Wheatley — as I shared the exciting results we had obtained — with a big smile on his face.

But one of the funniest presentations of these results was when I shared them at an international ION- GPS conference in Salt Lake City in 1993. There were several US Air Force officers — wearing their sharp-looking blue suits — in the audience, who were part of the GPS control segment and knew the GPS well. I could see their look of, "You can't do that!" on their faces. But when the Lord whispers, you can do that. These results are published in the proceedings of that conference.

Another friend of mine, who was at that same November 2000 meeting in Fountain Green, brought with him a young and very bright BYU EE grad student by the name of Gus German. For me this was a God-send. Gus and I ended up working together for 10 years on several interesting projects — starting off with SMD, and Gus moved to Fountain Green — making our working relationship most convenient.

We were able to obtain proof of concept for SMD in our little laboratory in our Fountain Green home, but nowhere else. The message seemed to be that it was not yet to be shared.

Gus helped me when we obtained the excellent results for ORNL using the data from six Loran-C stations in the Great Plains and contiguous areas of the United States. I did not have high hopes of ever coming close to the nanoseconds of error timing accuracies available from GPS having worked with Loran-C before and seeing 1 to 2 microsecond peak-to-peak errors in the Loran-C propagation delays. This was a thousand times worse than we needed.

As mentioned before, we obtained about 5 nanoseconds of stability and accuracy from real data and published these results.

Since the Lord prompted me to write this book a year ago, I have really enjoyed doing the research for it. But this chapter has been really fun to revisit and to review the several exciting experiences of discovery. I hope you feel some of the excitement that I have enjoyed. I believe, learning the truths of God give us a taste of the feelings in Heaven.

Having provided evidence for the existence of a fifth dimension — the Eternity-Domain — the domain where God and His angels do their work — the work of helping us fulfill our life's missions, now we are ready to learn how to best couple to that domain. The next chapter is exactly on that topic.

Chapter 22

It's About How Transcendental Time Couples Both to Mortality and to the Eternity-Domain, Opening the Door to Eternal Happiness, and How We Know God is Speaking

Now that the existence of a fifth dimension has been established scientifically — the Eternity-Domain — this chapter is about how we can couple to that domain. We will see the great importance of transcendental time and its very important purpose in our lives. As mentioned before, this is where God and His angels do their great work of orchestrating the plan of happiness for our past, present, and future. All is before the Lord; He knows the end from the beginning. Understanding and living God's plan brings peace, love, joy, and happiness into our lives. Without it life can be very empty. Fortunately, in His infinite goodness, everyone is born with a conscience (the light of Christ)[1] to know good from evil, which if followed leads us to His plan and the eternal happiness we all intrinsically seek. It is our choice each step of the way.

Some may like to think of this as emotional time, but transcendental time is a better descriptor. Transcendental time is how our body, mind, and spirit couple to the Eternity-Domain, which utilizes the full spectrum of our emotions. One can see that emotional time fits in this sense. I will use transcendental time since it better fits — in a broader sense — this important dimension of time.

---

1) John 1:9

The best way to think about transcendental time is to contrast it with man's time. Currently, man's time — described in Chapter 20 — is based on traditional science and is limited to the five senses and the four dimensions given to us by Einstein's theories here in this mortal sphere. GPS, for example, works in the three dimensions of space and the fourth dimension of man's time. Emotional or transcendental time, which is not restricted by man's time, allows us to move beyond space and time into the fifth dimension and couple to the Eternity-Domain.

It is interesting — as mentioned earlier in the book — that science has recently discovered brain cells in the heart and in the gut; yet the scriptures have contained those truths all along. Transcendental time is our opportunity to couple all of our emotions (body, mind, and spirit) — in harmony with eternal truth — toward a oneness with God. It is our great opportunity that a loving God has given us to commune with the infinite and to not be constrained by space and man's time; we call it prayer. It is no surprise that prayer is the most often repeated commandment in the scriptures. He wants us to come back to Him and to receive a fullness of joy.

When the disciples asked Jesus how to pray, one of the profound phrases he shared was to request: "Thy kingdom come; Thy will be done on Earth as it is in heaven." The kingdom of Heaven and the Lord's angels are perfectly organized to assist us in our mortal journey. The kingdom on Earth — made up of us mortals — is very imperfect, but as we follow this prayer admonition and listen to our conscience we can help His kingdom to prosper and ourselves on the path to perfection. The Savior summarized perfectly: "But seek ye first the kingdom of God, and His righteousness; and all these things shall be added unto you."[2]

We can never perfect ourselves. Christ is perfect and as we come unto Him, He will perfect us and the kingdom on Earth, which we can be part of as we come unto Him. This exciting process will happen as He comes to reign as King of kings and Lord of lords during His great and final millennial epoch — truly bringing peace on Earth. Then Jerusalem will be what is meant by its very name — the city of peace, and the word of the Lord will go forth from Jerusalem and the law will go forth from Zion as Isaiah prophesied.[3] We can look forward with great hope and anticipation.

---

2) Matt. 6:33
3) Isa. 2:3

Chapter 22

Opposition is a necessary part of the eternal plan of happiness. Without it we cannot grow or know good from evil. As part of that opposition, Satan and his angels were cast down to the Earth when Lucifer tried to destroy free choice — the agency of mankind.[4] Hence, we see the critical need as Abraham Lincoln so aptly said, "to listen to our better angels." We read that a "third part" fell from heaven and are here to tempt us and to try us and to attempt to make us miserable like unto them. Fortunately, they have no power over us unless we give it to them.

The Lord placed a veil over our memories so that during our mortal journey we would live by faith, which gives us the opportunity to learn how to listen to the " still small voice." This provides for our maximum growth in His perfect plan of happiness in the midst of opposition.

The Lord has provided His righteous angelic hosts in the spirit world to help us here in our mortal journey if we are willing to listen to the " still small voice." Learning to listen to the Spirit of Truth is the key to our eternal happiness. Raising up our children to learn how to listen is one of the greatest blessings we can give them as parents. One of the best books I know of outside the scriptures is the one mentioned earlier in the book by Boss and Boss, *Arming Your Children with the Gospel: Creating Opportunities for Spiritual Experiences*.

If we do not raise our children up in " light and truth," then we increase their vulnerability to the influence of Satan and the angels who fell with him to deceive them and us and to lead us down forbidden paths of sin and spiritual degeneration. For this and many other reasons, the family is the most important unit in time and in eternity. One can see why Satan is waging such a war to destroy the traditional family. That is where the truths of God are best taught and internalized by loving-caring parents.

So, how do we know it is God's voice that is speaking to our minds, our hearts, or our gut? First know that the fruits of the Spirit are love, joy, and peace…[5] God's truths are accompanied by a spirit of enlightenment — enlarging our soul — increasing our understanding — and perfect love

---

4) Revelations Chapter 12 and Isaiah Chapter 14
5) Gal. 5:22

"casteth out all fear."[6] Fear, worry, hate, envy, anxiety, etc. are tools of the devil to move us away from the Spirit and from learning the truths of God.

When we are coupled body, mind, and spirit (all of our emotions) to God through prayer, meditation, pondering, and in the quiet of our souls, we are coupled to the Eternity-Domain and are not limited by space and time. In this mode, we can commune with the infinite and potentially have access to all truth. At this portal we can ask, seek, and knock about anything, and a loving Father has promised us that we will receive, find, and have opened to us those requests made with a sincere heart.[7]

I know He is there to hear our prayer. Because He operates from the Eternity-Domain, He is not limited by space and time, and He knows all of our hearts and each of our names, and can and will answer all of our prayers in the way that is in our eternal best interest. Sometimes the answers are immediate; sometimes we need great patience, but know that every sincere prayer will be answered in the way and in the time that is best for us because He loves us with an infinite love. He sees the end from the beginning. He knows our needs and how to best help us have them properly met. All that He does is motivated by love and with the intent to help us find our way back home and into a fullness of joy.

A dear friend of ours, Dr. Joyce Brown, just came out with her new updated edition of her NDE book entitled *God's Heavenly Answers; Near—Death Experience Revealed*. It will be a great resource as well. She has received hundreds of letters from people thanking her for saving their lives as she helps them see their great worth and to know that God is there for them. I know His love is there for them; if they seek, they will find… answers and guidance in their lives regardless of the challenges.

Let us examine the hard question, "Why do people not get answers to prayer?" Here I will list what comes to my mind, my heart, and my gut:

- Pride: I don't need the Lord; I can do it; Look at me and what I have accomplished!

- Sin: So we don't feel like praying. He won't answer me; I am a sinner.

---

6) 1 John 4:18
7) Matt. 7:7-8; James 1:5-6

- Ingratitude: Not being grateful for all things — both the good and bad in our lives.

- Idolatry: We put other things before God — worshipping the works of men's hands.

- Resentments: Holding grudges and feeling offended; inability to forgive all.

- Too Busy: Caught up with life; no time for the Lord to pray and ponder His goodness.

- Lack of Faith: Seeing the evil, pain, and suffering, we doubt that God is there.

- Not Living The Law of Love: We ignore the first and great commandment to "Love the Lord thy God with all thy heart, and with all thy soul, and with all thy mind, and with all thy strength. And the second is like, namely this, Thou shalt love thy neighbor as thyself."[8]

The opposite of pride is humility, which for me is to be teachable — even by a child — open to and accepting truth from any source and in gratitude. An attitude of "I know and you don't" closes the door on humility and learning the truths of God. If we approach the throne of God with the feeling that He is all knowing and if we are humble and open in body, mind, and spirit, then His truths can flow and answers to prayer will come. Because of His love for us, He will force nothing on us — respecting our agency (right to choose good or evil). We need to have a deep desire to know Him and His marvelously-inspiring truths.

We are all sinners, but knowing that Christ's atonement covers them all is most warming and consoling to the soul. If only those without sin could pray, no one would be praying. "Forgive us our trespasses as we forgive those who trespass against us" is one of the main lines in the Lord's Prayer. One of the sweetest experiences of my life is when He has filled my soul with joy as I have asked forgiveness of my sins. This is one of the most meaningful answers to prayer, and it is there for us all. The infinite extent of the cleansing power of the atonement is beyond our comprehension. He invites all sinners to come unto Him.

---

8) Mark 12:29

The sin of child trafficking is one of the most horrible sins of our day. Those doing it have lost the Spirit and are headed for hell, but even they can be redeemed. Crying repentance seems hard, but in fact it is extended by a loving God as the only way to fullness of joy, and His hand is extended out to ALL.

An attitude of gratitude goes a long way, and is essential if we desire the heavens be opened to us. We often have a hard time being grateful for the bad things in our lives. God will not destroy our agency, and it is man's inhumanity to man that makes countless thousands weep. Great lessons are learned in the hardships of life. In every challenging experience, we can ask the question in humble prayer, "Lord, what lessons do you want me to learn from this experience?" He will answer and the lessons learned will be profound. By retiring each night with a prayer of gratitude for all that the day has brought into our lives and being most grateful for the atonement, then we are placing our spirits in a prime place to be open to the windows of heaven during the night so that the Lord can send His angels to instruct us in our missions in mortality; especially if we pray for that. It is good to get up and write down these impressions of the night and will give you a feeling of refreshment and enlightenment by His Holy Spirit.

Idolatry is a major problem in our society. Many worship the mind of man and the works of men's hands! A host of things can get between us and God, and this is idolatry. A lot of people look for the answers to their problems in science, in technology, in modern medicine, and fail to look to God. People are caught up with the things of this world: entertainment, sports, recreation, video games, preoccupation with texting, computer games, excess time on Facebook, etc. Not that any of these is bad in and of themselves, but they are way out of balance in our society and idolatrous. Then there are all the addictions that plague our society: pornography, drugs, alcohol, etc. which often keep a large segment of our population from praying and turning to God. All sins are in a very real sense idolatrous, and we see why the first two of the Ten Commandments are so written to help us overcome this most serious sin and all sins.

Resentment is a killer. Here I interject a story of profound significance. [9]

---

9) A more complete version of it can be found on our web site:
http://www.allanstime.com/Spiritual/BookReports/ Babbel_resentment.htm

Chapter 22

The following is a personal experience of Frederick Babbel found in the book he co-authored with his wife, June Babbel, *To Him That Believeth: Claiming Heaven's Blessings.*

> I was living in the Portland, OR, area; an urgent call came from a valued friend who had been bedfast for nearly a year. When I reached his home,. . . His wife was ironing his burial clothes.
>
> He told me that his family doctor had informed him. . . that his life was nearing its end and that it was now only a matter of a day or two, or perhaps a week, until he would expire. Then he remarked: "The strange thing is that the doctors still do not know what is wrong with me. Tonight I just felt that I wanted to visit with you before I prepare to meet my Maker."
>
> While continuing our conversation, I received a divine insight as to his problem. "Brother," I responded, "I believe I know what is wrong."
>
> He seemed startled, but genuinely interested, as he urged, "Please tell me."
>
> "You've had a number of very serious hurts and disappointments in your life," I said, "that have filled you with bitter resentment. Many of these have never been resolved."
>
> He seemed incredulous and somewhat apprehensive as he inquired, "What do you know about them?"
>
> "Not a thing," I replied, "unless you tell me about them. I only perceive that you have been deeply hurt many times. Yet you have never forgiven those who were responsible for these offenses."
>
> "Well, I must admit," he countered, "I have had some bitter experiences. But since I accepted the gospel, I believe that I could forgive those who were responsible if they asked for my forgiveness."
>
> "But that is not how the principle of forgiveness works," I said,

"When any serious grievance takes place, the Lord requires us to forgive the guilty party the moment the infraction occurs, if possible. . . .these negative feelings will finally consume and destroy [you]. This is what has been troubling you and what, even now, has brought you to the point of death."

My friend began to sob unashamedly. In the process he removed his nightshirt and showed me his bare back. I had never seen a back like this, not even in the concentration camps of Europe. Across his back were large crisscrossed scars that were scabbed over with ugly flesh. Some of them were so deep a person could almost lay his arm in them.

Then he related to me how his father used to come home occasionally in a mean, drunken stupor. His temper would flare up and he would take a heavy whip from the wall and flog whoever was within reach. This whip, a "cat o' nine tails," was leather with several strands. At the end of each strand was fastened a large brass ball with metal spikes that could tear the hide off an animal.

On one occasion my friend was the victim. Just 14 years old at the time, he was whipped into unconsciousness. How long he lay on the floor he did not know, but as he regained consciousness, he found himself lying in a pool of his own blood, with his back fairly torn to shreds. He managed somehow to crawl from his house, and he vowed he would never return.

"You've never forgiven your father for that flogging, have you?" I next inquired.

"No, I guess not," was his reply. "But if Dad were to ask for forgiveness, I think I could forgive him now."

"I'm concerned," I said, "that you still don't understand the underlying principle. You have had the divine responsibility of forgiving your father from the moment that you regained consciousness, so that the healing power of forgiveness could come into your own life and relieve you of this terrible burden. In doing so, you might also have started the process of healing for your father as well. But because you have continued to nurture this resentment, it has festered and grown until it is literally consuming you… and hastening you're untimely death."

When we finished talking, I invited him to sit upon a chair so I could give him a special blessing and outline for him what must be done. In the blessing he was instructed to get out of bed the following morning, take his wife, and drive to his father's home in North Dakota, with the assurance that his father was still alive…

About four are five weeks later my friend stopped his car in our driveway. As he stepped out of his car, I greeted him with, "Brother, you're a well man now, aren't you?"

"Yes," he responded, "I haven't felt this good in many years."

He. . . related his experiences. He told me about meeting his aged father, who was now in his '80s and nearly blind. When his father came to the door, he inquired in his usual gruff manner, "Who are you?"

My friend informed him that he was his son. Still rather brusquely, his father responded, "Well, what do you want now?"

My friend answered: "Dad, I have come home to ask for your forgiveness. For years I have held a bitter resentment against you for what you did to me when I was a young man. I had no right to feel resentment toward you. Can you forgive me for holding a grudge all these years?"

He said that his father looked stunned for a moment. Then he broke down and cried, threw his arms around his son, and sobbed, "Son, I'm the one who should have asked for your forgiveness, but I didn't have the courage. Can you forgive me?"

Then my friend added: "You know, we made a complete reconciliation. The spirit of peace and forgiveness flooded both of our lives... Today I am a happy, healthy man. I'm at peace with myself and with my Lord."

Within six months my friend was the third-highest sales producer for the large life insurance company he represented... More than thirty years later, as far as I am aware, he is still very much alive, enjoying life and serving his fellowman — this man who was doomed to die thirty years ago!

Indeed, resentment is a killer in many ways. We are often easily offended or to take offense, and unintentionally we offend others. It is critical to learn how to forgive all that we may be forgiven of all, and this needs to be an ongoing process. Doing so, as in this story, will work miracles in our lives and will definitely bring us closer to God. Only God can see our hearts; therefore, let us not pass judgment or criticize another. Christ set the perfect example in forgiving those who crucified Him.

How many of us are too busy to put God first in our lives? Setting priorities can be extremely important. We have work, school, friends, entertainment, eating, along with a task list of things we want to get done, and we often bounce around in our mortal sphere of space and time trying to deal with all of these activities and priorities in our lives and have no time for God. That can all change in a remarkable way if we will do all things with an eye single to the glory of God — continually rejoicing in His goodness and love. This is how we keep that profound commandment to pray always. To have a prayer in our heart at all times to do His will; this will help us tremendously in setting proper priorities. He can and will then bless us in our work, our school, with our friends, we will be entertained in better ways, our meals will be ideal — good food, with friends and loved ones, and not pressed for time so that we can visit and enjoy "breaking bread" together. The Lord will help us prioritize our tasks if we ask. Rather than being spread out and frustrated, we will be focused and productive — feeling good about our efforts before the Lord. If we set our own priorities, we almost never get everything done that we wanted to. If we ask the Lord to help us set them, then often we will have the fulfillment of getting all of them done, but we can count on being stretched so that we grow in the process.

Faith is a principle of action based on hope in our hearts to find and know the truth: about God and His perfect plan of happiness, about who He is, what are His characteristics and attributes for us to emulate, and that our path is in line with what we agreed with Him we would do during our mortal journey — that we may fully partake of the happiness He has designed for us. Such faith brings joy and peace to our hearts and leads us to know that He exists, that He loves each of us with a divine love and gave His Son to redeem us, and that He will instruct each of us on our missions in mortality if we will come unto His Son and seek, ask, and knock to know His will in our lives. The internet is a great tool, but the secularism that is there has destroyed the faith of many. The antidote for that is to spend more time in prayer and in the scriptures. Isaac Newton, the greatest scientist in history, spent more time studying the Bible than he did doing science. He said, "We account the Scriptures of God to be the most sublime philosophy. I find more sure marks of authenticity in the Bible than in any profane history whatever." And, "No sciences are better attested than the religion of the Bible." In their pride, modern scientists think to say, "Since the Bible doesn't agree with science, the Bible must be a myth." Newton would have said it just the opposite: "If science aligns with the Bible, then herein is truth." Agassiz, the greatest natural scientist of the 1800's, said, "It is the job of prophets and scientists alike to proclaim the glories of God." Ironically, Darwin — a contemporary of Agassiz — had great praise for Agassiz.

The Law of Love — as proven earlier — is the first law of heaven and is equivalent to the law of obedience when obedience is defined in a heavenly sense. The message of the *Torah* is a message of joy and love. Professor Dennis Rasmussen, who attended the Jewish Theological Seminary of America, summarized it well: "The *Torah* is a Law of Love. Obedience is a life of love. And always at the center of love, of peace, of joy, is the Holy One of Israel."

E. Douglas Clark in his book *The Blessings of Abraham... Zion...* beautifully points out that the fundamental message of Abraham is love and kindness. That example and heritage have perpetuated across the nations of the Earth. According to Professor Truman Madsen, current DNA studies show that 95% of the world's population tie back to Abraham, as stated in Clark's book. The ironies of the Lord make me smile. Abraham was promised that through his seed all nations of the Earth would be blessed; yet, he had to wait until he was 100 years old to have his birthright son! Now we have scientific evidence that this prophecy is being fulfilled.

Jesus gave us the first two and most important commandments, and they are based on love. Yet Satan has taken society in every other direction so that the ideal society — coming out of the law of love — is thwarted. Fortunately, living this law is much more powerful than we think and as Jesus profoundly shared in John 14, living it can bring us into the presence of God.

As mentioned in Chapter 17, we can gain an important perspective on how well we are living the teachings of Jesus from Michael H. Hart's book *The 100; the Ranking of the Most Influential Persons in History*. As you will remember, Hart lists Mohammad first, Isaac Newton second, and Jesus third. This is because Christians — for the most part — are not living the unique teaching of Jesus to " Love your enemies." Since Christians are not living it, Jesus' influence is minimized in that regard. In addition, current studies have shown that the most influential persons on current Christian ideology are St. Augustine and Martin Luther; Luther introduced the doctrine of saved by grace as documented by David W. Bercot in his book, *Will the Real Heretics Please Stand Up?*

Along the same lines, "Compassion" is the most frequently occurring word in the *Qur'an*. Each of its 114 chapters, with the exception of the 9th, begins with the invocation, "In the name of God, the Compassionate, the Merciful..." Yet what do we see from Islam today?

Atheists often point at all the heinous crimes done in the name of religion, often referring to the Christian crusades as an example, and say they want no part of it. What they don't understand is that if Christians and Muslims were to live the Law of Love taught in their scriptures, the world would be an amazing place to live.

As of 2014 the Muslims numbered about 2.08 billion, while the Christians numbered about 2.01 billion and with respective growth rates of 1.84 % per year and 1.32%. This difference is primarily a result not of their teachings but simply of their birth rates. Muslims are spreading into many nations and having lots of children at the same time.

Let us summarize the importance of the Law of Love:

Chapter 22

If the world lived the Law of Love:

- There would be no wars.

- There would be no crime.

- There would be no divorces and families would live happily together.

- There would be peace on Earth and good will among men.

- The resources of the Earth would be properly taken care of with no pollution

- There would be no abortions, and families and communities would work together in harmony and love to accommodate each other's needs.

- We would be living in an ideal society and the Lord and His angels would dwell among us helping us to organize and perpetuate the blessings of the Lord which would pour forth in rich abundance.

One may say this seems impossible, but the exciting promise is that such is coming, as will be shown in the next chapter, and emotional or transcendental time will be the operative time for that society.

It is interesting that genetically the children of Manasseh — the great grandson of Abraham — as well as some of the Asian peoples seem to have more of a tendency to operate in this emotional or transcendental time mode — meditating and listening to their hearts. There is inspiring scientific evidence that some of the native-American Hopewell Indian societies — living about 200 A.D. — had science and religion fully integrated and that it was a time of great peace.[10] It is also interesting that the DNA of the Hopewell society ties back to the descendants of Abraham.

If one reads the book 1491, you will be surprised to find documentation that the pre-Columbus American civilizations were greater than any others in the world at that time. The diseases brought by the European conquerors killed off massive amounts of these populations, and we tend to look at the surviving tribes as savages. The following DVD *Lost Civilizations of North America*, which I cited before shares how there were an estimated 200,000 mounds built in

---

10) See *http://www.allanstime.com/Spiritual/ Hopewell_Civilization_Great_Octagon/*

North America and now only about ½% of them remain. Americans treated the natives as savages and as if they had no rights or significant history. Now we see what a great mistake we made in treating them so.

Some may ask, "Does God play favorites? Does He have a chosen people?" This is often greatly misunderstood. He selects prophets and special people who commune with Him and then they share His messages. His design is that His truths be a blessing to all of His children. These prophets and special people are not better than any other peoples, but have accepted a responsibility in pre- mortal to help provide God's love and blessings for all of His children; thus, we have more copies of the Bible on this planet than any other book. And look at the blessings that come to those who *truly* live its teachings.

E. Douglas Clark, in his book *The Blessings of Abraham; Becoming a Zion People*, shares the perfect example of how Abraham reached out to all peoples and all nations to share the gospel message of joy, peace, and love. He knew of the Savior's coming (both first and second) and both lived and shared His teachings.

As we gain understanding of God's truths, we will see how He knows how to turn evil into good and that He will bring about — in due process — every promise He has made. In the next chapter, we will discuss His timing in how He is bringing those promises about. We will see that we live during the most exciting time in the history of the Earth.

Chapter 23

It's About Where We Are in God's Time

Here we are in man's time, Independence Day 4 July 2014. Are we Free? Ancient records now coming forth[1] show that Abraham knew more of astronomy than we do today — even with the Hubble telescope and its incredible eye to the sky — because he was taught by God in answer to his deep inquires. New records coming forth tell that he taught the Egyptians astronomy. Abraham discusses how one day with God is like a thousand of our years on Earth. The Apostle Peter apparently had access to Abraham's record — for he records the same time relationship in 2 Peter 3:8. So using this timing for God, it takes seven days of God's time to bring the Earth to perfection; one week of God's time is seven thousands of our years. Where are we in that perfection process? That is a very exciting question, and we will find the answer from God's word. We know we are living in the last days before the glorious millennial reign of Christ. We readily see the signs of the times and that He is hastening His work in His day. As we study God's time, it will help us know what the future holds and why God created the Earth in the first place as well.

In this regard Isaac Newton has a brilliant perspective. Our good friend, Professor Steven E. Jones felt to study the religious life and writings of Sir Isaac. Since this aspect of Newton's life is not published, Steve and his wife,

---

[1] See E. Douglas Clark's book *Blessings of Abraham*

Lezlee, and one of their daughters entered the Cambridge University library in June of 2001with pencil and paper to take notes from Newton's writings. Even today, the University restricts copying from Newton's writings on theology — no photocopies are allowed. All notes have to be hand-written on site. The following is mostly extracted from a very enlightening lecture on their findings given in our home 5 February 2010. He let me borrow copies of his slides and I took personal notes. From that and other information I share the following:

Sir Isaac Newton (1642-1727)

Newton's contributions to science have had a greater impact on science in the world than anyone else in history; yet, he wrote more on religion than he did of science — however, his religious beliefs have been largely suppressed. Religion was his quest — to know God. He was a great student of the Bible, but also of the writings of the early Christians. He not only found mistakes in translations available at his time, but proved that the Church of England was wrong in many aspects of their theology, which threatened him from having fellowship at Trinity College, and he was willing to lose this distinguished position rather than compromise his beliefs. He and his friends petitioned King Charles II, and the King set up the Lucasian Professor chair, which granted exemption for Newton and all subsequent holders of this chair to "all holy orders."

Alexander Pope wrote the following:

Nature and Nature's laws
lay hid in night;
God said, Let Newton be!
and all was light.

Newton's *Principia* contains more creativity and new science than any other book ever written. He continuously saw the hand of God in nature: "Whence is it that Nature doth nothing in vain; and whence arises all that Order and Beauty which we see in the World? . . .Was the Eye contrived without Skill in Optics, and the Ear without Knowledge of Sounds?"[2] "This most beautiful system of the sun, and planets, and comets, could only proceed from the

---

2) *Opticks*, Bk. 3, Query 28

Chapter 23

counsel and dominion of an intelligent and powerful Being. And if the fixed stars are the centres of other like systems, these being formed by the like wise counsel, must be all subject to the dominion of One"[3] This same truth inspired the prophet Alma as he said: ". . . yea, and all things denote there is a God; yea, even the Earth, and all things that are upon the face of it, yea, and its motion, yea, and also all the planets which move in their regular form do witness that there is a Supreme Creator."[4] We have a similar inspired statement in modern scripture: "The worlds were made by him; men were made by him; all things were made by him, and through him, and of him."[5] Further, Newton profoundly says: "This Being governs all things, not as the soul of the world, but as Lord over all. . .. The Supreme God is a Being eternal, infinite, absolutely perfect. . . and from his true dominion it follows that the true God is a living, intelligent, and powerful Being;. . . He is not eternity and infinity, but eternal and infinite; he is not duration or space, but he endures and is present. . .,"[6] and "Search the scriptures thy self & that by frequent (reading) and constant meditation upon what thou readest & earnest prayer to God to enlighten thine understanding - if thou desirest to find the truth. Which if thou shalt at length attain thou wilt value above all other treasures in the world by reason of the assurance & vigour it will add to thy faith, & steady satisfaction to thy mind which he only can know how to estimate who shall experience it."[7]

This expression of Newton's on how to learn truth reminds one of the following from the Prophet Joseph Smith, Jr., ". . .when you feel pure intelligence flowing into you, it may give you sudden strokes of ideas, so that by noticing it, you may find it fulfilled the same day or soon; (i.e.) those things that were presented unto your minds by the Spirit of God, will come to pass; and thus by learning the Spirit of God and understanding it, you may grow into the principle of revelation, until you become perfect in Christ Jesus."[8]

In a world going awry caused by a people turning from God with ever increasing pornographic "distortions," Newton has an elegant and simple solution: "The way to chastity is not to struggle directly with incontinent

---

3) *Principia*, Bk3, pp. 543-546
4) Alma 30:44
5) Doctrine and Covenants 93:10
6) *Principia*, Bk3, pp. 543-546
7) *Yahuda* Manuscript collection, Cambridge University Library, 1.1 Also: Newton Project, Yahuda MS 1.1
8) *Teachings of the Prophet Joseph Smith*, p151

thoughts but to avert the thoughts by some employment, or by reading, or by meditating on other things, or by conversation."[9] This reminds one of a modern scripture: "Look unto me in every thought; doubt not, fear not." [10] AND "Therefore whosoever heareth these sayings of mine, and doeth them, I will liken him unto a wise man, which built his house upon a rock: And the rain descended, and the floods came, and the winds blew, and beat upon that house; and it fell not: for (he) was founded upon a rock (Christ)."[11] "And whoso treasureth up my word, shall not be deceived, for the Son of Man shall come, and he shall send his angels before him with the great sound of a trumpet, and they shall gather together the remainder of his elect from the four winds, from one end of heaven to the other."[12]

Newton rejected the creeds. Prof. Jones extracted the following 12 summary points from Isaac's hand written notes about the Godhead:

1. The Son confesses the Father greater than him, calls him his God

2. The Son acknowledges the original prescience of all future things to be in the Father only. (E.g., when the Father gives Book to the Son, in Revelation)

3. The Son in all things submits his will to the will of the Father, which would be unreasonable if he were equal to the Father.

4. The Son in several places confesses his dependence on the will of the Father.

5. The union between him and the Father is like that of the saints — one with another. That is in agreement of will and counsel.

6. It is a proper epithet of the Father to be called Almighty. For by God Almighty we always understand the Father. Yet this is not to limit the power of the Son, for he doth whatsoever he seeth the

---

9) Yahuda MS. 18.1, fol. 2v; quoted in Frank E. Manuel, *The Religion of Isaac Newton*, Clarendon Press, 1974, p. 13.
10) Doctrine and Covenants 6:36
11) Matthew 7:24-25
12) Joseph Smith-Matthew 37

Father do; but to acknowledge that all power is originally in the Father & that the Son hath no power in him but what he derives from the Father.

7. When, after some heretics had taken Christ for a mere man & others for the supreme God, St John in his Gospel endeavored to state his nature so that men might have from thence a right apprehension of him and avoid those heresies & to that end calls him the word or ΛΟ'ΥΟΣς (LOGOS): we must suppose that he intended that term in the same sense that it was taken in the world before he used it when in like manner applied to an intelligent being. For if the apostles had not used words as they found them how could they expect to have been rightly understood? Now the term before St John wrote, was generally used in the sense of the Platonists, when applied to an intelligent being, & the Arians understood it in the same sense, & therefore theirs is the true sense of St John.

8. The Word itself was made flesh and took upon him the form of a servant.

9. It was the Son of God which He sent into the world & not (an ordinary) human soul that suffered for us.

10. The (word) God is nowhere in the scriptures used to signify more than one of three persons at once.

11. Whenever it is said in the scriptures that there is but one God, it is meant of the Father.

12. The word God put absolutely without particular restriction to the Son or Holy Ghost doth always signify the Father from one end of the scriptures to the other.[13]

Joseph Smith's vision of the Father and the Son agrees exactly with Newton's findings: "...When the light rested upon me I saw two Personages, whose brightness and glory defy all description, standing above me in the air. One of them spake unto me, calling me by name and said, pointing to

---

13) Spelling modernized

the other—This is My Beloved Son. Hear Him! My object in going to inquire of the Lord was to know which of all the sects was right, that I might know which to join. No sooner, therefore, did I get possession of myself, so as to be able to speak, than I asked the Personages who stood above me in the light, which of all the sects was right (for at this time it had never entered into my heart that all were wrong)—and which I should join. I was answered that I must join none of them,. . . that all their creeds were an abomination in his sight;. . ." just as Newton had found.[14]

David W. Bercot documents, from the early Christian writings, how the creeds came about and why they are an abomination in the sight of God in his book, *Will the Real Heretics Please Stand Up*. Bercot, like Newton, made a great study of these writings. We find a very different Christianity amongst the saints in the meridian of time than exists, generally, in the world today.

In fact, Newton's studies led him to finding mistakes in Bible translations. As a relevant example, consider 1 John 5:7, "For there are three that bear record in heaven, the Father, the Word, and the Holy Ghost: and these three are one." About it, Newton says: " In Jerome's time and both before and long after it, this text was never thought of... Now it is in everybody's mouth and accounted the main text for the business" that is, supporting Trinitarian dogma. Newton continues: "Let them make good sense of it who are able; for my part, I can make none..."Newton concluded that this verse was added to the Bible by someone in the Dark Ages, and many scholars today agree; e.g., *Unger's Bible Handbook*: "Verse 7 (of I John 5) is not in the oldest and best manuscripts and should be omitted." It is evident that the creeds were having their effect on the translators.

Newton has profound insights on the apostasy of the primitive church. In his treatise on the Book of Revelation, Newton writes: "If you now compare all with the Apocalyptic visions and particularly with the flight of the Woman in the Wilderness and the reign of the Whore of Babylon, they will very much illustrate one another. For these visions are as plain as if it had been expressly said the true Church shall disappear and in her stead an Idolatrous Church reign in the world. It is to be observed that this Apostasy was to be general. . . This I gather from (Revelation and) . . . the working of iniquity in St. Paul's time set on foot by Simon Magnus, Manander, . . .

---

14) Joseph Smith-History:17-19

Chapter 23

Basilides, Carpocrates, Corinthus and other heretiques like these..."

Newton's studies revealed a quartet of false teachings:

- Creation ex nihilo (Latin: out of nothing); God uncreated, man created ex nihilo.
- Jesus did not (or does not now) have a tangible flesh body.
- Tri-une deity, uncreated and incomprehensible.
- Man cannot bridge the gap and become like this Deity

As Newton unveiled these false teachings, he discovered the truth regarding each of them. Here again, we have remarkable agreement with the truths that came to the Prophet Joseph Smith as part of the restoration, which was also foretold by Newton:

- Earth formed from existing materials, not ex nihilo. "We will take of these materials, and we will make an Earth. . ." (Abraham 3:24) It is interesting that we have 'ex nihilo' again in the "Big-bang" theory as science tries to figure out the creation without God.
- Jesus' spirit came to Earth in a tangible flesh body, and now he has a resurrected body of flesh and bone (and so does the Father).
- Three distinct members of the Godhead, comprehensible, man in God's image.
- Man CAN bridge the gap and become like this Deity.[15]

Newton warned of the false teachings of his day: "Look about thee narrowly, lest thou shouldst in so degenerate an age be dangerously seduced and not know it. . .. (Amidst) so many religions . . . there can be but one true and perhaps none of those that thou art acquainted with. . ." — in remarkable agreement with what Joseph Smith was told in his first vision.

Newton saw the great apostasy of the "dark ages" as well as previous apostasies. Then he foresaw the glorious restoration: "And there appeared unto them Moses & Elias & they were talking with Jesus - And (the disciples)

---

15) see Philippians 3:21

asked him saying why say the Scribes that Elias must first come. And he answered & told them Elias verily cometh first & restoreth all things. . .[16] Jesus said unto them (his disciples) "Elias shall first come & restore all things,"[17] "I will lay the Land most desolate & the pomp of her strength shall cease, & the Mountains (i.e. Cities) of Israel shall be desolate. . .[18] Jerusalem shall become heaps. . . But in the last days it shall come to pass that the Mountain of the house of the Lord shall be established in the top of the Mountains & it shall be exalted above the hills[19] and above all other temples.[20] "So in Daniel 2, the New Jerusalem extending its dominion over the Earth is represented by a great mountain which filled the whole Earth." "For as often as mankind has swerved from (truth) God has made a reformation. When the sons of Adam erred & the thoughts of their heart became evil continually, God selected Noah to people a new world. & when the posterity of Noah transgressed & began to invoke dead men, God selected Abraham & his posterity. & when they transgressed in Egypt, God reformed them by Moses. & when they relapsed to idolatry & immorality, God sent Prophets to reform them & punished them by the Babylonian captivity." "& when they that returned from captivity mixed human inventions with the law of Moses under the name of traditions... upon outward acts & ceremonies, God sent Christ to reform them, & when the nation received him not, God called the Gentiles. & now the Gentiles have corrupted themselves; we may expect that God in due time will make a new reformation."

In Summary:

- Newton is acknowledged as one of the greatest scientists and thinkers who ever lived. Rejecting Aristotle as "the" authority, he discovered the Law of Universal Gravitation, much about light and optics, and described the three fundamental Laws of Motion. He also invented Calculus.

- Newton was a devout Christian — wrote more about religion than he did about science.

- Based on his own studies of the Bible and early Christian-era teachers, he concluded that the Trinity doctrine was false and apostate.

---

16) Mark 9.4, 11, 12, 13
17) Matthew 17.11, 12, 13
18) Ezek 33.28
19) Isiah 2:2, Mica 4:1,2
20) Newton interprets the other temples to be "Idol-temples"

- He therefore refused the rites of the Anglican Church, and almost lost his position at Cambridge University and would have had the King not interceded. He also refused "last rites" at the time of his death

- Newton concluded that a general Apostasy had occurred and that a future Restoration of the Lord's Church would occur.

As a scientist, I am extremely impressed with Newton's work scientifically, but more importantly, as a devote Christian, I am impressed with Newton's religious contributions. I find his religious insights to be even more important than his scientific contributions. (Thank you, Professor Jones, for pulling together these outstanding materials on the religious insights of Sir Isaac Newton.) They are profound, and we can see that they agree remarkably well with the truths coming forth with the restoration through the Prophet Joseph Smith.

Harold Bloom in his book *The American Religion* points out that the Mormon Church today has changed from when it was organized. Can these changes be documented within the Church's history, and if there are changes what are they and are they significant?

In 1990, I was asked to represent The ^IChurch of Jesus ^IChrist of Latter-day Saints as part of a comparative religion debate held at the University of Colorado. I prepared a handout for the students — knowing they would not be able to remember all that was shared, and an updated version of this handout is on our web site.[21] There is a chart in that handout which addresses the prognostication of the famous historian, Count Leo Tolstoy, which he made in 1892 in regard to what he called "The American religion" — referring to the Mormon Church. He said:

The Mormon people teach the American religion; their principles teach the people not only of heaven and its attendant glories, but how to live so that their social and economic relations with each other are placed on a sound basis. If the people follow the teaching of this Church, nothing can stop their progress — it will be limitless. There have been great movements started in the past but they have died or been modified before they reached maturity. If mormonism is able to endure, unmodified, until it reaches the third and fourth generation, it is destined to become the greatest power the world has ever known.[22]

---

21) http://www.allanstime.com/Spiritual/lds_synopsis.htm
22) LeGrand Richards, *Marvelous Work and a Wonder*, pp 412-14

In the above link is a chart, which illustrates that the Count was basically right on target. However, there is more to the story.

The chart covers the period from when the Church was organized in upstate New York in 1830 with six members up to the year 2,000. The membership now numbers over 15 million and is worldwide. I plotted four parameters: membership, the number of missionaries serving, the number of copies of the Book of Mormon distributed per year (Along with the Bible the Book of Mormon is part of our scriptures), and the number of languages into which the Book of Mormon has been translated. All four of these parameters show increasing slopes over that 170 year period in agreement with the Count's prognostication.

A couple of years ago I looked at these same four parameters over the last decade — from the year 2000 through 2009. I was surprised to see the slopes on all four turn down over this last decade. One is immediately prone to ask why? Have things changed as Harold Bloom indicates? And upon studying this change in the context the word of the Lord (Scriptures) as we look at what is happening in the world we will see some significant reasons why. This will be brought to a very important conclusion at the end of the chapter and that will tie into the chapter title in a fundamental way. We will see God's timing in the purpose for the Earth's existence as well as where we are in that timing.

In considering God's timing for the Earth and His purposes for creating it, one may be inclined to study the history of the world and extrapolate to anticipate the future. There are two big problems with this approach. First, every historian is limited to writing his or her own history through his or her own paradigm, which is always an imperfect view. Some historians have gone through major paradigm shifts as they have gathered the data for their version of history. Second, extrapolation is prone to be wrong because God often works through cataclysmic events — like Noah's flood. In the days of Peleg, the Earth was divided.[23] As we look forward to the Second Coming of Christ, we anticipate major cataclysmic events in that regard.

I have chosen a different approach in looking at the Earth's history and purposes and projecting forward. In this chapter, I have tried to share as

---

23) Genesis 10:25; 1 Chronicles 1:19

best I can what I believe to be the Lord's perspective on this most important topic — to look through His eyes as best I can. His stated purpose for creating the Earth is, "For behold, this is my work and my glory — to bring to pass the immortality and eternal life of man."[24] I find this scripture most exciting. He created the heavens and the Earth to bring us back into His presence to receive a fullness of joy and with a glorious resurrected body like Christ's made possible because of His infinite atonement for us. This our Father-in-Heaven does because of His infinite love for us.

The historian's approach is bound to be flawed because of mortal imperfections. I have more hope in my approach because God is perfect. However, as you can well imagine, this approach is no small challenge. How do you look through God's eyes?

In pursuing this challenging approach, I have used what I have learned in my own scientific research with the Unified Field Theory as I feel I have been taught from on high as explained in Chapter 21. I have used the scriptures coupled with my own meditation, thoughts and prayers — using the concepts that were shared in Chapter 22 to ascertain and know the truth. I have also used as resource several good books written by people I trust and who also believe in the scriptures. I take the position — like Sir Isaac Newton — that if it aligns with the scriptures, then that is a good sign of its truth and validity. This is contrast to the last century where if science disagreed with the Bible then the Bible was put in the category of mythology.

As one studies the scriptures, we learn first and foremost that God loves us and has created this Earth for us to enjoy and to learn some critical lessons in our journey to be perfect like He is perfect.[25] We could not have learned or gained those experiences had we stayed in our pre-mortal sphere. One of the fundamental purposes of the Earth is a place where we could come to gain a body. His entire plan of happiness for His children is centered on free-choice. He will force no one to heaven. We are all spirit children of God as declared by the Apostle Paul in Hebrews 12:9 and in Acts 17:29 — born in a pre-mortal sphere of existence, and there we chose to follow our Heavenly Father's plan for us. Those who didn't were cast out of heaven and became the devil and his angels and were not privileged to gain a physical body.

---

24) Moses 1:39
25) Matthew 5:48

Though some may not feel it here, we all rejoiced in the opportunity to come to Earth, to gain a body, and to gain Earthly experience. We knew that a veil would be placed over our memory of our pre-mortal existence so that we could grow in our spiritual development by having to live by faith. In gaining Earthly experiences we knew there would be opposition — good and evil, and we knew we would make mistakes and commit sins. Hence, there was a need for a Savior to pay for our sins. When a heavenly law is broken, the law of justice requires a payment.

We read from the 4th chapter of the Book of Moses a very important scripture in this regard.

> Satan (said) Behold, here am I, send me, I will be thy son, and I will redeem all mankind, that one soul shall not be lost, and surely I will do it; wherefore give me thine honor. But behold, my Beloved Son, which was my Beloved and Chosen from the beginning, said unto me — Father, thy will be done, and the glory be thine forever. Wherefore, because that Satan rebelled against me, and sought to destroy the agency of man, which I, the Lord God, had given him, and also, that I should give unto him mine own power; by the power of mine Only Begotten, I caused that he should be cast down; And he became Satan, yea, even the devil, the father of all lies, to deceive and to blind men, and to lead them captive at his will, even as many as would not hearken unto my voice.

Satan's plan was one of force; Heavenly Father's plan is one of choice. We read, "For God so loved the world, that He gave His only Begotten Son, that whosoever believeth in Him should not perish, but have everlasting life."[26] Through the atonement, Jesus extends to us the law of mercy which can satisfy the law of justice. We chose to follow Heavenly Father's plan in pre-mortal — hence we are here — and now we get to choose again whether or not to follow Him — to receive His grace and to be saved by the law of mercy.

Our Savior has provided a way both by His perfect example and teachings and by His infinite atonement that we may have a sure path to return with a glorious resurrected body — like our Savior's — to our Loving Heavenly

---

[26] John 3:16

Father. There we can receive a fullness of joy with our loved ones in His presence. It is our choice. If we will come unto Him, He can turn all of our imperfections into perfection. He is the way, the truth, and the life. [27] There is no other way, and Satan is doing his best with his angels to convince us of lots of other paths filled with worldly pleasures designed to satisfy the lusts of the flesh. In contrast, in due process by keeping our personal desires kept within the bounds the Lord has set we can experience the fullness of joy we all seek as well as the greatest happiness in this life.

As we strive to look at history through the Lord's eyes, we see that He works with His children through covenants and commandments, and the greatest commandment is to love God with all of our hearts, with all of our souls, with all of our strength, and with all of our minds; and our neighbors as ourselves.[28] Having covenant people and covenant lands make it look like God is partial, but we need to realize that He uses these covenant relationships to bless all of His children. We also need to remember that in our pre-mortal we were both happy with and agreed to the time and place of our birth and our Earthly missions.[29]

We learned there that the path we chose was best for us in time (mortality) and in eternity. Like with Jeremiah,[30] the Lord told him that he was chosen before he was born, and so He gives us prophets,[31] prophetesses and wise teachers,[32] and He establishes covenant peoples and covenant lands to best help us in our mortal journey[33] — knowing that many will use their agency (free-choice) to try and thwart the work of God, but in the end we will clearly see that His work cannot be frustrated and we will rejoice in what He has done, is doing, and will do for us. In the commandment to love we find the greatest joy in this life and it leads to a fullness of joy in the world to come. Yet, how many of us turn from living that most fulfilling and joyful of all the commandments, which, if lived, would solve all the world's problems?

I have been also impressed by the work of Timothy Ballard, who ties ancient covenants to modern covenants. His book *The Covenant: One Nation*

---

27) John 14:6
28) Luke 10:27
29) Acts 17:22-31
30) Jer. 1:5
31) Amos 3:7
32) Alma 30:10
33) Abraham 2:9-10

Under God — America's Sacred and Immutable Connection to Ancient Israel is amazingly insightful in this regard. I have gotten to know Timothy personally and can vouch for him as a person of great integrity. He spends all his professional time trying to stop the horrible child trafficking problem now plaguing this planet, which is now the fastest growing criminal activity in the world. Child trafficking must be one of the most offensive sins in the eyes of the Lord — because of His love for little children. It should be most offensive to us as well, and it is good for us to get behind those who are fighting it like Tim Ballard, who, by-the-way, was featured on the Glenn Beck show and even hosted one of his shows.

Part of the following is both inspired by and taken from Ballard's book. I have labeled my write up here:

The Land of the Free — Prepared by the Lord
— How a Utopian society (Zion) will come out of it —

Here we define Zion as the "Pure in Heart" with no poor among them; a people of one mind and one heart. The world hopes for such a utopian society. While they hope, Jesus is doing everything necessary to bring it about. Such a society (Zion) will come as sure as night follows day, and we can be part of it. It is our choice. This land of the free (America) was prepared by the Lord as His consecrated place for Zion to be established in the last days. The foundations of liberty, which are essential in a Zion society, were laid out by our inspired founding documents — except for the issue of slavery, which Lincoln later resolved. Tim Ballard's second book addressed how that important step came about; it is entitled *The Covenant, Lincoln, and the War (The Covenant)*.

These foundations of liberty and the moral fabric upon which they were founded are eroding rapidly as our government has moved away from the Declaration of Independence and our Constitution. Furthermore, the minds of modern Americans, for the most part, do not appreciate the incredible price paid for liberty to give us this "land of the free." Over the entrance of the Norlin Library in Boulder, Colorado, is a famous quote by George Norlin: "Who knows only his own generation remains always a child." Let us learn as well from some very important documented messages of American history. The better we can view history from the Lord's perspective, the

Chapter 23

better we will know who we are and how we can help the cause of TRUTH to move forward toward a Zion society and to bless the lives of all who are willing to receive this most important message of joy and gladness in the midst of the storms of life.

Interestingly, as carbon dated by scientist, about 200 AD there was a very large civilization here in America that had science and religion fully integrated, and they were living in peace with no wars among them. Their remains are still evident in a golf course in Ohio. We can and should learn from them if we desire to have harmony between science and religion today. We have visited the great octagon in Ohio and it is amazing.[34] The DVD *Lost Civilizations of North America* has some nice pictures of it as well. It seems they were way ahead of us today as they had science and religion in harmony and lived in national peace and prosperity — a model society that we could definitely learn from. This ancient civilization lost that magnificence in civil strife after living in peace for over 200 years; let us learn from that lesson as well. The Book of Mormon tells the religious dimension of their story. These ancient civilizations prophetically knew this was a land of promise and a land of Liberty above all other nations and that the United States of America would be raised up here — they saw in vision the founding of this land of liberty. And they knew that it would be the place where the Lord would build up Zion in preparation for His Second Coming.

Pastor John Robinson, speaking to the Pilgrims just before they set sail for America prophetically proclaimed: "Now as the people of God in old time were called out of Babylon civil, the place of their bodily bondage, and were to come to Jerusalem, and there to build the Lord's temple... so are the people of God now to go out of Babylon spiritual to Jerusalem (America)... and to build themselves as lively stones into a spiritual house, or temple, for the Lord to dwell in.. For we are the sons and daughters of Abraham by faith."

Native American Indian, Squanto, was raised up by the Lord and stayed with the Pilgrims their first year, teaching them how to raise corn, to hunt, and to fish. As William Bradford said of him that he was a "special instrument sent of God" to save their lives. Similarly, Indian chief Massasoit welcomed the Pilgrims as friends and extended providential care.

---

34) I have cited the following link before as I was deeply moved by what is there:
http://www.allanstime.com/Spiritual/Hopewell_Civilization_Great_Octagon/index.html

Then we have the famous *Mayflower Compact* written in 1620: "In the name of God, Amen... Having, undertaken for the glory of God, and Advancement of the Christian faith. . . a Voyage to plant the first colony. . . do by these Presents, solemnly and mutually in the presence of God and of one another, covenant and combine ourselves together into a civil Body Politick for our better Ordering, and Preservation, and Furtherance of the Ends aforesaid."

Puritan Jonathan Edwards prophetically proclaimed, "God presently goes about doing some great thing in order to make way for the introduction of the church's latter-day glory — which is to have its first seat in, and is to rise from (this) new world." They named one of their communities "Salem," which translated from the Hebrew means "peace." Samuel Sewall of Salem declared that America was to be the host nation of the "New Jerusalem." He further believed that the American Indians were "Israelites unawares." Current DNA studies tie many of the Hopewell Indians to being descendants of Israelites.[35]

In 1631, Roger Williams, as part of the Massachusetts Bay Colony, left his First Baptist Church ministry proclaiming, "There is no regularly constituted church on Earth, nor any person qualified to administer any church ordinances; nor can there be until new apostles are sent by the Great Head of the Church for whose coming I am seeking."

In 1755, during the French and Indian War, George Washington, who headed up the battle at Monongahela, was fighting against one of the great Indian chiefs, Grand Sachem, where Washington later explained, ". . .death leveling my companions on every side of me. . ." Over half of his troops were killed. ". . .by the miraculous care of Providence, that protected me beyond human expectations; I had four bullets through my coat, and two horses shot from under me, and yet escaped unhurt." As the Father of our Country, the Lord had yet a great work for him to do.

Fifteen years later, this same Chief met Washington years before Washington became famous and said, "I am a chief and ruler over many tribes. . . I have traveled the long and weary path of the wilderness road that I might once again look upon the young warrior (Washington) of the great battle. (During the battle) I called my young men and said (pointing

---

35) see DVD, *Lost Civilizations of North America*

to Washington), 'Mark you tall and daring warrior! He is not of the redcoat tribe, he is of the Long knives. He has an Indian's wisdom. Let your aim be certain and he dies.' I, who can bring the leaping squirrel from the top of the highest tree with a single shot, fired at this warrior more times than I have fingers. Our bullets killed his horse, knocked the war bonnet from his head, pierced his clothes, but 'twas in vain; a Power mightier far than we shielded him from harm. The Great Spirit protects that man and guides his footsteps through the trails of life. He will become the chief of many nations, and when the sun is setting on the remaining few of my people and the game has departed from our forests and streams, a people yet unborn will hail him as the founder of a mighty empire. I have spoken."

Washington responded, "Our destinies are shaped by a mighty Power, and we can but strive to be worthy of what the Great Spirit holds in store for us... If I must needs have such lot in life as our Red Brother presages, then I pray that the Great Spirit give unto me those qualities of fortitude, courage, and wisdom possessed by our Red Brother. I, the friend of the Indian, have spoken."

In the year of the famous Declaration of Independence, 1776, Thomas Paine succinctly said, "We have it in our power to begin the world over again. A situation, similar to the present, hath not happened since the days of Noah until now." Then we have the immortal words in the Declaration of Independence inspired by the pen of Thomas Jefferson, "We hold these truths to be self-evident, that all men are created equal, that they are endowed by their Creator with certain unalienable Rights, that among these are Life, Liberty and the pursuit of Happiness." At the conclusion of the eight-year war, which logically we should not have won being outnumbered 10 to1 and lacking most of the necessities to fight the strongest military force on the planet, George Washington, the General of our Army, simple summed up why we won the Revolutionary War, "Divine intervention."

After the war Washington declared:

> May the same wonder-working Deity, who long since delivered the Hebrews from their Egyptian oppressors, planted them in a promised land, whose providential agency has lately been conspicuous in establishing these United States as an independent nation,

still continue to water them with the dews of Heaven and make the inhabitants of every denomination participate in the temporal and spiritual blessing of that people whose God is Jehovah.

Washington's mother upon her death bed reportedly said to her son, "Go, George, fulfill the high destinies which Heaven appears to have intended for you." Abigail Adams said of Washington, "Mark his majestic fabric. He's a temple sacred from his birth and built by hands divine." Washington said, "It is impossible to rightly govern the world without God and the Bible."

The feeling at that time is profoundly evidenced in a speech given by the President of Yale University, Ezra Stiles, in 1783 — the year America officially gained independence:

> God determined that a remnant should be saved... recovered and gathered... from the nations whither the Lord had scattered them in his fierce anger... and multiply them over their fathers — and rejoice over them for good, as he rejoiced over their fathers. Then the words of Moses... of Abraham shall be nationally collected, and become a very distinguished and glorious people under the Great Messiah, the Prince of Peace. He will then make them "high above all nations which he hath made in praise, and in name, and in honor," and they shall become "a holy people unto the Lord" their God.

Now, given the devolution of American morals and the secular influence on religion, we cannot even imagine such a talk being given by the President of any university as America has turned from the God of this land, the Great Jehovah.

A vital historic perspective most do not know is that the Lord prepared this land of liberty for the restoration of the fullness of His gospel, for the establishment of Zion, and to prepare a people for His glorious Second Coming. As Sir Isaac Newton so profoundly deduced, the Lord's Church and Kingdom had been lost during the Dark Ages, and there was a fundamental need for them to be restored in this land of the free. In the early 1800s, living in Sainsbury, Connecticut was a man named Robert Mason. In his community, he was known as Father Mason because of

Chapter 23

his saintly nature and great faith. He had a vision of this restoration of the fullness of the gospel, which he shared with the young lad, Wilford Woodruff:

> . . .the voice of the Lord came to me saying, "Son of man, thou hast sought me diligently to know the truth concerning my Church and Kingdom among men. This is to show you that my Church is not organized among men in the generation in which you belong; but in the days of your children the Church and Kingdom of God shall be made manifest with all the gifts and the blessings enjoyed by the Saints in past ages. You shall. . . not partake of its blessings before you depart this life. You will be blest of the Lord after death because you followed the dictation of my Spirit in this life.

Then to Wilford he said, "I shall never partake of this fruit in the flesh, but you will and you will become a conspicuous actor in the new Kingdom." Wilford Woodruff became one of the Twelve Apostles as part of the restoration and later the fourth President of the Church.

In 1820, Joseph Smith, Jr. had his enormous epiphany, when the Father and the Son appeared to him commencing the dispensation of the fullness of times in preparation for the Lord's coming again to the Earth. Using experimentation, science uses the five senses to ascertain truth, which truths change with time as more is learned. Joseph opened to the world the sixth sense to know the truths of God, which are eternal and do not change. Joseph put the Lord to the test when he read in James 1:5, "If any of you lack wisdom, let him ask of God, who giveth to all men liberally and abraideth not; but let him ask in faith, nothing wavering." He found this scripture to be literally true by experimentation with the sixth sense. Thus began the "restoration of all things" prophesied by Peter and Paul, and opened the door to all of us to bask in the light of truth.[36] The enormity of this "First Vision" of the Prophet Joseph is yet to be appreciated by the world. In the fullness of the gospel restored is the vital truth that each of us can come to know the Lord on a personal basis as Joseph did, which is eternal life — the greatest of all the gifts of God.[37]

---

36) Acts 3:18-21; Eph. 1:10, respectively
37) John 17:3; Doctrine and Covenants 14:7; 93:1

In the same year (1820), Thomas Jefferson had the following perspective. The contributions of Jefferson to this land of liberty are enormous.[38]

> I hold the precepts of Jesus, as delivered by himself, to be the most pure, benevolent, and sublime which have ever been preached to man. I adhere to the principles of the first age; and consider all subsequent innovations as corruptions of this religion, having no foundation in what came from him. . . if the freedom of religion, guaranteed to us by law in theory, can ever rise in practice under the overbearing inquisition of public opinion, truth will prevail over fanaticism, and the genuine doctrines of Jesus, so long perverted by His pseudo-priests, will again be restored to their original purity. This reformation will advance with the other improvements of the human mind, but too late for me to witness.

As a humorous aside, President John F. Kennedy one time had several Nobel Prize laureates with him dining in the White House. He said to them, "There is about as much intelligence in this room now as there was when Thomas Jefferson dined here alone."

Starting in 1830, the Lord had Joseph Smith, as a prophet of God, organize the Church, as it was in ancient times, with apostles and prophets and as several had seen in vision. One of the most beautiful dimensions of the restoration through the Prophet Joseph Smith is to understand history in the context of Heavenly Father's grand plan of happiness for His children, and why He created the Earth and the inhabitants thereof in the first place. Those who study history from a secular point of view only, as is traditionally done, miss much in true historical perspective and the significance of many events that change the course of history as part of Father's grand plan, such as when the Savior came to Earth the first time. The restoration also brought great understanding as to the importance of "free choice" (agency) being the heart of Father's plan with the natural and necessary consequence of "opposition in all things" — explaining pleasure and pain, good and evil, etc. (2 Nephi 2 explains this doctrine profoundly). Chapter 2 of this book deals with the reason for opposition in detail and why we have pain, suffering and evil in the world.

---

38) See the following link enumerating some of his incredible accomplishments: *http://www.allanstime.com/Spiritual/jefferson.htm*

Chapter 23

With the background of the restoration, one can appreciate a grand historic chiasmus — covering 5,000 years of history in one sentence from the Savior in which He said, "...the first shall be last; and the last shall be first." [39] This chiasmus is explained in 2 Nephi 13:42 and in Jacob 5. The first is the House of Israel, whose children were given prophets and the gospel, the record, and history of which comprises much of the Old Testament. When the House of Israel rejected the prophets and the Savior, the gospel was taken from them and given to the Gentiles,[40] who became the last — the record and history of which comprises much of the New Testament. With the Gentiles' rejection of the prophets and teachings of the Savior, as evidenced by the Great Apostasy and the Dark Ages, the Lord prepared this "Land of Liberty," as a place for the restoration of apostles and prophets and the pure teachings of the Savior to this Gentile nation — making the last becoming the first as part of the "restitution of all things" that will precede the Second Coming.[41] The records and history associated with these events are the Book of Mormon, the Doctrine and Covenants, and the Pearl of Great Price.

In regard to this epoch, Jefferson further stated,

> The religion-builders have so distorted and deformed the doctrines of Jesus, so muffled them in mysticisms, fancies and falsehoods, have caricatured them into forms so monstrous and inconceivable as to shock reasonable thinkers... Happy in the prospect of a restoration of primitive Christianity, I must leave to younger athletes to encounter and lop off the false branches which have been engrafted into it by mythologists of the middle and modern ages.

So Newton and Jefferson both saw the need for a restoration. In a very similar sense, David W. Bercot has shown in his 1989 book, *Will the Real Heretics Please Stand Up*, that St. Augustine has had more influence on the teachings of current Christianity than any other person including the Savior. Bercot, as a contemporary investigator, did his research totally independent of Newton and Jefferson, but they arrive at many of the same conclusions — even though Newton and Bercot are separated in their lifetimes by about 300 years. Newton had the advantage that he was conversant in Latin and Greek and could better compare ancient texts.

---

39) Matt 19:30
40) Matt. 21:43
41) Acts 3:19-21

Now, the Lord gives us a very important sign: "when the Gentiles shall sin against my gospel, and shall reject the fulness of my gospel...," "...then will I remember my covenant, which I have made unto my people, O house of Israel..." and "I will bring the fulness of my gospel... unto them."[42] Then the first will become the last, completing the last major step in this grand chiasmus in preparation for the Second Coming, and taking us into and through the millennium — thus completing the 5,000 years of this pivotal chiastic structure.

A major part of this last step is the preparing the Bride to meet the Bridegroom, as beautifully portrayed in Revelations 19. This will be the Zion society (a pure-in-heart people of one mind and one heart with no poor among them) that are prepared to meet the Lord at his coming. The records of this last epoch are yet to come forth, but we can be as sure of it as we are of the occurrence of that which has transpired. Indeed, historically, we now live in one of the most exciting times ever. Prophecy is more accurate than history books, because the former comes from the Lord. We can be just as sure that Babylon the great will fall (the wickedness of the world) in our day as described in Revelations 17 and 18. Our choice is to be Christ-centered and to come totally out of Babylon to be His Bride, or to be otherwise and to fall with "the whore...the mother of harlots" and be destroyed. Read the books *Visions of Glory* by John Pontius, *A Greater Tomorrow, My Journey Beyond the Veil* by Julie Rowe, and *Beloved Bridegroom* by Donna Nielsen if you want to be filled with hope and get an excellent idea of what is ahead going into the millennium. Julie Rowe has a second book coming out soon on how to best prepare for the days ahead — Spring Creek Book Company.

Now, as mentioned earlier, over this last decade, we have seen a downturn in the rate of growth of The Church of Jesus Christ of Latter-day Saints as the influence of the world has crept in; we know the Lord knows what is going on and that this is His Church restored by Him to be a vehicle to bring us to Christ and to prepare for His glorious return. The Lord used the Prophet Joseph Smith to bring forth the fullness of the gospel and a "marvelous work and a wonder" as prophesied in Isaiah 29. As prophesied by the Savior, "when the Gentiles... shall reject the fullness of my gospel..., then... I will bring the fullness of my gospel... unto them (O house of Israel)." At that day, which is now, "If the Gentiles will repent and return unto me, saith the Father,

---

42) 3 Nephi 16:10

behold they shall be numbered among my people, O House of Israel." As the first becomes the last, the Lord will work an even greater "marvelous work and a wonder" as also prophesied by the Savior in 3 Nephi 21. All the prophets looked forward to our day and we are privileged to be part of it. Nephi prophesied of a great division,[43] and the Lord said, "upon my house shall it begin,. . ."[44] The purging will be from top to bottom as prophesied by Zenos,[45] as the wicked are destroyed and the "humble followers of Christ" are pulled out if they are not deceived by the precepts of men.[46]

We live in the most exciting time in the history of the Earth. The Lord tells us that "…our forefathers have awaited with anxious expectation (looking to the) last times,.. (when) nothing shall be withheld." You can see why all of the prophets looked forward to our day.[47] We have every reason to be filled with hope and love — anticipating the glorious return of our Lord and Savior. Satan — knowing it is his last hour — is filling the world with hate, fear, anxiety, envy, pride, sexual perversions, depression, discouragement (What can I do?), despondency, obesity, and dumbing us down so that we cannot see how he is deceiving the masses. Thus we see a world spiraling down in wickedness. This spiral is a massive sign of the times.

In glorious contrast, the humble followers of Christ are spiraling up in love, joy, and peace in the midst of the storms of life, because we know Him who is coming and that the purging times ahead are a necessary part of saving our constitution and the principles of liberty and freedom and the preparing a pure-in-heart people (Zion) who will be ready to meet Him as the Bride meets the Bridegroom. Satan has no power to destroy such a bright hope when we are totally focused on Him, who is our Redeemer, and are filled with His love.

Fortunately, the invitation to come unto Christ is independent of denomination. Everyone is born with the Spirit of Christ to know good from evil as part of Heavenly Father's perfect plan, and the Lord looks on the heart. If we choose to do good, we will be led to Him and a fullness of the gospel.

As the first becomes the last, we will see many exciting things. As

---

43) 2 Nephi 30:10
44) Doctrine and Covenants 112:25
45) Jacob 5:61-66
46) 2 Nephi 28:14
47) Doctrine and Covenants 121:26-32

previously mentioned, the downturn in the slope of the growth rate of The Church of Jesus Christ of Latter-day Saints is but a sign of the times as we come to the end of the Gentile era and we move into the era of the gathering of the pure in heart — the elect — from the four corners of the Earth. The influence of secularism (Babylon) on Christians in general is quite apparent. Darwinian evolution has almost totally infiltrated our education system at all levels, creating an enormous bias against God and Christian-Judeo morality. We started out as a Christian nation, but have moved far from that as a result of the ever increasing secular influence in our society. Too many are worshiping the mind of man and the works of men's hands rather than God.[48] Sports and games have become our American Sabbath activities. Who has the most toys seems to be one of the main games, and pornography is a major addiction problem and is destroying families.

The Lord will turn all of this retrogression around as the first becomes the last. The retention of new converts will be 100% as the pure in heart are gathered from the four corners of the Earth into the fold of God, and a Zion (utopian) society will come out of it as the Bride is prepared to meet the Bridegroom.

Moroni saw this coming and proclaimed to Joseph Smith, when he told him about the Book of Mormon coming forth, that the day would come when those who would not heed the voice of the Son of God would be cut off at His coming.[49] The equivalent negative transpose of this scripture is important to appreciate: "Those who want to be with Him at His Coming will hearken to His voice." Given that the great deceiver is having so much success, the scripture in JS-M 37 is of enormous value:

> And whoso treasureth up my word, shall not be deceived, for the Son of Man shall come, and he shall send his angels before him with the great sound of a trumpet, and they shall gather together the remainder of his elect from the four winds, from one end of heaven to the other.

In other words, during this purging time that is coming, as we focus totally on Christ and His teachings, we are promised that we will not be deceived. And furthermore He says: "Come unto me, all ye that labour and are heavy laden, and I will give you rest. Take my yoke upon you, and learn

---

48) Doctrine and Covenants 1:14-16
49) JS-H 40

of me, for I am meek and lowly in heart; and ye shall find rest unto your souls."[50] And this He will do in the midst of the purging ahead if we totally come unto Him. He will gather us as a hen gathers her chicks under her wing as is promised in Doctrine and Covenants 29:1-2, and this is a sure promise for those who come unto Him.

Moroni further warned us of our day in the Book of Mormon[51] that "when," not "if," we see "secret combinations" come among us, we need to "awake to (our) awful situation" that these may not get "above" us. The New World Order (NWO), promulgated by Satan, in its many facets fits Moroni's description in the detail, and yet most do not have a clue. Many in high places are part of the NWO as is much of the media. As Moroni says, they are caught up with "pride" and the NWO is about "control," "power," and "gain." To use force was Satan's plan. The US constitution and the principles of liberty and freedom upon which this country are built are in opposition to the NWO agenda. The NWO is doing its best to take down the constitution and to take away our God-given liberty.

As an example, in 1910 seven men gathered incognito in the ruse of a duck hunt on Jekyll Island off the coast of Georgia. These seven men represented 1/4 of the world's wealth. There they designed the Federal Reserve System. Three years later it was (with illegal legislation) put in place in the US government — giving the big banks control of our money system. This is well documented in the book by G. Edward Griffin, *A Creature from Jekyll Island*. The Federal Reserve — coming out of that ruse — is neither "federal" nor "reserve," but was designed by the big bankers to control us. Guess what?

Another devastating example of NWO activity is 9/11. Now, it has been scientifically documented that the Twin Towers could not have been brought down by the airplanes hitting the Towers. In addition, Building 7 fell at near free-fall velocities with no airplane hitting it, as would happen in a controlled demolition. The Twin Towers came down the same way. When super-thermite was found at a level of 0.1% in the dust of all three buildings, this gave a direct explanation of why the buildings fell as they did. This finding is well documented on the web site 9-11 studies. It seems that the NWO folks wanted the American public to believe that our war attacks on

---
50) Matt. 11:28-30
51) Ether 8:18-26

Iraq and Afghanistan were defensive, when in fact they were offensive with many other hidden agendas. In large measure their massive deceit perpetrated by "secret combinations" has been successful, but it will be exposed as the TRUTH becomes known by the seekers of truth.[52]

As one sees the world spiraling down, one may lose heart, but this discouragement is exactly what the adversary wants us to feel. When we know who is really in charge, we take heart; our hope and faith build in Him who will allow the wicked to destroy the wicked as they dig their own pit. He will pull out the wheat from among the tares. As the scriptures so profoundly teach us, we will not "be cut-off at His coming" if we hear and heed His voice. We will hear and heed His voice as we love Him with all of our heart, soul, mind, and strength (Mark 12:30) and our neighbors as ourselves. On these two commandments hang all the law and the prophets.

Let us help as many as we can to come to know of the Father's love and of His Beloved Son's (the Bridegroom) and to drink deeply of this love, which is most desirable to the soul and will bring a fullness of joy. There is not a more important message for the human family. How grateful we should be for the privilege of living in this incredible epoch as we prepare for the Lord's glorious coming.

> And then cometh the New Jerusalem (in this land of the free); and blessed are they who dwell therein, for it is they whose garments are white through the blood of the Lamb; and they are they who are numbered among the remnant of the seed of Joseph, who were of the house of Israel. And then also cometh the Jerusalem of old; and the inhabitants thereof, blessed are they, for they have been washed in the blood of the Lamb; and they are they who were scattered and gathered in from the four quarters of the Earth, and from the north countries, and are partakers of the fulfilling of the covenant which God made with their father, Abraham.[53]

The Lord has given us specific signs for our day. We have seen the sign of the transition from the LAST becoming the FIRST to this century where the FIRST is becoming the LAST because of how this — the USA — as a

---

52) See *http://www.allanstime.com/Spiritual/Christ_is_Coming_How_can_we_Best_Prepare.htm*
53) Ether 13:10-12

Chapter 23

Gentile nation is rejecting the God of this Land, Jesus Christ. In 3 Nephi 21 the Savior gives the ancient inhabitants of this land another sign: " And verily I say unto you, I give unto you a sign, that ye may know the time when these things shall be about to take place — that I shall gather in, from their long dispersion, my people, O house of Israel, and shall establish again among them my Zion;" Then in the next six verses he explains that it will be when the Book of Mormon record goes back to the descendants of father Lehi, who came here around 600 B.C. And that is happening as we write this.

On the book's web site is an article entitled *And It Shall Come To Pass =Jehovah.* [54] There you will find some 25 pages of scriptures and information around this phrase telling of our day and the events leading up to and including the Second Coming of our Lord. The scriptures associated with the phrase "And it came to pass" discussed in Chapter 3 and in the article on the book's web site entitled *And It Came To Pass=El=God,*[55] along with the scriptures shared in the above article compiled together appear to be a book of scriptures within our scriptures designed to come forth now, triggered by the work of Professor Leonora Leet.

In this article, we have shown using the deductive method of scientific reasoning that the phrase "And it shall come to pass" validates the harmony between all of the scriptures, both ancient and modern, just as was ascertained for the phrase "And it came to pass." As we study the phrases in this article, we see they are very important in helping us know of and best prepare for the exciting days ahead as we will see the Lord do His divine work of preparing a pure in heart people (His Bride) to meet Him at His Coming. It is so exciting that we now have this information to help us be prepared to be part of the Bride.

As we know from the birth of Jesus, the Lord also gives signs in the heavens. In this regard, Val Brinkerhoff, in his book, *Seven Heavenly Witnesses of the First and Second Coming of Jesus Christ*, documents some of these heavenly signs. (available from *www.digitalegend.com*) Specifically, he documents the significance of the four blood moons as they are called, which occur during 2014 and 2015 on specific Hebrew feast days.

About 1906 in a work called the *Pseudo-Philo* the writings of Zenos were discovered. The Prophet Zenos was documented as living before Isaiah and after

---
54) *http://ItsAboutTimeBook.com/and-it-came-to-pass-jehovah-science-and-religion/*
55) *http://ItsAboutTimeBook.com/science-and-religion-it-came-to-pass/*

Moses. He is quoted in the New Testament by Jesus and Paul, and his incredible allegory is shared by the Prophet Jacob in the Book of Mormon (Jacob Chapter 5). The allegory covers 5,000 years of the earth's time-frame and discusses also the FIRST-LAST, LAST-FIRST grand chiasmus. Studying it tells you also that we are at that last transition in preparation for the coming of the Lord.

Taking Val's work and Zenos' allegory along with Moroni's profound tying of the coming of the New Jerusalem and the restoring of the Old as prodigious events to FIRST-LAST, LAST-FIRST chiasmus as well, we see that we are indeed at that juncture.

Jesus further shares: "For in that day, for my sake shall the Father work a work, which shall be a great and a marvelous work among them; and there shall be among them those who will not believe it, although a man shall declare it unto them."[56]

Let us rejoice in this great epoch in which we live as we come totally out of Babylon and center our lives in Christ that we "shall not be deceived,"[57] and rejoice with as many as will "hear (and heed) His voice" as we become part of the family of Christ — His Bride. He will bring forth a "new heaven and a new Earth" at the beginning of the millennium, as the Earth goes from telestial (this current mortal sphere) to terrestrial and receives its paradisiacal glory (millennial), as well as at the end when the Earth goes from terrestrial to celestial and the Savior delivers His kingdom unto the Father.[58]

There we will receive all that the Father hath and a fullness of joy in the order of heaven, which is eternal family. His is the perfect plan of Happiness. Let us rejoice in it and share it at every opportunity — a message of gladness in a world going awry.

Now that we can see where we are in God's time table, in the next chapter we will tie this information into God's "wheel of time" and "one eternal round." In all of this grandness we will see the love of God manifest. It is so exciting.

---

56) 3 Nephi 21:9
57) JS-M 37
58) Rev. 21:1; Ether 13:9; Doctrine and Covenants 29:22-23

Chapter 24

It's About the Wheel of Time — One Eternal Round

I had a fascinating discussion with the Catholic priest after the comparative religion debate held at the University of Colorado in 1990. I had been asked by The ^lChurch of Jesus ^lChrist of Latter-day Saints to represent our faith. As we were chatting, he said, "May I ask you a difficult question?" I said, "Of course!" He queried, "Is it true that Mormons believe they can become a god?" I responded with a question, "Did the Savior ask us to be perfect like our Father-in-Heaven is perfect?" He paused, and then said, "I have never thought of it in that way before."

Often, the biggest limitation we place on ourselves is our limited thinking. I have found it imperative to trust in the Lord and not the arm of flesh, which is a most important lesson we learn from the center chapter and the center verse of the KJB.[1] We saw in Chapter 1 how the Lord placed that miraculously between the shortest and longest chapters in the KJB — showing how the Lord is in the details of our lives throughout the ages to bless us and to show His love.

If we take the Lord's word as truth with total trust and then open our minds, our hearts, our souls and our bodies to the truth, then truth and enlightenment will come, and it will be delicious, as we commune with God

---

1) Ps 118:8

in His infinite capacity to love and share the mysteries of godliness. He will bless us in all dimensions of our beings and it will bring great joy to the soul. In due process, it will allow us to partake of the fruit of the tree of life, which is the most desirable to the soul and will bring the greatest joy in this life and in the life to come — to be with Him and to be like Him in eternal family relationships; this is the great promise of eternal life.

The Savior told the Prophet Joseph Smith, "The Spirit of truth is of God. I am the Spirit of truth, and John bore record of me saying: He received a fullness of truth, yea even of all truth; And no man receiveth a fullness unless he keepeth his commandments. He that keepeth his commandments receiveth truth and light until he is glorified in truth and knoweth all things."[2] What an exciting promise the Lord makes to us! And the Apostle Paul promises, "Eye hath not seen, nor ear heard, neither have entered into the heart of man, the things which God hath prepared for them that love him."[3]

The Lord tells us through Isaiah that, "...my thoughts are not your thoughts, neither are your ways my ways saith the Lord."[4] As mortals, we cannot comprehend the glory, the majesty, the extent, the wonder, and the joy of the Lord. How much God loves us we also cannot comprehend and what He has in store for those who love Him and keep His commandments. Loving Him and our neighbors is the path to the fullness of joy. On these two commandments "hang all the law and the prophets."[5] The greatest joy we can partake of on Earth is only an infinitesimal part of what He has in store for those who come unto His Beloved Son and partake of the full blessings of the gospel through the atonement that His grace offers to each of us if we will so choose to hearken to His voice.

The Apostle Paul tells us of three degrees of glory in the resurrection.[6] The Prophet Joseph Smith and Sidney Rigdon together saw them in vision and describe them.[7] Those who were present when Joseph had this vision said his countenance glowed and he looked almost transparent. We have in them the mouth of two witnesses and in Paul the third of these degrees of glory. They are called the glory of the celestial, the glory of the terrestrial, and the glory of

---

2) Doctrine and Covenants 93:26-28
3) 1 Corinthians 2:9
4) Isa. 55:8
5) Matthew 22:36-40
6) 1 Corinthians 15:40-44
7) Section 76 of the Doctrine and Covenants

the telestial Kingdoms. As Paul says, they are like unto the sun, the moon, and the stars in glory, and that "as one star differs from another star in glory, so also is the resurrection of the dead." As Paul also says, all are resurrected.[8] And as the Apostle John says, we are judged according to our works[9] — the first being faith in the Lord Jesus Christ; the Prophet Joseph Smith taught that faith is the principle of action in all intelligent beings. Faith is naturally followed by repentance motivated by love as we change our behavior to align with that of our perfect Savior — being changed by Him as we come unto Him.

We further read that we cannot even comprehend the glory of even the lowest kingdom, the telestial kingdom, which is where the thieves, whoremongers, etc. go in due process. Then how great is the celestial kingdom?

As mentioned before, the Lord works through cataclysmic events: the six days of creation of the Earth were cataclysmic; the fall of Adam and Eve was cataclysmic; the flood was cataclysmic; when the Earth was divided that was cataclysmic; there were major cataclysmic events around the atonement; the end of the world (of wickedness) will be cataclysmic — we will have a new heaven and a new Earth; and the end of the millennium will be cataclysmic as our Savior presents this terrestrial sphere to the Father, and we will again have a new heaven and a new Earth as it becomes the celestial kingdom. "The meek shall inherit the Earth." And "the pure in heart shall see God."

Now let us tie these celestial promises of our infinitely loving Heavenly Father to the Earth, on which we live and move and have our being, to our astronomical relationships. These are questions Abraham asked and got answered in his day — coming to know more of the heavens than our current world astronomical community knows today because Abraham, the Father of nations, was taught directly by God.

Our solar system is about 27,000 light years from the center of the Milky Way Galaxy. Einstein's theory says that mass cannot travel faster than the speed of light. According to Albert, mass becomes infinite as it reaches light speed, and so according to his theory it would take much more than 27 thousand years to move our Earth from the center to where we are now. Let's see what God can do.

---

8) 1 Corinthians 15:13-24
9) Rev. 20:12

In God's Eternity-Domain and in the new UFT there is a very different perspective, and we have experimental evidences for same. Tele-transport of matter as well as telepathy are not limited by the speed of light in God's equations — contrary to current theory. ESP, for example, has been shown to not be limited by the velocity of light, as mentioned before in the Electric Universe talk by Dr. Dean Radin. His team's research work is fascinating.

We have in Heavenly Father's perfect plan of happiness a grand chiasmus for the Earth. We and our Earth were given birth in celestial realms and will return there perfected by the atonement of our Beloved Savior. After the creation and our spiritual birth, the Earth was moved to a terrestrial realm for the Garden of Eden part of its journey. The fall of Adam and Eve and the flood were both cataclysmic events during the Earth's journey through space. Now we find ourselves in this telestial realm in our current solar system configuration. With the Second Coming of Christ, we will move into His millennial reign and back into a terrestrial realm. That thousand years (one day with the Lord) is the time needed to perfect us and the Earth through the atonement and then the Savior will present us and the Earth back to the Father into celestial realms of glory if we have chosen to follow Him. We read the description of those in celestial glory given to the Prophet Joseph Smith and to Sidney Rigdon as follows:

> These are they who are just men made perfect through Jesus the mediator of the new covenant, who wrought out this perfect atonement through the shedding of his own blood. These are they whose bodies are celestial, whose glory is that of the sun, even the glory of God, the highest of all, whose glory the sun of the firmament is written of as being typical.[10]

As mentioned before, the scriptures tell us that one day with God is a thousand years with man. We know that the number seven is significant in the Lord's arithmetic. It is the number of completeness or perfection. We see in the six thousand years of the Earth's mortal journey since the fall of Adam and Eve and then with the Sabbath millennium comes the perfecting of the Earth in its seven thousand years mortal journey. Hence, as mentioned before, it will take one week (seven of the Lord's days) to bring about the perfection of the Earth.

---

10) Doctrine and Covenants 76:69-70

Chapter 24

We can think of ourselves on this giant space ship called Earth, which was created as we read in the scriptures nigh unto Kolob, which I believe is in the center of our galaxy. When the Andromeda Galaxy was discovered, three colors were apparent. The center is white, the next band out is red, and the outer band is blue. Is it a coincidence that the United States' flag colors are red, white, and blue as this great land of the free provided the turf for the restoration of the fullness of the gospel to help us in our journey back to celestial realms? We have a similar color distribution in our Milky Way Galaxy.

The Lord then moved our space ship Earth from the center of our galaxy (the white region) to the red band of our galaxy during the Garden of Eden epoch. With the fall, the Lord then moved the Earth to the blue band, where we are now. As part of the Second Coming of Christ as we move to a new heaven and a new Earth we will move back to the red band during the millennium. The Light of Christ will replace the sun as our light and we will no longer have the sun and our solar system as it is now. We read how the stars will fall from heaven as part of the Second Coming. Indeed, we are traveling in the Lord's way from the blue band of our galaxy to the red band. Then at the completion of the millennium, the Lord brings us home to the center of our galaxy to be with our infinitely loving Father-in- heaven — having perfected us and the Earth through His atonement.

To be with Him and to be like Him is His work and His glory; this is eternal life. In His infinite love, He also provides immortality to all who chose with Christ to follow the Father's plan in our pre- mortal epoch. This means that all will be resurrected to a degree of glory with the exception of those who turn against God and follow Satan. They go to outer darkness as part of the final judgment.

So we see a big wheel turning in our Galaxy in a few weeks of God's time with one week for the Earth's mortal journey and perfection. And as Abraham, Moses, and Paul all share, God is the creator of worlds — even without number to us. Modern astronomers estimate there could be millions of worlds like ours in just our galaxy. They estimate there are about 10^{11} (100 billion) stars in our galaxy. On a clear starry night — with no city lights to bother — like in Fountain Green where we live, you can see about forty thousand stars. Essentially all the stars you can see are in our Milky Way Galaxy. This means that for every star you can see, there are about ten million that you cannot see, but are in our galaxy. Astounding!

Then current astronomical estimates are that for every star in our galaxy there is another galaxy. More astounding! And these galaxies are all nominally in a plane. In some near-death experiences and visions of others they have seen plane upon plane upon plane. (See the books *Visions of Glory* by John Pontius and *Taught by Christ* by Ralph V. Jensen) It seems God's work goes on forever and ever in all directions without limit. Man cannot comprehend the majesty of God, and this is who we pray to and He hears every prayer because He is not limited by space and time as we are unless we open up to transcendental time. Incredibly more astounding!

Then to know we are His offspring and can become like Him — all knowing, all loving, and filled with a fullness of joy, which does nothing but grow in magnitude as more and more of us come unto Him. In the scriptures it is called "One Eternal Round." A synopsis of this concept can be found in an article on the book's web site.[11]

As we examine the title of this book, *It's About Time*, we have seen its very broad implications and applications. Now we would like to ask ourselves with the understanding gained, "What is time?" As we attempt to answer it in the next and concluding chapter, we will again gain some fascinating insights and perspectives. These insights and perspectives will manifest God's love for us and how extensive His love is.

---

11) *http://ItsAboutTimeBook.com/the-time-table-of-the-lord-spiritual-science/*

Chapter 25

It's About Time

When I retired from the lab in Boulder, Colorado, I named my new consulting company Time Interval Metrology Enterprise (TIME). Time interval is not time. However, when we talk about time, it is usually around the concept of the ordering of events — measured in days, months, years, as for history from some agreed upon origin like the birth of Christ, which is still a time interval. Or, we measure how far we are into a day in hours, minutes, and seconds for our daily activities. These man-made intervals are not time either. A common scriptural phrase to describe such time intervals is from "time to time."

The time interval for the swing of a pendulum is called its period. The clock's frequency (designated by the Greek letter ν "nu") is the reciprocal of its period, $\nu = 1/\tau$. As we discussed in Chapter 20, the time of a clock is simply kept by adding up these intervals, τ, from some counting origin. The more stable the intervals from one cycle to the next the better the clock. As we have shared, time interval or frequency can be measured with more precision and accuracy than any other physical quantity. Is plus or minus one second in 16 billion years good enough? An atomic clock, as described in Chapter 20 uses as its "pendulum" the vibration frequency of the electromagnetic signal of a photon associated with some quantum transition in some appropriate atom or molecule.

It's About Time

Calendars and clocks are all time-interval devices. They do not give us time. If someone asks you, "What time is it?" You look to your watch or wall clock and tell them without hesitation. Are you telling them time? What you share is the accumulated-time intervals unique to that clock since that clock or watch was synchronized with some other timekeeping device. What you share will never be in exact agreement with official time as defined by your country or the world. Your hope is that it is approximately correct. For most of society, a few minutes of error is probably good enough. However, our high-tech society is now needing worldwide microsecond accuracy, which GPS can amazingly provide.

The official time for the United States in Boulder, Colorado, is not a physical clock. It is some optimal-weighted average of the clock signals from several atomic clocks. These readings are in a computer- time-scale algorithm to give a best estimate of what we conceptualize as ideal time. This ideal time is metaphysical; it is an idea in our heads, and we cannot access or measure it!

As astronomers saw the order in the heavens, they devised ways to use those regular periodic motions for timekeeping purposes. Astronomy was probably the first application of science. Can the heavens give us time? With what we know now, we observe irregularities in the periods of all cyclical-heavenly bodies. They are like our clock problem; they all keep different and variable rates with a big variety of accuracies and precisions.

The dictionary definitions of time are mostly about time-interval, and nowhere have I found a good definition of time. I believe that is the case because time is metaphysical. In the book, we have shared man's time (physical time and how it is generated), God's time (in the Eternity-Domain), spiritual time (which I have called transcendental time because of its coupling of us to God), and we could come up with many other kinds of time: psychological time, biological time, emotional time, meal time, time of birth, time of death, and the list goes on ad infinitum. But none of these are time in the abstract sense that we usually think about, and which scientist — in their text books — write down as an independent variable. Is it? Does it exist?

Perhaps the title of this book should be "It's About Clocks," since we can't get hold of time. I like the metaphysical definition of time given by my good friend, Chauncey Riddle, "time seems to be the possibility of change."[1] The flow of time, as many like to think about it, is in some sense a measure of the changes in our existence — the growth of plants, the changes in the seasons, and our human aging process. As my wife and I enter our 80th orbit around the sun, we are very well aware of that one! Getting old is not for the timid!

I believe the word "change" is significant, which is the key word in Chauncey's definition of time. In the atomic-clock world, which provides official time for our nation and the world — the Allan variance is the optimum estimator of clock-rate "change" for the ideal noise in atomic clocks. As mentioned above, the smaller the change the better the clock.

The best definition of time that I can come up with is, "time is the ordering of events to bring about the work of the Father, which is to bring to pass the immortality and eternal life of man."[2] This definition has the short coming that events are not time. Perhaps we can think about the "wheel of time" and the concept of "one eternal round" discussed in Chapter 24 as God's clock, which He orchestrates from the Eternity-Domain — outside of space and time as we view it.

In the Father's work, faith in the Lord Jesus Christ and repentance are key elements. We live in the "Dispensation of the fullness of times" as Father brings about His timing in seeing that the gospel message is heard by all of His children. We have His promise. Note in this quote the phrase "For it shall come to pass in that day..." — giving us a direct indicator of a very important future event — key to His Coming:

> And then cometh the day when the arm of the Lord shall be revealed in power in convincing the nations, the heathen nations,... of the gospel of their salvation. For it shall come to pass in that day, that every man shall hear the fulness of the gospel in his own tongue, and in his own language, through those who are ordained unto this power, by the administration of the Comforter, shed forth upon them for the revelation of Jesus Christ.[3]
>
> And because iniquity shall abound, the love of many shall wax

---

1) Riddle, *Think Independently*, p 79
2) Moses 1:39
3) Doctrine and Covenants 90:10-11

cold. But he that shall endure unto the end, the same shall be saved. And this gospel of the kingdom shall be preached in all the world for a witness unto all nations; and then shall the end [of the world] come.[4]

And I saw a new heaven and a new Earth:. . .[5]

Change is also associated with both repentance and with sin. We can use our agency, which has its sacred origins in our intelligences in the eternities, to choose repentance — change in our direction toward the characteristics of God by the power of the atonement through which we can overcome all sin and without which we cannot. Here the law of grace and mercy has full sway in our lives. Or we can choose sin — change in the other direction. Here the law of justice has full sway.

If we choose to use our time to make those changes in our lives to conform to the commandments of God, then we are the beneficiaries of the atonement and God's promised blessings. Then we are using our time in mortality to bring about the greatest joy in time and in eternity, which is life with God, life like God, and eternal family. He wants us to use our time to lay up these treasures in Heaven where "neither moth nor rust doth corrupt, and where thieves do not break through nor steal."[6]

If ye abide in me, and my words abide in you, ye shall ask what ye will, and it shall be done unto you. Herein is my Father glorified, that ye bear much fruit; so shall ye be my disciples.[7]

In the dictionary, space is defined both in terms of distance (the space between two people, for example) and in terms of time interval (the space of time needed to do something, for example). In our studies of the new UFT, space can be dissolved and two people can feel close when physically far away from each other.

Einstein's equations involve the four dimensions of space and time. His equations and theories are limited by the velocity of light, and the application

---

4) Matthew 24:12-14
5) Revelations 21:1
6) Matthew 6:20
7) . John 15:7-8

Chapter 25

of his equations for GPS are always associated with the changes in time over an interval as the clocks on board the satellites are resynchronized periodically.

In mortality, we typically think that we are constrained in our freedoms by our space and time — the size of our property and the length of our lives, for example. But in fact, our freedoms are dependent on coming to Christ. Only in Him can we be "free indeed."[8]

In a more general sense, quoting again from my friend's book: "Space seems to be the possibility of existence. Where there is no space, nothing can exist. If something does exist. It is because there is space there for it to exist."[9] We believe there is no such thing as immaterial matter. Even spirit is matter, but is more fine or pure, and can only be discerned by purer eyes.[10] Science has recently proven that the mind operates outside of the brain.[11] These findings are totally consistent with the new Unified Field Theory (UFT) research we have performed as reported in Chapter 21 and with the diallel-field-line structure. In this research, we have learned that God is not constrained and neither need we be by space and time as is typically viewed in the world. We described in Chapter 22 how to transcend space and time as we utilize transcendental time. Translated beings can move to wherever instantaneously, as an example. We know and have proven that prayer works. We do have a direct link to God in His time frame — the Eternity-Domain.

In summary, this new UFT, which we believe is not new, but as old as God, our prayer channel is a diallel-field line. This field exists throughout the universe and interconnects everything. Lynne McTaggert, in her excellent book, *The Field, The Quest for the Secret Force of the Universe*, describes several experiments showing examples of these diallel-field lines. She does not call them that, however, as diallel-field lines are unique to our work. We have conducted seven experiments to date — showing their existence and some of their characteristics. We feel we are on the edge of an enormously important piece of God's physics. We have given illustrations of how all four of the force fields in traditional physics are carried by these diallel-field lines. By way of review, these force-fields are gravity, electromagnetic forces, and the strong and weak nuclear forces. We believe that if these forces were viewed properly,

---
8) John 8:36
9) Riddle, *Think Independently*, p 79
10) Doctrine and Covenants 131:7
11) Sheldrake, *Science set Free*

from God's perspective, they would be viewed very differently than we view them today in traditional physics.

The diallel-field lines are much richer than just the four-force fields. As described in Chapter 21, seven is an important number in the Lord's arithmetic; it stands for completion or perfection. As is known, there are seven basic electron shells that provide the fundamental quantum states for the electrons making up all of the known elements. These, nominally, have spherical symmetry. The diallel-field lines, nominally, have cylindrical symmetry and seven communication channels and carry all four of the known force fields. In completeness of this diallel structure, there are seven spectral bands in these communication channels. They may be divided up in generic terms as follows:

1. the communication band (TV, radio, satellite, etc.);

2. the molecular band (coming from quantum transitions outside the nucleus); most of the visible light from the sun is in this band;

3. the nuclear band (coming from quantum transitions inside the nucleus) — a nuclear blast is a manifestation of this band;

4. the creation and/or annihilation band (where $E = mc^2$ as matter and light convert with the emission and/or absorption of cosmic rays);

5. this is nominally the gravitational-field band;

6. this is the band used for mind-to-mind, mind-to-plant or animal or another object, mind-to and from-God, ESP, etc.;

7. this is the band used by our great-loving Creator to bring about the grand harmony we see in the heavens and the Earth.

Bands five, six and seven are not limited by space and time and the velocity of light.

We feel like our few experiments are like baby-steps into the giant woods of God's infinitely-extending physics by which, motivated by pure and perfect love, He brings about His work of bringing to pass the immortality and eternal life of His precious sons and daughters, who we are. This is how He uses His time, as defined above, which is outside of our perceived space and time, with the goal of eternal lives for all who choose to come unto His Beloved Son. Because Christ is perfect and worked out a perfect atonement — fulfilling the will of the Father — He can perfect us — in spite of our imperfections. His grace and mercy are offered to all. The only ones excluded from receiving a fullness of joy in His eternal realms are those who choose to not come unto Him and to partake of all that He offers to give. I love the lines penned by Eliza R. Snow in the fifth verse of the Hymn, *How Great the Wisdom and the Love*:

How great, how glorious, how complete,
Redemption's grand design,
Where justice, love, and mercy meet
In harmony divine!

If we truly understood the majesty of God, we would let nothing get between us and Him. To be filled with His love and to do His will — bringing honor and glory to His Holy Name — would be our utmost desire. Again, prayer is our doorway and grand opportunity to come to know Him and His majesty. In that feeling, we would want the whole world to know Him as well. I am writing this book to help in that process for I feel a great love for all of His children and want them to know our great and loving God that all may partake of the joy that fills my soul as I write this in honor of Him whom I serve. It is my prayer that you will open your heart to the marvelous truths of God and partake of His blessings by hearkening to His voice and coming unto Him.

Conclusion

The vast majority of the human population has been deceived by a false postulate introduced into how science is done and how histories are written during the last century. This assumed postulate or guideline for how things should be done is commonly referred to as "naturalism," which in a summary sense excludes God or things spiritual. A very important message of this book is showing, scientifically, that this postulate is false. The ramifications are enormous. The spiraling down of morality and the increase in worldliness are direct evidences of the devastating effects of this false postulate. In contrast to this degeneration of society, we have shown that there is every reason to hope.

Jesus Christ, the Savior said, "I am come that they might have life, and that they might have it more abundantly."[1] I love this scripture as it has been fulfilled in my life to such a large extent. One of my other favorite scriptures is the prophet Lehi's counsel to his sons given some 588 years before Christ after they had come here to the Promised Land — North America. In his aged years Jacob (Israel) gave each of his sons a blessing.[2] In the blessing the Lord gave Joseph, under Israel's hands, the Americas as a place for Joseph's seed. Lehi is a descendant of Joseph.[3] We read from 2 Nephi Chapter 2:24-28 the following:

---

1) John 10:10
2) Genesis Chapter 49
3) Further details of that blessing can be found in 2 Nephi Chapter 3.

> But behold, all things have been done in the wisdom of him who knoweth all things. Adam fell that men might be; and men are, that they might have joy. And the Messiah cometh in the fulness of time, that he may redeem the children of men from the fall. And because that they are redeemed from the fall they have become free forever, knowing good from evil; to act for themselves and not to be acted upon, save it be by the punishment of the law at the great and last day, according to the commandments which God hath given. Wherefore, men are free according to the flesh; and all things are given them which are expedient unto man. And they are free to choose liberty and eternal life, through the great Mediator of all men, or to choose captivity and death, according to the captivity and power of the devil; for he seeketh that all men might be miserable like unto himself. And now, my sons, I would that ye should look to the great Mediator, and hearken unto his great commandments; and be faithful unto his words, and choose eternal life, according to the will of his Holy Spirit;

I believe these scriptures; they are a reality in my life, and I continually bless the name of the Lord for His countless blessings in my life. I have learned that all things testify of Him when we have hearts to feel, ears to hear, and eyes to see.

I have written this book in token of gratitude to my God, hoping that everyone who reads it will feel closer to God. I have shared light and truth within my mortal limitations as best I can with pure intent to serve you with this goal in mind. In those areas where I am not an expert, I have drawn on the best experts in the world to the best of my knowledge.

In that regard and as we anticipate the oneness of a Zion society – the ideal society of God – we see much of society caught up with fear and worry about the future. The Lord tells us that "perfect love casteth out all fear." (1 John 4:18; Moroni 8:16) Fear and worry add to stress, which is one of our main killers and causes of disease. Dr. Lissa Rankin says that the immunity system shuts down when we are in a stressful state. We need to be in a relaxation mode to help our immunity system heal our bodies and resist disease. With the Ebola scare, this drives many into fear and worry, which is then counterproductive to resisting diseases. There are lots of natural remedies and ways to ward off Ebola and other pandemics. For example, read the article on the book's web site, *Protect Yourself During A Pandemic,*[4] and gain understand of the body's divine design to be healthy.

In the Savior's last great High Priest's prayer,[5] he prayed that we all may be one as He and the Father are one. I believe that oneness comes as we are born again of body, mind, and spirit and come to know the Father and the Son. Life eternal is the greatest of all the gifts of God, and in that prayer Jesus defines life eternal as knowing the Father and the Son.[6] When we know them we will know how to love as they love us. And then we will have the ideal society — living the first two greatest commandments. Abundance, joy, and eternal life will be the rewards of all those who focus on the Lord and His word. We have the sure promise of not being deceived when we so focus and treasure Him and His word.[7]

Many believe man is an evolved creature. I believe man has devolved from the state of our glorious first parents. As we are privileged to learn from the truths of God, from the scriptures, and from those who have had irrefutable near-death experiences, we can clearly see the degraded state of modern society as they have turned from God. My hope is that with this book, I can help in some way to turn that devolution trend around into helping us evolve toward God. This is the evolution I promote.

In contrast to the world view, as an example, I extract the following glorious Truths from an NDE of a dear friend of ours; whom I know and trust.

Exquisite Eve

Ralph V. Jensen has authored the book, *Taught by Christ,* about his near-death experiences. He is a humble, compassionate, and sensitive person, and my wife and I count it a great blessing to be in his circle of friends. We met him about a year ago in the occasion for him to share with a small group of friends some of the incredible details of teachings he received from the Savior in his NDEs.

Ralph is very much like me. He loves to take things apart to see how they work, and sometimes like me, when the thing is put back together there are

---

4) http://ItsAboutTimeBook.com/physical-spirituality-health-pandemic-protection/
5) John Chapter 17
6) John 17:3
7) JS Matt 1:37

parts left over! But you learn a lot in the process. Also, like me, he is very inquisitive — desiring to learn all he can in any given experience. So when he went across the veil with the Savior by his side showing him the things of eternity, his mind was racing with questions, and they got answered. The Lord knows the thoughts of our hearts and all the answers! So what an awesome experience he had, even though the pain of his heart attacks was excruciating? They were so bad, the doctors attending said he should not have survived, but the Lord had other plans for Ralph; among other things he needed to write a book and share these remarkable experiences.

One that he shared with us that first evening we met was soul-changing for me. When he was across the veil with the Savior, he was thinking who is more important, the man or the woman? The Savior responded to his thought question with a question, "Can a man make a baby?" The husband has the sacred privilege of planting the seed, but his wife has the divine nature to grow that baby within her womb and then nurture it after birth — making mother-child love the nearest to God's love. Consider the perfect shape of mother Eve created by God and placed in the Garden of Eden. Is there anything more beautiful than her shape from head to toe? She is exquisite!

Ralph learned from Christ that the Genesis story of Adam and Eve is symbolical. They were brought to the Garden of Eden from elsewhere. He was not privileged to share the details of that process, but described it as being perfectly logical and inspiring as God follows the laws of heaven in bringing about this great event. He saw that when Eve (the mother of all living) walked into the garden, everything was in awe and reverence. All the animals, the birds, the beautiful flowers there, all the plant life, and even the Earth itself paused in silent reverence because of her beauty and who she represented.

It is the woman who has within her divine makeup the ability to give birth to a potential god and goddess. Ralph's sharing what he witnessed lifted my view of mothers to see them more nearly as God sees them — helping Him bring about celestial beings in His perfect plan of happiness.

Since Ralph shared this glorious perspective, I have looked on the fairer gender with much more nearly celestial eyes. The female beauty and divine design inspires my heart at a new level. Her role is fundamental to God's work to bring to pass the immortality and eternal life of families that can be

Conclusion

together forever with Him. The birth of a baby brings Him additional joy to add to His fullness, which He wants to share with us. So the perfect plan of happiness works so that we can be like our Heavenly Parents as we come unto Christ, who is the way, the truth, and the LIFE

Ever since Satan realized it was he, not Eve, who was deceived in the Garden, he has been trying to denigrate her daughters. Let us help the Lord preserve their virtue, loveliness, and sacred role. President Nathan Eldon Tanner — after whom our youngest son is named — said the following in the October 1973 General Conference of the Church of Jesus Christ of Latter-Day Saints:

> A mother has far greater influence on her children than anyone else, and she must realize that every word she speaks, every act, every response, her attitude, even her appearance and manner of dress affect the lives of her children and the whole family. It is while the child is in the home that he gains from his mother the attitudes, hopes, and beliefs that will determine the kind of life he will live, and the contribution he will make to society.

Is there anything more powerful in the world than a mother's love? It is the closest to God's love.

How blessed I am to have a wonderful wife and mother of our children. Both my father and mother had a great impact for good on my life.[8]

For those who want to know Jesus better, I highly recommend you watch the DVD *Matthew*, and then read the book written by the actor, Bruce Marchiano, *In the Footsteps of Jesus*, whose life was changed forever as he came to know Jesus in an intimate way in the process of acting the part. He portrays a Jesus filled with love and joy the likes of which you have never seen before. In his preparations for the film, Bruce was deeply touched by a book he was divinely led to: *Jesus Man of Joy: Finding Meaning for Your Life through Knowing God* by Sherwood Eliot Wirt.

It is exciting to see true religion and true science coming into harmony in this century. The Lord has promised:

- God shall give unto you knowledge by the... Gift of the Holy Ghost...

---

[8] An excellent interview with Ralph is here: *http://www.youtube.com/watch?v=yjg0wBBTlMA*

- That has not been revealed since the world was until now.

- Which our forefathers have awaited with anxious expectation to be revealed in the last times,

- A time to come in the which nothing shall be withheld… and all things testify of Him.

- And I believe that the Unified Field Theory and the full and beautiful dimensions of the diallel-field-line structure will be brought to full fruition in the years ahead; it is the Lord's physics.

- The grand harmony of true science and religion will be a massive blessing to all the inhabitants of the earth during and after the purging of the Earth.

As the Lord sets His hand again the second time as He has promised[9] to bring about the "restitution of all things,"[10] we will see new science and new religious truths brought forth in harmony. The gospel will go to all nations and the righteous will be gathered from the "four corners of the Earth."

In summary, as I have shared before, we live in the most exciting time in the history of the earth. All of the prophets looked forward to our day. We have every reason to be filled with hope and love — anticipating the glorious return of our Lord and Savior. Satan — knowing it is his last hour — is filling the world with hate, fear, anxiety, envy, pride, sexual perversions, depression, discouragement (What can I do?), despondency, obesity, and dumbing us down so that we cannot see how he is deceiving the masses. Thus we see a world spiraling down in wickedness. This is a massive sign of the times.

In glorious contrast, the (few) humble followers of Christ are spiraling up in love, joy, and peace in the midst of the storms of life, because we know Him who is coming and that the purging times ahead are a necessary part of returning to the liberty and freedoms vouchsafed by the United States founding documents (US Constitution and Declaration of Independence) and the preparing a pure in heart people (Zion) who will be ready to meet Him as the Bride meets the Bridegroom. Satan has no power to destroy such a bright hope when we are totally focused on Him, who is our Redeemer and are filled with His love.

---

9) Isaiah 11:11; 2 Nephi 25:17; Jacob 6:2
10) Acts 3:19-21

Conclusion

The Apostle John and the prophet Mormon tells us what we need to do to receive these enormous blessings:

> Beloved, let us love one another; for love is of God; and everyone that loveth is born of God, and knoweth God. He that loveth not, knoweth not God; for God is love. In this was manifested the love of God toward us, because that God sent his only begotten Son into the world, that we might live through him. Herein is love, not that we loved God, but that he loved us, and sent his son to be the propitiation for our sins. Beloved, if God so loved us, we ought also to love one another. No man hath seen God at any time, except them who believe. If we love one another, God dwelleth in us, and his love is perfected in us.[11]

> Wherefore, my beloved brethren, pray unto the Father with all the energy of heart, that ye may be filled with this love, which he hath bestowed upon all who are true followers of his Son, Jesus Christ; that ye may become the sons of God; that when he shall appear we shall be like him, for we shall see him as he is; that we may have this hope; that we may be purified even as he is pure. Amen.[12]

I feel so blessed to be the recipient of the truths of the Restoration of the fullness of the gospel of Jesus Christ. I know that Joseph Smith, Jr. is the prophet called of God to head this last dispensation in preparation for the coming of our Lord and Savior, Jesus Christ. In Joseph's vision of the Father and the Son, he validated Newton's conclusions from the scriptures and from early Christian writings. The heavens have been opened and the truths of the Restoration Newton saw are going forth to bless all nations and to bring them to a true knowledge of the Father, the Son, and the Holy Ghost — to know of their love. The Spirit has witnessed to my soul that Heavenly Father loves each of us with an infinite love, and thus He gave us His Beloved Son to work out the infinite and perfect atonement that brings perfection to Father's perfect plan of happiness — to reach out to us in our imperfections — and opens the door to every soul to receive that fullness of joy if they but choose to come unto the Father through the Son, for He is the way, the truth, and the life.[13] This I know and testify, in the name of Jesus Christ.

---

11) JST 1 John 4:7-9
12) Moroni 7:48
13) John 14:6

We finished writing the first edition of this book on July 4th, 2014; INDEPENDENCE DAY — symbolic, I believe, of "Ye shall know the truth, and the truth shall make you free."[14] It is only by truly coming to Christ that we can be free as individuals and as nations. I pray that we may do so and look forward to His glorious millennial reign as King of kings and Lord of lords when, indeed, that shall come to pass. How glorious it will be.

---

14) John 8:32

Glossary

BIPM
　Bureau International des Poids et Mesures (International Bureau of Weights and Measures)

BPA
　Bisphenol A chemical; in plastics, look for BPA-free.

CD
　Compact (audio) disc

D&C
　Book of Doctrine and Covenants

DDT
　Dichlorodiphenyltrichloroethane insecticide; has been banned in the USA

DNA
　Deoxyribonucleic acid

DVD
　Digital Video Disc

EE
　Electrical Engineering

FDA
　Food and Drug Administration

GPS
　Global Positioning System

ID
　Intelligent Design

ION
　Institute of Navigation

IRS
: Internal Revenue Service

JST
: Joseph Smith Translation: The Lord instructed Joseph to go through the King James edition of the Bible and to fix some translation errors and to provide some missing text.

KJB or KJV
: King James edition of the Bible

LDS
: Latter-day Saint – referring to the members of the Church of Jesus Christ of Latter-day Saints as a noun; as an adjective, something associated with the Church.

LDS Scriptures
: Bible, Book of Mormon, Doctrine and Covenants, and Pearl of Great Price

Mormon
: Nickname for a member of the Church of Jesus Christ of Latter-day Saints; Mormon was an ancient prophet-historian who lived in America (311-385 AD)

NBS
: National Bureau of Standards for the USA

NDE
: Near Death Experience

NIST
: National Institute of Standards and Technology (USA)

NWO
: New World Order

ORNL
: Oak Ridge National Laboratory

PCT
: Pacific Crest Trail

Glossary

PVC
Poly(vinyl chloride) polymer

SLC
Salt Lake City, Utah USA

SAR
Search and Rescue

TAI
Time Atomic International (International Atomic Time)

UFT
Unified Field Theory

UTC
Universal Time Coordinated

WWV
United States standard time and frequency and broadcast station (Ft. Collins, Colorado)

WWVB
United States standard time and frequency and broadcast station at 60 kHz (Ft. Collins, Colorado) – the source for time for the "so called" millions of atomic clocks that are commonly available

Index

A

Abinadi 61, 119

Abraham 8–9, 32, 39, 43, 57, 59, 78, 108, 260, 323, 331, 333–335, 341–342, 347, 349, 352, 360, 365, 367

Accountability 21

Adam viii, 15, 21, 37, 39, 49, 55–56, 68, 101, 111, 154, 157, 223, 342, 365–366, 378, 380

Adamic language 55–56, 60–61, 69, 72, 128

addiction xv, 81, 83, 86, 102, 204–205, 209, 358

Afrikaans 51–54, 62–63

Agassiz, Louis 9–12, 34, 148, 155–157, 331

Augustine, St. 50, 250, 258–259, 332, 355

Allan, Collin iii–v, xii, xiv, xvi, xviii–xx, 184, 266, 275, 288, 290–293, 309–310, 371

Allan, David W. iii–v, xii, xiv, xvi, xviii–xx, 184, 266, 275, 288, 290–293, 309–310, 371

Allan, Dean iii–v, xii, xiv, xvi, xviii–xx, 184, 266, 275, 288, 290–293, 309–310, 371

Allan, Edna iii–v, xii, xiv, xvi, xviii–xx, 184, 266, 275, 288, 290–293, 309–310, 371

Allan, McKaylee iii–v, xii, xiv, xvi, xviii–xx, 184, 266, 275, 288, 290–293, 309–310, 371

Allan, Nathan iii–v, xii, xiv, xvi, xviii–xx, 184, 266, 275, 288, 290–293, 309–310, 371

Allan, Sterling iii–v, xii, xiv, xvi, xviii–xx, 184, 266, 275, 288, 290–293, 309–310, 371

Allan, Sylvester iii–v, xii, xiv, xvi, xviii–xx, 184, 266,

275, 288, 290–293, 309–310, 371

Allan variance xviii–xx, 266, 275, 288, 291, 293, 309–310, 371

Alma 60, 63, 95–97, 99, 106, 112, 119, 122–123, 125, 337, 347

Almighty 18, 338

AMA 185

America xi, xv, xxii, 13–14, 21–22, 43, 51, 60, 63, 71, 75, 110, 153, 155–158, 166, 175, 185, 194, 198, 241, 249, 255, 259–260, 331, 333–334, 348–350, 352, 377

American diet 164–167, 192–193, 195

American Religion, The 43, 343

amino acids 187–188

Amondawa people xxviii

Amos 60, 347

And it came to pass viii, 53, 55–61, 64, 69–70, 95, 119, 169, 180, 361

And it shall come to pass viii, 57–58, 60–61, 69–70, 120–121, 123, 361

And now it came to pass 69–72

anti- Christ 21, 155, 245

Angel 28, 43, 97, 125, 167

angel 28, 43, 97, 125, 167

Angels xxvi, xxix, 27–33, 40, 77, 97, 124, 148, 175, 236, 248, 299, 303–304, 320–323, 326, 333, 338, 345, 347, 358

angels xxvi, xxix, 27–33, 40, 77, 97, 124, 148, 175, 236, 248, 299, 303–304, 320–323, 326, 333, 338, 345, 347, 358

Apollo 13 201–202

Ashby, Neil 279

Arab 78

Archaeology 260

Aristotle 208, 342

Index

Art of Listening ix, 126

Atheism 7, 11, 21, 35, 105, 149–152, 155, 176

atheism 7, 11, 21, 35, 105, 149–152, 155, 176

Atheist xiii, xxv, 11, 41, 153, 155

atheist xiii, xxv, 11, 41, 153, 155

Atheists xiii, xxii, xxv, 11, 25, 41, 149–150, 154–155, 245–247

atheists xiii, xxii, xxv, 11, 25, 41, 149–150, 154–155, 245–247

atomic clocks xvii, xxi, 4, 234, 268, 271–274, 277–278, 280–285, 287–289, 293–295, 297, 314, 370–371

atomic clock iii, xv, xx, xxv, 183, 273, 276–278, 284, 288–290, 314, 369

atonement viii, 5, 15–17, 21, 25, 33–34, 37, 42, 50, 59, 64, 68, 78, 85, 88–89, 91, 96, 98, 101, 103, 106–107, 110–112, 114–115, 154–155, 175, 201–205, 225, 255, 259, 325–326, 345–346, 364–367, 372, 375, 383

attributes 33, 41, 79, 89, 124, 301, 331

Aurelius, Marcus 208

B

Babbel, Frederick & June 327

Babylon 5, 21, 123, 207, 255, 268, 340, 349, 356, 358, 362

Back to Eden 167–168

Baker, Amy 206

Ballard, Timothy 347–348

baptism 95, 103

baptism of fire 95, 103

Barack Obama and the Enemies Within 243

Barnes, James A. 275, 277

Bates, William H. 191

Batmanghelidj, Fereydoon 181–182, 184, 197, 212

Beck, Glenn 243, 348

Béchamp, Antoine 198

belief xiii, xxv, 5, 20, 41, 152–153, 157, 167, 188, 195, 217–218, 302, 305

Beloved Son 25, 34, 68, 78, 87–88, 90, 104, 106, 109, 124, 231, 255, 340, 346, 360, 364, 375, 383

Bement, Delta 181, 223

Benson, Ezra Taft 98–100, 208, 215

Benson, Herbert 98–100, 208, 215

Bercot, David W. 14, 50, 115, 249–250, 257–259, 262, 332, 340, 355

Berlinski, David 149–152

Bible xv, xxv, xxx, 5, 9–15, 20, 37, 42–44, 49–50, 57, 59–61, 63–72, 78, 80, 108, 146, 154–155, 157, 238, 243, 249–250, 257, 259–260, 267, 331, 334, 336, 340, 342, 344–345, 352

Big Bang 7

Biology 147–149, 151, 195, 198, 217–218

Biology of Belief 195, 217–218

BIPM xx, 274, 281, 283–286, 296

Bishop xviii, xxiii, 52, 55, 62–63, 73, 254

Blue-jets 312–317

blood 25, 98, 100, 120, 136, 174, 182, 184–186, 188–191, 195–197, 201, 209, 225, 229, 233, 237, 254, 328, 360–361, 366

blood moons 361

Bloom, Harold 43, 343–344

body iii, xx, xxvii–xxix, 4, 19, 24–25, 27, 29, 39, 44, 78, 80, 83–85, 87, 103, 108, 112, 115–116, 121, 125–126, 128, 144–145, 163–164, 166–168, 171–172, 175, 181–192, 194–195, 197–199, 203–204, 207, 212, 214–219, 221, 224–225, 229–230, 233, 254, 261, 293, 301, 304, 321–322, 324–325, 341, 345–346, 350, 379

Bohm, David 302

Index

Bohman, Fred 194

Book of Abraham 8–9

Book of Ether 56

Book of Mormon viii, xv–xvi, 44–45, 51–57, 59, 61–63, 65–67, 69–72, 95, 97, 101, 209, 259–260, 344, 349, 355, 358–359, 361–362

Book of Moses 346

Boulder, Colorado xi–xii, xvii–xix, xxi, xxiii, xxv, 175, 177, 180–181, 183, 223, 237, 268, 272, 283, 286–287, 290, 292–293, 297, 314–315, 348, 369–370

Born of God 81, 88, 91–92, 95–101, 383

Boss, Wayne R. & Leslee S. 93, 323

brain xxix, 28, 49, 85, 103, 106, 121–122, 167, 174–176, 185–190, 204, 221–234, 302–303, 322, 373

Brain Longevity 221–222, 225–226

Bride 180, 255–256, 356–358, 361–362, 382

Bridegroom 77, 180, 255, 356–358, 360, 382

Brinkerhoff, Val 361

Brook, Michael 195, 199, 217–218

Brown, Hugh B. 177

Brown, Joyce 324

Brown, Victor L. 83, 85

Brummer, Johannes P. 52, 63

Burns, Robert 105

BYU (Brigham Young University) iii, xiv, xvi, xxi, 314

C

Cambridge University 148, 336–337, 343

Cancer 164, 172, 184, 189, 192, 211–212

Cancer is a Fungus: A Revolution in Tumor Therapy 211

Capernaum 230

Caplan, Mariana 128

Cataclysmic 5, 344, 365–366

Catholic xiv, 43, 363

Celestial 15–16, 108, 113, 126–127, 203, 225, 261, 285, 362, 364–367, 380

Chapman, Sandra Bond 223, 229–232

Chandler, Keith 199

charity xxx, 40, 79, 110, 115–116, 122, 124, 126, 181, 242

chiasmus 96–97, 355–356, 362, 366

chlorophyll molecule 114, 237

Christ viii, xv–xvii, xxiv, xxvi, 5, 15–16, 21, 23–24, 26–27, 31, 33–35, 37, 40, 42–43, 50, 57–59, 64–65, 68, 76, 79–80, 88–89, 91–94, 96–98, 100–103, 105–108, 110–111, 113–117, 121–122, 124–125, 127, 136–137, 150, 154–155, 159, 172, 180, 201–209, 214, 225, 243, 245–248, 250–252, 254–255, 259–260, 262, 266–268, 321–322, 325, 330, 335, 337–339, 342, 344–345, 356–358, 361–362, 365–369, 371, 373, 375, 377, 379–384

Christ-centered xv, xxiv, 89, 102, 110, 113, 207, 356

Christ-centered home 89, 113

Christian iii, xv, xxiv–xxv, 12, 14, 57, 68, 75, 87, 135–136, 154, 156, 175, 229, 249, 251–252, 255–259, 262, 300, 332, 340, 342–343, 350, 358, 383

Christian Nation 14, 175, 249, 256, 262, 358

Christianity xxv, 14, 35–36, 42, 50, 161, 204, 249–250, 255, 259, 340, 355

Church of Jesus Christ of Latter-day Saints xv–xvii, 57–58, 68, 113, 204, 356, 358, 381

Clark, E. Douglas 8, 43, 52, 331, 334–335

Clement 252

Clovis culture 157

Clovis point 156–157

Clingo, Karie vi 195–196

Index

Colbert, Jeannette vi

Coldwell, Leonard 211–212, 221

communism 241, 244

Constantine 257

Constitution 14, 170, 239–240, 243, 249, 348, 357, 359, 382

Constitutional principles 244

correct time 271, 295

cortisol 221, 226

Crawford, Myron xxiii

create evil 35–36, 105

creation 5, 7, 15, 68, 105, 115, 127, 149, 154–155, 158, 170, 231, 290, 306, 341, 365–366, 374

creationism 149–150

creationist 10, 307

D

Dahlen, Marie 191, 195

Daniel 60, 169–170, 342

Danielson, Magnus xix

Dark Ages xxi, 14, 259, 340–341, 352, 355

Darwin, Charles 9–10, 21–22, 148–149, 155–156, 176, 331

Darwin's Doubt 22, 148–149

daughter of Zion 118, 168

Day, Lorraine 191, 212

Dawkins, Richard 11, 148–152, 154, 246

Dead Sea scrolls 259–260

deceit 76, 123, 147, 246, 360

deceiver 20, 34, 76, 243, 358

Declaration of Independence 14, 249, 348, 351, 382

Deduction 5-6, 13

defense 152, 165, 239, 241–243, 301

degrees of glory 364

dehydration 185–191

dementia 164

depression 184, 187, 189, 216, 357, 382

Devil's Delusion 149–151

diabetes 164, 166, 184, 189, 192

Diamond, Marian 176, 227–228

distorted love 74–75, 78, 82, 94

divine love 25, 75, 331

Divine Providence 117

devil 16, 33, 40, 149–151, 245, 324, 345–346, 378

diallel-field lines xxix, 104, 144, 299–304, 306–309, 311–315, 317, 373–374

DNA 8, 11, 147, 156, 331, 333, 350

Doctrine and Covenants 19, 32, 38–40, 57, 61, 69, 108, 110, 114, 157, 225, 237, 265, 337–338, 353, 355, 357–359, 362, 364, 366, 371, 373

D&C 77, 90, 98, 106–107, 110, 112–117, 120–125, 127–129, 165–168, 171, 179–180, 203, 208, 225, 248

Drummond, Henry 79–81

Dufty, William 191

Duke, James A. 114

Duncan, Kirk and Kim 231

Dutch Reformed Church 51–52, 54

E

Earth xi, xiii, xxvii, 4–5, 13, 16, 19–21, 24–26, 28–31, 33, 40–41, 43, 59, 61, 68, 77, 89–90, 98, 103, 106–107, 111, 119–120, 125, 127, 129, 138, 143, 156–157, 167–169, 171, 178–180, 194, 202–203, 207, 214, 216, 219, 231, 236, 242–243, 251, 254–256, 261–263, 265–266, 268, 272–273, 278, 283–288, 294–295, 300, 306–309, 311–313, 315, 317, 322–323, 331, 333–335, 337, 341–342, 344–346, 350, 353–354, 357–358, 360, 362, 364–367, 372, 374, 380, 382

Earthbound 17, 29–30

Earth's spin rate 272–273

east xvi, 39, 142, 146, 157, 178

East Asia 157

Eden 21, 23, 39, 76, 167–168, 366–367, 380

education xiv, xxii, 7, 21, 45, 62, 148, 246, 358

Edwards, Jonathan 350

Egyptian 8, 52–56, 351

Einstein, Albert xxii, xxv, xxvii, 4, 7, 18, 227, 232, 280, 297, 299–300, 302, 322, 365, 372

Eisenhower, Dwight D. 208, 240, 243

El viii, 58–60, 69, 272, 361

Electric Universe 7, 245, 247, 302, 366

Enderlein, Günther 198

energy xii, 7, 74, 94, 111, 116, 123, 154, 194, 225–226, 233, 235–237, 240, 267, 273–274, 295, 306–309, 311–313, 315–317, 383

Enoch 50, 169, 171, 265

Ensign 62, 100

ephemeris 273, 286

ESP 302, 306, 366, 374

Essen and Perry 273–274

estrogen 172–175

eternal joy 78, 115

eternal life xxvi, xxix, 16, 25, 34, 39–40, 68, 74, 81, 87, 89, 95, 99, 103–104, 107, 109–111, 114–115, 124, 128–129, 155, 209–210, 230, 234, 254, 266, 345, 353, 364, 367, 371, 375, 378–380

eternity iii, xi, xxii, xxv–xxvii, xxix–xxx, 4, 18–19, 37, 73–74, 82, 91–92, 104, 112–113, 116, 127, 138, 155, 159, 225, 229, 297, 301, 307, 312, 320–324, 337, 347, 366, 370–373, 380

Eternity-Domain xi, xxii, xxv–xxvii, xxix, 4, 18–19, 104, 297, 301, 307, 312, 320–322, 324, 366, 370–371, 373

Evangelicals 67, 257

Eve 15, 21, 39, 49, 55, 68, 154, 156–157, 365–366, 379–381

evil 4, 16, 21, 24, 29, 35–37, 39–40, 42, 79, 81, 88, 105–107, 115, 118, 125, 153, 205, 207–208, 245, 250–251, 253, 255, 321, 323, 325, 334, 342, 346, 354, 357, 378

evolution, law of 5–7, 9, 21, 55, 58, 147–151, 155–157, 159–161, 176, 247, 358, 379

evolution theory of 5–7, 9, 21, 55, 147–149, 155, 160, 176

E

Expelled: No Intelligence Allowed 150–151

experiment 5, 9, 276, 278, 290, 294, 309–318

Exquisite Eve 379

Eyring, Henry B. (LDS leader) -Henry B. Eyring 100, 314

Eyring, Henry (scientist) - Henry Eyring 100, 314

Ezekiel 60

F

faith xiv, xx, xxiii, 19–20, 24–25, 33–34, 37, 40–42, 49, 65, 71, 75, 85, 87–89, 95–96, 100, 102–103, 106, 110, 116, 122, 124, 126, 132–133, 138, 140, 149, 177, 180, 201–202, 204, 234, 242, 248, 254, 258–263, 302, 304, 323, 325, 330–331, 337, 346, 349–350, 353, 360, 363, 365, 371

Index

family i, vi, xi–xii, xiv, xvii, xx, xxiii, xxx, 18, 25–26, 30–31, 65–66, 73, 75, 82, 85, 89–90, 94, 101–102, 112–113, 121, 127, 129, 133, 135, 138, 141, 144–146, 155, 157, 164, 167, 175, 177, 181, 183, 209, 216, 236, 256, 305, 323, 327, 360, 362, 364, 372, 381

fall of Adam 15, 21, 68, 101, 111, 365–366

father (parent or founder) xiv, 8, 14, 21, 56, 63, 71–72, 81, 95, 97, 101, 119, 127, 132, 213, 249, 253, 258, 260, 299, 328–329, 346, 352, 357, 360–361, 381, 382

Father (God, Heavenly Father) see also *Heavenly Parents* 5, 24–26, 34, 37–38, 41, 74, 78, 82, 104, 108–109, 111, 124, 127, 135, 138, 155, 177, 207, 225, 231, 266, 345–347, 354, 357, 365–366, 383

Federal Reserve 359

Fields, Jonathan 214–215, 228

F

fifth-dimension xxii, 19, 299

flood 5, 20, 59, 68, 154, 157–158, 203, 300, 344, 365–366

foreordained 33–34, 106

Fountain Green, Utah 55, 73, 237, 309, 316, 318–319, 367

four dimensions xxii, xxvii, 4, 18, 104, 116, 121, 124, 126, 297, 299, 322, 372

food ix, 83, 101, 139, 144, 163–173, 175, 187–190, 192–194, 207, 234, 330

FDA 166–167, 188

fractals 277

Franklin Institute in Philadelphia 18, 32

free choice 16, 31, 33, 37, 41, 105–106, 277, 323, 354

freedom 44, 76, 95, 103, 107, 136, 153, 171, 204, 236, 240, 242, 244, 255, 291, 354, 357, 359

G

galaxy xxviii, 202, 365, 367–368

Gamble, Foster 237

Garden of Eden 21, 23, 76, 366–367, 380

Gautschi, Paul 167

Genesis 15, 39, 56, 59–60, 78, 84, 265, 344, 377, 380

Gentile 78, 116, 355, 358, 361

Gus German, Gus 319

gift 52, 70, 86, 88, 90, 92, 95, 104, 107, 151, 258–259, 267, 381

Gilder, George 147–148

GPS iii, ix, xi, xx–xxi, xxvii, 4, 134, 268–269, 271–272, 274, 278–282, 285–286, 290, 295–297, 299, 318–319, 322, 370, 373

Global Positioning System xi, 274

GMT 287

Gödel, Kurt 19

God's love xxx, 4, 12, 26, 28, 36, 45, 103–105, 116, 154, 156, 158, 216, 248, 334, 368, 380–381

God's Time i, iii, v, xi, xix, xxii, xxvii, xxix, 16, 269, 335, 362, 367, 370

Goble, Edwin G. 260

gospel xvi, 16, 25, 32, 34, 37, 42, 56–57, 64, 91, 93, 98, 102–103, 112–113, 116–117, 127–128, 159, 243, 253, 259–260, 323, 327, 334, 339, 352–353, 355–357, 364, 367, 371–372, 382–383

Gould, Stephen Jay 148

Grand Sachem 350

gravitational xxvi, 306, 308, 310, 312, 374

gravity xiii, xxvi, 219, 312–313, 373

great and abominable church 120

Green Pharmacy 114

Guiding Life Force 215

H

Index

Halford, Don xx

Handel 11, 176, 178, 180

happiness iii, xv, xxvii, xxix, 15, 17, 20, 24, 28, 30, 32, 37, 39–41, 64, 73–74, 79–81, 84, 88, 90, 98, 103–108, 111–113, 126–127, 138, 155, 159, 176, 185, 203, 210, 215–216, 219, 224, 233, 245, 255, 267, 321, 323, 331, 345, 347, 351, 354, 362, 366, 380–381, 383

Hormone Imbalance 172

harmony vi, viii, xi, xiii, xx, xxiii, xxv, xxx, 1, 3, 5, 7, 9, 13–14, 22, 61, 69–70, 72, 81, 89–90, 102–103, 112, 149, 154, 156–158, 171, 175–176, 187, 192, 202, 218, 306–307, 322, 333, 349, 361, 374–375, 381–382

Harrison, Colleen C. 101

Harrison, John 101–102, 279–280

Harrison, Philip A. 102

Hart, Michael 255, 352

Harvard 10, 85, 93, 148, 215

healing 31, 82, 102, 114, 128, 145, 164, 181–182, 199, 201, 204–205, 212–219, 228, 328–329

health iii, xxvii, 114, 125, 164–167, 171–172, 175, 184–189, 191–199, 216–219, 223, 228, 233, 237

heart i, vi, xii, xv, xxiii, xxix–xxx, 4, 7, 15, 22, 26, 32, 37–38, 43–44, 59, 66, 71, 74–75, 78, 81, 83, 86, 89, 97, 100–102, 105–106, 108, 110–111, 113, 116–118, 121–128, 138, 144, 164, 169, 171, 177, 180, 184, 189–190, 192, 194, 199, 208, 213–214, 229–230, 233, 243, 248, 252, 266–267, 274, 300, 305, 322, 324–325, 330, 340, 342, 348, 354, 356–361, 364–365, 375, 380, 382–383

heart attack 199, 233

heaven 3, 14–18, 20, 27–28, 30–31, 33, 42, 59, 68, 73–75, 77, 89–91, 96, 107, 109–110, 113, 116, 120, 122, 127, 169, 178–179, 202, 207, 215–216, 224–225, 231, 247, 251, 253, 261, 265–268, 320, 322–323, 326–327, 331, 338, 340, 343, 345, 352, 358, 362–363, 365, 367, 372, 380

Heavenly parents 15, 25, 32, 106–107, 114, 121, 127, 224, 381

Hebrew 12, 52–53, 56–57, 61, 350, 361

hemoglobin molecule 114, 237

hen viii, 117, 121, 359

high fructose corn syrup 165

HFCS 165–166

Hinckley, Gordon B. 67–68

Hinze, Brent 24

Hinze, Sarah 17, 19, 24, 27, 32

Hodge, Cliff 279

Holy Ghost 15, 40, 43, 64–65, 95, 98, 110, 122, 124–125, 172, 248, 254, 267, 339–340, 381, 383

Holy Sabbath 103

Holy Spirit 40, 50, 65, 77, 96, 101, 104, 124, 247, 326, 378

Hopewell Indians 350

Hosea 61, 169

Hugo, Victor xxvi

human nature 80, 100, 208–209

humble 13, 23, 28, 41, 65, 75, 101, 117, 121, 126, 242, 325–326, 357, 379, 382

humor 24, 233

Hutchison, John 240

hydration 139, 181–182, 185–186, 188–189, 197, 212

I

ideal meal 164–165, 167

idolatry 41, 43, 325–326, 342

IEEE xviii–xix

Ilibagiza, Immaculée 115, 256

immune xii, 192, 233, 237

immunity 164, 198, 378

induction 5–6, 13, 24, 32

inspiration xi, xviii, xxi, 44, 51, 57–58, 61, 63–64, 70, 181, 202, 283, 299, 304

intimacy 82–86, 209, 233

integrity xxii, 6, 22, 24, 43, 64, 71, 99, 124, 150, 153, 155, 213, 241, 259, 284, 348

intelligence 21, 34, 49, 116, 121, 149–152, 214, 224–225, 234, 275, 301, 337, 354

intelligent design (ID) 11, 147, 149–151, 156

International Bureau of Weights and Measures (BIPM) xx, 274, 281

international time 280, 282, 285, 314

internet xi, xv, 23, 75, 83, 156, 215, 245–247, 286, 303–305, 331

IRS 240

Isaiah xxviii, 33, 35, 44, 51, 60, 105, 118–119, 134–135, 179–180, 208, 228, 322–323, 356, 361, 364, 382

Islam 50, 332

J

Jehovah viii, 24, 58–61, 69, 117, 179, 255, 262, 352, 361

Jefferson, Thomas 150, 166, 351, 354–355

Jekyll Island 359

Jensen, Ralph V. 23, 368, 379

Jeremiah 17, 23, 60, 76, 118, 169, 347

Jerusalem 13, 118–119, 168, 265, 322, 342, 349–350, 360, 362

Jew 78, 150, 152

Job i, xvii, 10, 25, 32, 34, 76, 85, 105–106, 150, 155, 180, 218, 237, 268, 273, 281, 283, 302, 331

Joel 61

Jonah 50

Jones, Steven E. 335, 338, 343

Jude 56

K

Kabbalah 57, 69

Kabbalistic 58–61

Kaptchuk, Ted 215

Kastleman, Mark 82–85, 102, 204, 209

Kennedy, John F. 354

Kessler, Fred i

Khayyám, Omar 36

kibbutz 265

killer xii, 216, 221, 326, 330

 killed xxv, 36, 68, 117, 237, 259, 333, 350–351

King Benjamin 88, 101, 114

King James xxx, 11, 63–64, 72

King of kings 13, 154, 203, 255, 263, 268, 322, 384

kingdom of God 16, 81, 99, 135, 207, 322, 353

kingdom of heaven 96, 109–110, 216, 253, 322

Klobuchar, John A. (Jack) 281

Kolob 202, 367

L

Lake Tahoe 141, 145–146

Lambson, Celeste i, iv

Lambson, Kevn i, iv

land of promise 234, 349

Index

land of the free 234, 239, 260, 348, 352, 360, 367

language xxviii, 15, 52–56, 60–64, 69, 72, 94, 128, 146, 209, 227, 303, 371

laughter xii, 233

Levine, Judah xix, 287

law i, xiii, 7, 18, 38, 40, 68, 73, 76, 104, 107–109, 114, 116–117, 122, 124, 159–160, 171, 179, 225, 243–244, 252, 259, 265–267, 322, 325, 331–333, 342, 346, 354, 360, 364, 372, 378

law of heaven 73, 107, 109, 116, 122, 265, 331

Law of Love 109, 114, 122, 266, 325, 331–333

leap seconds 285, 287

learn iii, xi–xii, xv, xxiii–xxv, xxvii, 4, 14, 16–18, 25–28, 30, 32–33, 38, 41, 56, 74–75, 78, 81–82, 87–88, 93, 104, 106, 108–110, 113–116, 121, 126, 134, 142, 145, 153, 158, 164, 185, 191, 202, 219, 223, 228, 232, 242, 247, 266, 271, 303, 320, 323, 326, 330, 337, 345, 348–349, 358, 363, 379–380

Lee, Harold B. 209

Lee, Ranae i, 57, 209, 299, 316

Leet, Leonora 57–61, 69, 361

Left to Tell ix, 115, 256

Lehi 70, 361, 377

Lewis, C. S. 20, 37, 42, 105, 203

liberty xxii, 14, 16, 40, 234, 239, 241–244, 348–349, 351–352, 354–355, 357, 359, 378, 382

Life After Life 19

life-before-life 18–19, 24, 32

Life Before Life 17, 32

light xv, xx, xxiii, 3–4, 22, 26, 28, 30, 35, 37, 51, 64, 75, 86, 102, 114–115, 122, 133, 152, 163, 171, 176–178, 180, 191, 195, 208, 214–216, 219, 230, 232, 236, 247–248, 251, 261, 278, 284, 286, 295, 301, 303, 305–306, 308, 311, 313, 316–317, 321, 323, 336, 339–340, 342, 353, 364–367, 372, 374, 378

Lighthill, Michael James 7

Lincoln, Abraham 323, 348

Lipton, Bruce 195, 199, 217–218

listening ix, xii, 35–36, 110, 126, 191, 214, 228–229, 300, 333

LOGOS 339

loneliness 82–83, 93, 144, 216

Loran-C xxi, 319

Lorbeck, Jeff 300, 316

Lord of lords 154, 203, 255, 263, 268, 322, 384

Lost Civilizations of North America 13, 158, 333, 349–350

Loudon, Trevor 243–244

love implies obedience 108

Love your enemies 251, 332

Lucifer 33, 323

Luke 56, 112, 116, 138, 209, 230, 237, 252–253, 347

Luther, Martin 50, 258–259, 332

Lyons, Harold 273

M

Magnetometer 240

Malheiro, Beverly xiv

Mandelbrot, Benoît B. 277

Man's Inhumanity to Man 105, 326

Man's Time iii, xix, xxii, xxvii, xxix, 175, 267–269, 271–272, 299, 322, 335, 370

Manasseh 333

Mapleton, Utah i, xiv, xvii

Marchiano, Bruce 381

Index

Mark xxviii, 56, 59–60, 65–67, 82, 102, 116, 161, 204–205, 209, 325, 342, 351–352, 360

Martyr, Justin 63, 252

Mason, Robert 352

Massasoit 349

Matthew 6, 16, 39, 42, 56, 75, 135, 197, 224, 251, 256, 261, 338, 342, 345, 364, 372, 381

Maxwell, Neal A. 38

May, Wayne N. i, iv, xii–xiii, xxii–xxiii, xxviii–xxix, 4–6, 10, 18, 20, 23, 27–28, 34–35, 38–41, 45, 50, 52, 62, 64, 74, 77–80, 82, 86–91, 93–95, 98, 100–101, 104, 106, 111, 114–116, 121, 123–124, 128–129, 134–135, 138–140, 146, 150, 153, 158, 165, 169–170, 172, 175–176, 184, 190, 192, 197, 204, 208, 214, 223–224, 241–243, 247, 251–252, 254–256, 258, 260–261, 263, 267, 271, 277, 281, 289–290, 297, 300, 302, 304–306, 311–312, 321, 330–331, 333–334, 337, 342, 344, 346, 351, 359–361, 363, 374–375, 378–379, 383–384

Mayflower Compact 350

Message of Gladness 102, 179–180, 362

Mezuzah 123

McClean, Matt 196–197

McKay, David O. 98, 100, 206, 208

McTaggert, Lynne 235, 373

Measure(ment) iii, xi, xxi, 9, 277, 288–292, 296, 318

Mediator 16, 40, 225, 366, 378

medicine xii, 36, 185, 195, 198–199, 213–217, 219, 233, 326

Meldrum, Rodney 260

Messiah 11, 16, 23, 40, 56, 176, 178, 180, 256, 352, 378

metaphysical xiii, 370

Meyer, Stephen C. 22, 147–151, 156

Micah 60

mighty prayer 117, 121

millennium i, 5, 170, 215, 255, 268, 356, 362, 365–367

military 71, 240–243, 272, 279, 281, 286, 318–319, 351

Militia 240

Mills, Roy 17–18, 24–26, 224

mind iii, xiii, xxvii, 4, 11, 15, 18, 27, 34, 49, 67, 78, 83–85, 97, 103, 105, 108, 111, 115–116, 121–126, 129, 135, 138, 148–149, 151, 153, 161, 163–164, 167–168, 171–172, 175–178, 180, 195, 198–199, 204, 207, 213–219, 221, 224–227, 229, 231, 234, 245, 252, 262, 266, 300–302, 305–306, 317, 321–322, 324–326, 337, 348, 354, 356, 358, 360, 373–374, 378–380

Mind Over Medicine 199, 213, 216

miracles 131, 140, 181, 216, 218, 301, 330

Miracles of Mind 301

Mijnhardt, Felix 51–52, 54–55, 62–63

modified Allan Variance xix, 291, 293

Moffet, Thomas 195

Mormon viii, xiv–xvi, xxiv, 43–45, 51–57, 59, 61–63, 65–67, 69–72, 95, 97, 101, 107, 110–111, 115, 176, 209, 259–260, 343–344, 349, 355, 358–359, 361–362, 383

Moroni 16, 40, 43, 56, 61, 71, 79, 87, 102, 105, 107, 110–111, 115–116, 124, 172, 180, 267, 358–359, 362, 378, 383

mortal xi, xxiii, xxv, xxvii–xxix, 13, 17, 19, 23–27, 32, 38, 41–43, 49, 64, 74, 82, 89, 91, 104–106, 114, 116, 121, 124, 127, 138, 141, 160, 203, 215–216, 224, 268, 275, 299, 322–323, 330–331, 334, 345–347, 362, 366–367, 378

mortal sphere of existence 24, 38, 41, 299, 345

Montagu, Ashley 93–94

mortality xi, xxvi, 24–26, 30, 34, 38, 113, 321, 326, 331, 347, 372–373

mother xiv, 15, 24–28, 66, 74, 94, 127, 135, 156, 169, 189, 191, 196, 207, 304, 352, 356, 380–381

Mother in Heaven 15

Index

Moses 5, 13–14, 33, 39–40, 44, 57, 74, 104, 107, 111, 120–122, 124–125, 129, 169, 171, 179, 265, 341–342, 345–346, 352, 362, 367, 371

Mount Ararat 20

Mt. Hood 131–133, 137, 139, 141–142, 145

Mount Moriah 59, 78

Mozart 176, 228

MSG 167

Muhammad 255

Musha, Professor 234

Muslims 332

N

Nain 230

National Bureau of Standards xi, xvii, 275

NBS 271, 273, 275, 282–283

National Guard 133, 140

National Institute of Standards and Technology xi, 275

NIST xix, 271, 274, 283, 286, 290, 293, 296–297

National Physical laboratory 274

nature xxv, 18, 20, 36, 57, 73, 75, 80, 100, 148, 176, 192, 208–209, 217, 227, 245–246, 275–278, 292, 295, 302, 307, 315, 336, 339, 353, 380

near-death experience (NDE) xxix, 18, 42, 153, 176, 307, 368, 379

neighbor xxiii–xxiv, 35, 79, 90, 108, 115, 121, 124, 126, 128, 138, 167, 255, 299, 325

Nelson, Lee i, 57, 217, 219

Nephi 15–16, 33, 38–40, 53–54, 61, 70, 72, 83, 90, 96, 100, 110–112, 117, 119, 121, 209, 223, 225, 248, 259, 265, 267, 354–357, 361–362, 377, 382

New Jerusalem 265, 342, 350, 360, 362

409

New Scientific Approach 17

New Testament 11, 56–57, 63, 355, 362

New World Order 243, 245–246, 359

NWO 247, 359

New York iii, v, xvi, 44, 93, 151, 165, 181, 344

Newbold, Brian 62

Newell, Lloyd 92–93

Newton, Isaac xxi, xxvi, 7, 9–10, 68, 150–151, 155–157, 255, 331–332, 335–343, 345, 352, 355, 383

Nibley, Hugh 8

Nichols, Beverly 100, 209

Nielsen, Donna 356

Nightingale, Florence 198–199

Noah xiii, 5, 20, 59, 68, 76, 78, 108, 119, 157, 203, 342, 344, 351

Noah's Ark 20

Noah's flood 5, 20, 68, 157, 203, 344

Norlin Library 13, 348

North America 13, 22, 156–158, 333–334, 349–350, 377

nutrition 164–165, 197

O

Oak Ridge National Laboratories xxi

Oaks, Dallin H. 84

obedience xxv, 49, 73, 107–109, 125, 252, 257–258, 261, 305, 331

obesity 164, 166, 173, 186, 357, 382

offspring of God 26, 121

Ohio Earthworks 22

Old Testament 54, 56, 255, 260, 355

Oliveri, Rachel i, iv

one eternal round xxvii, 362–363, 368, 371

opposition 16, 33, 40, 75, 107, 225, 300, 323, 346, 354, 359

Origen (theologian) 254

Origin of Species 148, 155

Owen, Mary 131, 133–134, 137, 139

Owen, Ruth 131, 133–134, 137, 139

Owen, Bruce and Shelli 131, 133–134, 137, 139

P

paradigm xii, xxiii, 5, 8, 10, 15, 21, 149, 189–191, 217, 258, 260, 307, 344

Pasteur, Louis 198–199

Paul 5, 15, 25, 37, 56, 89, 92, 101, 111–112, 167–168, 203, 254, 257, 259–260, 267, 315, 340, 345, 353, 362, 364–365, 367

Pearl of Great Price 57, 355

Peckham, Gene 55

Peleg 5, 344

pendulum xxvi, xxviii, 272–273, 294, 308, 310–313, 369

perfect love 26, 42, 79, 101, 107, 323, 374, 378

perfect plan xxvii, xxix, 24, 29, 37, 39, 64, 105–107, 111, 126–127, 203, 224, 267, 323, 331, 357, 362, 366, 380–381, 383

Peter 10, 16, 21, 33, 56, 64, 68, 103, 106, 126, 248, 257, 260, 301, 335, 353

pH Balance 195, 197

philosophy iii, xv, xxi, 7, 36, 44, 68, 151–152, 331

photosynthesis 114, 237

physical iii, xvii, xxvi–xxvii, xxix, 18, 24–25, 27, 29–30, 39, 75, 83–84, 121, 152, 159–160, 165, 183, 198, 201, 210, 212, 216–218,

224–225, 227, 233, 272–274, 290, 345, 369–370, 379

physics xiii–xiv, xvi–xvii, xx, xxvi, 160, 274, 281, 289, 296, 299–300, 314, 317, 373–374, 382

Pickering, Wayne 194

Pilate 250

Pilgrims 63, 349

Pollan, Michael 165–166

Pontius, John 51, 62, 356, 368

Pontius, Terri 62

Pope, Alexander 206, 336

pornography xv, xxii, 75–76, 78, 82–85, 101–102, 204–205, 209, 326, 358

power of prayer 140, 228

pray xvii, 31, 49, 65, 67, 75, 90, 100, 111, 113, 115–116, 123, 131, 153, 177, 205, 244, 250–252, 256, 263, 267, 303, 306, 316, 322, 325–326, 330, 351, 368, 383–384

prayer i, xviii, xx, xxvii, 12, 14–16, 25, 31, 41, 65, 90, 92, 100, 104, 116–117, 121–122, 132–133, 137–138, 140, 153, 161, 177–178, 202, 212, 225, 228–229, 248, 300, 305, 316, 322, 324–326, 330–331, 337, 368, 373, 375, 379

pre-mortal existence 346

Principia 336–337

Prodigal Son 80

progesterone 172–175

Promised Land 351, 377

prophet 6, 8, 10, 12, 17, 19–20, 23, 32, 37, 42–62, 104, 108, 110, 117–124, 155–157, 163–170, 178, 240, 252, 256, 259–260, 265–266, 331, 334, 337, 341–342, 347–366, 382, 386

prophetess 10, 15, 42, 64

Protestants 65, 67

Providence 117, 350

prozac 188

Pseudo-Philo 361

Q

quantum states 104, 306, 374

Qumran society 259

quantum transitions 271, 273, 306, 313, 374

R

Rabbi 66

Radin, Dean 302, 306, 366

Rankin, Lissa 195, 199, 212–214, 216–218, 228, 378

Rasmussen, Dennis 331

recovery program 204–205, 209–210

red-sprites 312–313, 317

repentance 41, 50, 66–68, 85, 88–89, 102–103, 128, 262, 305, 326, 365, 371–372

resentment 326–327, 329–330

Rheaume-Bleue, Kate 192–194

Riddle, Chauncey C. i, iii, xix, xxi–xxii, 26, 159, 371, 373

Ridenhour, Lynn 67

Rimmasch, Paul 5

Ritchie, George 19

Roberts, Frances J. 86

Robinson, John 349

Roman 250

Rowe, Julie 356

Rubáiyát 36

rubidium 281–282

Russo, Aaron 240

S

SA degradation 318

Sabbath 103, 255, 358, 366

salvation 77, 86, 119, 166, 179, 251, 257–259, 262, 371

Satan 10, 26, 33, 50, 76–78, 83, 85, 89, 94, 107, 110, 113, 116, 120, 155, 163, 168, 202–204, 243–244, 254–255, 268, 323, 332, 346–347, 357, 359, 367, 381–382

saved 53, 110, 120, 137, 178, 181, 248, 257–260, 267, 332, 346, 352, 372

Savior xxvi, 6, 15–16, 20, 24–26, 33–34, 37, 39, 42, 49, 56, 59, 68, 79, 82, 86, 88, 90, 95, 98, 100–101, 106–108, 111–113, 116–117, 124–125, 128, 136, 154, 175, 203–205, 225, 243, 250, 253, 255, 259–260, 262, 266, 268, 274, 300, 305, 322, 334, 346, 354–357, 361–366, 377, 379–380, 382–383

Schweitzer, Albert 36

science and religion iii–iv, xi, xiii, xxiii, xxx, 3–5, 7–10, 13–14, 19–22, 156, 158–161, 333, 349, 382

scientific method vi, xii, xxi, xxvii, 1, 3–5, 9, 14, 17, 19, 22, 34, 42, 44–45, 149, 300–301

Science Set Free 7, 246, 373

scientist iii, xii, xxiv–xxv, 5, 8–10, 12, 18, 148, 155–156, 246, 314, 331, 343, 349, 370

scriptures xi, xv–xvi, xxix–xxx, 3, 10, 19–20, 24, 31, 38, 42, 51, 55, 57, 59–61, 63–64, 68–70, 72–73, 78–79, 99, 101, 107, 114, 117, 121, 154, 157, 168, 171, 178, 205, 229, 248, 255, 260–261, 303, 322–323, 331–332, 337, 339, 344–345, 360–361, 366–368, 377–379, 383

Second Coming 58, 60–61, 206, 255, 265, 344, 349, 352, 355–356, 361, 366–367

secret combinations 359–360

selfishness 113

sermon on the mount 109, 251, 253

Index

service 79, 113, 128, 178, 216, 241, 285, 304

seven viii, xii, xvii, xxii, 4, 15, 56, 59, 103, 142, 176, 212–213, 227–228, 234, 300, 306, 335, 359, 361, 366, 373–374

Sewall, Samuel 350

sex xxii, 83–84, 209, 216

Sheldrake, Rupert 7, 225, 245–247, 302, 373

Signature in the Cell 22, 147, 156

Simoncini, T. 211

sing 88, 118–119, 121, 175–176, 179

singing 97, 118–120, 148, 153, 175, 248

SMD 318–319

Smith, Emma 8–9, 32, 37, 43–45, 51, 53–55, 57, 61, 64, 67, 70–71, 124, 165, 170–171, 178, 245, 259, 337–341, 343, 353–354, 356, 358, 364–366, 383

Smith, George A. 8–9, 32, 37, 43–45, 51, 53–55, 57, 61, 64, 67, 70–71, 124, 165, 170–171, 178, 245, 259, 337–341, 343, 353–354, 356, 358, 364–366, 383

Smith, Hyrum 8–9, 32, 37, 43–45, 51, 53–55, 57, 61, 64, 67, 70–71, 124, 165, 170–171, 178, 245, 259, 337–341, 343, 353–354, 356, 358, 364–366, 383

Smith, Joseph Jr. 8–9, 32, 37, 43–44, 51, 53–55, 57, 61, 64, 67, 70–71, 165, 170, 178, 259, 337–341, 343, 353–354, 356, 358, 364–366, 383

Snow College 62

Snow, Eliza R. 375

Sobel, Dava 279

Soul's Remembrance 17–18, 25, 224

solar home 167, 237

Solomon 266

Spectrum of Love 79

spiritual time xxvii, xxix, 370

Squanto 349

Stake President 177, 180

Stein, Ben 21, 149–152

Sternberg, Robert 232

Stiles, Ezra 352

still small voice 16, 25, 40–41, 45, 51, 142, 323

stress xii, 181, 184, 186, 188–189, 198, 212, 215–216, 221, 226, 233, 378

Stress Management 226

strength 28, 32, 78, 86–87, 108, 111, 116, 122–123, 128, 138, 154, 170–172, 179, 182, 188, 226, 263, 302, 325, 342, 347, 360

Stromberg, Gustaf 18–19, 32

Strong, Jeanne iii, xiv, 27, 43, 52, 76–77, 86, 100, 110, 193, 198, 207, 212, 229, 237, 242, 373

sugar 166, 186, 188–189, 191–192, 196

Supreme Creator 337

T

Tanner, Nathan Eldon 381

Tavella, Patrizia xviii–xix

technology iii, xi, xxi–xxii, 8, 157, 215, 236, 246, 275, 297, 308, 313, 318, 326

TED 215, 246–247

telestial 108, 362, 365–366

temper 79–81, 328

terrestrial 362, 364–366

Tertullian 252, 254

test 51, 87, 107, 183, 215, 236, 315, 353

The Everlasting Decree 260

The Field xvii, 169, 225, 235, 295, 301, 373

Index

The Way 15, 37, 77, 113–115, 135, 246, 248, 347, 381, 383

Think Independently 371, 373

This Land 119, 234, 239, 255, 260, 348–349, 352, 354, 360–361

Thrive 237

TAI (Time Atomic International) 281, 284

time and eternity xxx, 74, 82, 91, 116, 138, 225

Time, definition xix, 370-371

time variance 293

Torah 331

touching xxx, 93–94, 138, 183, 230

Tower of Babel 55–56

Townes, Charles 294–295

toxins 163, 174

Tracy, Ann Blake 188

transcendental time iii, xxii, xxvii, xxix, 269, 321–322, 333, 368, 370, 373

Tuttle, Boyd i

Twain, Mark 59–60

Twelve 187, 250, 252–253, 353

Twin Towers 239–240, 310, 359

Tyndale, William 11, 63–64

tyranny 236

U

UN xxv, 68, 243

unconditional love 26, 154

Unified Field Theory xxii, xxvi–xxvii, 19, 103, 144, 235, 240, 299, 345, 373, 382

UFT xxii, xxvii, 4, 19, 35, 103, 237, 299–300, 302, 307–309, 312–313, 315, 318, 366, 372–373

United States of America xi, xxii, 249, 260, 349

universe vi, xi, xiv, xx, xxvi, xxix, 7, 12, 39, 57, 78, 81, 103–104, 136–137, 151, 161, 175, 224, 245, 247, 301–302, 307, 366, 373

Universal Intelligence 214

University of Alaska 312–313

University of Colorado xvii, 13, 297, 343, 363

University of Oklahoma 312

Universal Time Coordinated xx, 290

UTC xx, 201, 281, 283–287, 290, 293, 314

UV spectrometer 314–315, 317

V

velocity of light 306, 317, 366, 372, 374

vitamin D 168

vitamin K2 ix, 192–193

Voss, Richard F. 277

Washington, George 150, 350–351

W

waste places 119

watchmen 118–119

water 20, 95–96, 135, 158, 172, 181–187, 189–191, 194, 197–198, 203, 212, 237, 352

Waters, Mike 59, 169, 236

Webster's, Noah 76, 79, 108

Weil, Andrew 194

Welch, John W. 97

Wheatley, Charles III (Chuck) 318–319

Wheel of Time xxvii, 269, 362–363, 371

White, Burton xvii, 85, 98, 137, 141, 143, 178, 276–277, 289, 292, 294–295, 308, 354, 360, 367

Wilcox, Keith W. 42, 177–178

Will the Real Heretics Please Stand Up 14, 50, 115, 249, 257, 332, 340, 355

Williams, David 172–175, 350

Williams, Roger 350

Wineland, Dave 297

wisdom iii, xxi, 14, 18, 26, 29, 31–32, 49, 63, 65, 125, 144, 154, 161, 165–167, 170–171, 201, 226, 266, 284, 304, 307, 351, 353, 375, 378

Woodruff, Wilford 10, 12, 353

WWV 290

WWVB 290

Y

Yale University 181, 352

yeast 195–197

Young, Robert O. 191, 195, 198

Yudkin, John 191

Z

Zechariah 61, 118

Zenos 50, 357, 361–362

Zion 83, 90, 98, 118–122, 124–125, 168–171, 179–180, 207–208, 248, 255–256, 265–267, 322, 331, 334, 348–349, 352, 356–358, 361, 378, 382

Zion society 121, 171, 255, 265–266, 348–349, 356, 378